VOLUME SEVEN HUNDRED AND NINE

METHODS IN
ENZYMOLOGY

Time-Resolved Methods
in Structural Biology

METHODS IN ENZYMOLOGY

Editors-in-Chief

ANNA MARIE PYLE

*Departments of Molecular, Cellular and Developmental
Biology and Department of Chemistry
Investigator, Howard Hughes Medical Institute
Yale University*

DAVID W. CHRISTIANSON

*Roy and Diana Vagelos Laboratories
Department of Chemistry
University of Pennsylvania
Philadelphia, PA*

Founding Editors

SIDNEY P. COLOWICK and NATHAN O. KAPLAN

VOLUME SEVEN HUNDRED AND NINE

Methods in ENZYMOLOGY

Time-Resolved Methods in Structural Biology

Edited by

PETER MOODY
*Leicester Institute for Structural & Chemical Biology
& Department Molecular & Cell Biology
University of Leicester,
Leicester, LE1 7RH, United Kingdom*

HANNA KWON
*Leicester Institute for Structural & Chemical Biology
& Department Molecular & Cell Biology
University of Leicester,
Leicester, LE1 7RH, United Kingdom*

Academic Press is an imprint of Elsevier
50 Hampshire Street, 5th Floor, Cambridge, MA 02139, United States
525 B Street, Suite 1650, San Diego, CA 92101, United States
125 London Wall, London, EC2Y 5AS, United Kingdom

First edition 2024

Copyright © 2024 Elsevier Inc. All rights are reserved, including those for text and data mining, AI training, and similar technologies.

Publisher's note: Elsevier takes a neutral position with respect to territorial disputes or jurisdictional claims in its published content, including in maps and institutional affiliations.

No part of this publication may be reproduced or transmitted in any form or by any means, electronic or mechanical, including photocopying, recording, or any information storage and retrieval system, without permission in writing from the publisher. Details on how to seek permission, further information about the Publisher's permissions policies and our arrangements with organizations such as the Copyright Clearance Center and the Copyright Licensing Agency, can be found at our website: www.elsevier.com/permissions.

This book and the individual contributions contained in it are protected under copyright by the Publisher (other than as may be noted herein).

Notices
Knowledge and best practice in this field are constantly changing. As new research and experience broaden our understanding, changes in research methods, professional practices, or medical treatment may become necessary.

Practitioners and researchers must always rely on their own experience and knowledge in evaluating and using any information, methods, compounds, or experiments described herein. In using such information or methods they should be mindful of their own safety and the safety of others, including parties for whom they have a professional responsibility.

To the fullest extent of the law, neither the Publisher nor the authors, contributors, or editors, assume any liability for any injury and/or damage to persons or property as a matter of products liability, negligence or otherwise, or from any use or operation of any methods, products, instructions, or ideas contained in the material herein.

ISBN: 978-0-443-31456-8
ISSN: 0076-6879

> For information on all Academic Press publications
> visit our website at https://www.elsevier.com/books-and-journals

Publisher: Zoe Kruze
Editorial Project Manager: Saloni Vohra
Production Project Manager: James Selvam
Cover Designer: Gopalakrishnan Venkatraman

Typeset by MPS Limited, India

Contents

Contributors		*xi*
Preface		*xv*

1. The growth of microcrystals for time resolved serial crystallography   1
Alexander McPherson

1.	Introduction	1
2.	Properties of macromolecular crystals	2
3.	Microcrystals for time resolved serial crystallography	4
4.	Screening and optimization	5
5.	Supersaturation, nucleation and growth	6
6.	Matching crystallization methods to the presentation of crystals to the X-ray beam	10
7.	Creating a state of supersaturation	12
8.	Precipitants	21
9.	Visualization	23
	References	23

2. Use of fixed targets for serial crystallography   29
Sofia Jaho, Danny Axford, Do-Heon Gu, Michael A. Hough, and Robin L. Owen

1.	Introduction	30
2.	Common aspects of SSX and SFX: optimising the experiment	32
	2.1 Preparing for the experiment	32
	2.2 Moving from standard MX to serial experiments	33
	2.3 Sample preparation	34
	2.4 Anaerobic fixed target SSX and SFX	37
	2.5 Fixed target SSX beamline hardware: I24	38
	2.6 Fixed target SFX beamline hardware: SACLA	40
3.	Data collection	40
	3.1 Setup and data collection strategies	40
	3.2 Fast-feedback from data acquisition	41
	3.3 Dose	42
	3.4 Choice of X-ray energy	44
4.	Record keeping, log and beamtime logistics – best practice	46
	4.1 Iterative optimisation of experiment design	48

4.2	Unit cell as a metric for reaction progression: contrasting examples of crystallin and myoglobin	48
4.3	Exploring radiation damage and X-ray driven reactivity using fixed-target SSX	51
5.	Summary	53
	Acknowledgements	53
	References	53

3. Sample efficient approaches in time-resolved X-ray serial crystallography and complementary X-ray emission spectroscopy using drop-on-demand tape-drive systems 57

Jos J.A.G. Kamps, Robert Bosman, Allen M. Orville, and Pierre Aller

1.	Introduction	58
	1.1 Complementary spectroscopy	62
	1.2 X-ray emission spectroscopy	65
	1.3 XES setup components	67
2.	Materials	69
	2.1 Sample delivery	69
	2.2 XES, detector and analyzer crystals	71
3.	Methods	74
	3.1 Preparing microcrystal slurries	74
	3.2 Reaction initiation scheme	78
	3.3 Data collection/beamtime	84
	3.4 Future developments	93
	References	96

4. Sample delivery for structural biology at the European XFEL 105

Katerina Dörner, Peter Smyth, and Joachim Schulz

1.	Introduction	106
2.	Scientific instruments at the European XFEL	107
	2.1 The Single Particles and Biomolecules & Serial Femtosecond Crystallography (SPB/SFX) instrument	108
	2.2 Structural biology at other instruments at the European XFEL	109
3.	Sample delivery for SFX via liquid jets	109
	3.1 Introduction	109
	3.2 Sample properties for sample delivery via GDVN	114
4.	High viscosity extrusion for SFX experiments	116
	4.1 Introduction	116
	4.2 Crystal properties and viscous media choices	118

Contents vii

5. Testing of sample delivery — 120
6. Fixed-target sample delivery — 122
7. Drop-on-demand sample delivery — 124
8. Summary and outlook — 125
References — 126

5. Experimental approaches for time-resolved serial femtosecond crystallography at PAL-XFEL — 131

Jaehyun Park and Ki Hyun Nam

1. Introduction — 132
2. NCI experimental hutch — 134
   2.1 Beamline instruments — 135
   2.2 Optical lasers — 136
3. Sample environments for TR-SFX at the NCI experimental hutch — 137
   3.1 Sample chamber — 137
   3.2 Sample injectors — 141
   3.3 Fixed target scanning — 144
4. Sample preparation — 146
   4.1 Sample preparation laboratory — 146
   4.2 Sample preparation — 146
5. Experimental setup for TR-SFX at NCI experimental hutch — 149
   5.1 Procedure — 149
6. Data processing — 153
7. Safety considerations and others — 155
   7.1 Preliminary discussion for experiment — 155
   7.2 Laser — 156
   7.3 Hutch access — 156
   7.4 Data access — 156
Acknowledgments — 156
References — 157

6. Time-resolved IR spectroscopy for monitoring protein dynamics in microcrystals — 161

Wataru Sato, Daichi Yamada, and Minoru Kubo

1. Introduction — 162
2. Overview of instrument — 163
3. Experimental procedure — 165
   3.1 Instrument preparation — 165
   3.2 Alignment of the pump beam — 165

3.3 Collection of a background spectrum	166
3.4 Sample preparation	167
3.5 Sample packing within the FTIR cell	167
3.6 Selection of the sample-containing wells	168
3.7 Adjustment of the pump irradiation timing and rapid-scan measurements	168
3.8 Setting measurement conditions	170
3.9 Data analysis	170
4. Results and discussion	171
4.1 Observation of CO photolysis in CcO	171
4.2 Observation of P450nor intermediate	173
5. Conclusion	174
Acknowledgments	175
References	175

7. Multiplexing methods in dynamic protein crystallography   177

Margaret A. Klureza, Yelyzaveta Pulnova, David von Stetten,
Robin L. Owen, Godfrey S. Beddard, Arwen R. Pearson, and
Briony A. Yorke

1. Introduction	178
1.1 Laue and XFEL X-ray sources	179
2. Theory	180
2.1 Principles of multiplexing measurements	180
2.2 Error and the crystallographic multiplexing advantage	183
3. Experimental design	189
3.1 Combining HATRX with existing pump-probe techniques	190
3.2 S-matrix selection	192
3.3 Temporal encoding	193
3.4 Data collection	198
4. Data processing	199
5. Conclusion and outlook	201
Acknowledgments	201
References	202

8. Processing serial synchrotron crystallography diffraction data with DIALS   207

James Beilsten-Edmands, James M. Parkhurst, Graeme Winter, and
Gwyndaf Evans

1. Introduction	208
2. Methods	211

2.1 Model parameterisation using a multivariate-normal distribution	212
2.2 General impact of the RLP distribution	214
2.3 Parameter estimation via maximum likelihood methods	215
2.4 Integration and partiality estimation	217
3. DIALS command-line tools for processing SSX data	218
3.1 Indexing with dials.ssx_index	219
3.2 Integration with dials.ssx_integrate	219
3.3 Data reduction with dials.cosym and dials.scale	220
3.4 Associated tools from cctbx.xfel	221
4. Xia2.ssx: an automated processing pipeline	221
4.1 Integration and reduction of SSX data with xia2.ssx	222
4.2 Standalone data reduction with xia2.ssx_reduce	225
4.3 Merging in groups based on metadata	225
5. Demonstration and evaluation on example data	227
5.1 Processing anomalous SSX data: AcNiR above the K-edge	227
5.2 Processing SSX dose-series: photoreduction of Fe(III) FutA	233
6. Discussion	239
Appendix A. Block matrix inversion	240
Appendix B. Derivatives of the log-likelihood and Fisher Information matrix	241
References	242

9. Time-resolved scattering methods for biological samples at the CoSAXS beamline, MAX IV Laboratory — 245

Fátima Herranz-Trillo, Henrik Vinther Sørensen, Cedric Dicko,
Javier Pérez, Samuel Lenton, Vito Foderà, Anna Fornell,
Marie Skepö, Tomás S. Plivelic, Oskar Berntsson,
Magnus Andersson, Konstantinos Magkakis, Fredrik Orädd,
Byungnam Ahn, Roberto Appio, Jackson Da Silva,
Vanessa Da Silva, Marco Lerato, and Ann E. Terry

1. Introduction	246
2. Technical design of CoSAXS	248
2.1 Optical design	249
2.2 Experimental hutch	250
2.3 Vacuum vessel	252
2.4 Data acquisition and processing	252
3. SUrF—combined SAXS with UV–vis/Raman/fluorescence	254
3.1 Design of the multiprobe platform SUrF	255
3.2 Time-dependent acid-induced unfolding of BSA with SUrF	259
4. Time-resolved SAXS studies using microfluidic chips	265
4.1 Microfluidic chip design	267

4.2	Following surfactant-induced structural changes in a microfluidic chip	269
5.	Structural kinetics investigated with stopped-flow	275
5.1	Structural changes induced by addition of Ca^{2+} ions	276
6.	Time-resolved X-ray solution scattering, TR-XSS	282
6.1	Design of the TR-XSS experimental setup	283
6.2	Thermal response of lysozyme to laser induced temperature jumps	285
6.3	Laser-induced activation of ATP binding in adenylate kinase	289
7.	Future perspectives	290
Acknowledgments		292
References		292

Contributors

Byungnam Ahn
MAX IV Laboratory, Lund University, Lund, Sweden

Pierre Aller
Diamond Light Source, Harwell Science & Innovation Campus; Research Complex at Harwell, Rutherford Appleton Laboratory, Didcot, United Kingdom

Magnus Andersson
Department of Chemistry, Umeå University, Umeå, Sweden

Roberto Appio
MAX IV Laboratory, Lund University, Lund, Sweden

Danny Axford
Diamond Light Source Ltd, Harwell Science and Innovation Campus, Didcot, Oxfordshire, United Kingdom

Godfrey S. Beddard
School of Chemistry, University of Edinburgh, David Brewster Road; School of Chemistry, University of Leeds, Woodhouse Lane, Leeds, United Kingdom

James Beilsten-Edmands
Diamond Light Source Ltd, Harwell Science and Innovation Campus, Didcot, Oxfordshire, United Kingdom

Oskar Berntsson
MAX IV Laboratory, Lund University, Lund, Sweden

Robert Bosman
Diamond Light Source, Harwell Science & Innovation Campus, Didcot; University Medical Center Hamburg-Eppendorf (UKE), Hamburg, Germany

Jackson Da Silva
MAX IV Laboratory, Lund University, Lund, Sweden

Vanessa Da Silva
MAX IV Laboratory, Lund University, Lund, Sweden

Cedric Dicko
Division of Pure and Applied Biochemistry, Lund University, Lund, Sweden

Katerina Dörner
European XFEL, Schenefeld, Germany

Gwyndaf Evans
Diamond Light Source Ltd; Rosalind Franklin Institute, Harwell Science and Innovation Campus, Didcot, Oxfordshire, United Kingdom

Vito Foderà
Department of Pharmacy, Faculty of Health and Medical Sciences, University of Copenhagen, Universitetsparken, Copenhagen, Denmark

Anna Fornell
Division of Biomedical Engineering, Department of Materials Science and Engineering, Science for Life Laboratory, Uppsala University, Uppsala, Sweden

Do-Heon Gu
Diamond Light Source Ltd; Research Complex at Harwell, Harwell Science and Innovation Campus, Didcot, Oxfordshire, United Kingdom

Fátima Herranz-Trillo
MAX IV Laboratory, Lund University, Lund, Sweden

Michael A. Hough
Diamond Light Source Ltd; Research Complex at Harwell, Harwell Science and Innovation Campus, Didcot, Oxfordshire, United Kingdom

Sofia Jaho
Diamond Light Source Ltd; Research Complex at Harwell, Harwell Science and Innovation Campus, Didcot, Oxfordshire, United Kingdom

Jos J.A.G. Kamps
Diamond Light Source, Harwell Science & Innovation Campus; Research Complex at Harwell, Rutherford Appleton Laboratory, Didcot, United Kingdom

Margaret A. Klureza
Institute for Nanostructure and Solid State Physics, University of Hamburg, HARBOR, Hamburg, Germany

Minoru Kubo
Graduate School of Science, University of Hyogo, Kouto, Kamigori, Ako, Hyogo, Japan

Samuel Lenton
Department of Pharmacy, Faculty of Health and Medical Sciences, University of Copenhagen, Universitetsparken, Copenhagen, Denmark

Marco Lerato
MAX IV Laboratory, Lund University, Lund, Sweden

Konstantinos Magkakis
Department of Chemistry, Umeå University, Umeå, Sweden

Alexander McPherson
Department of Molecular Biology and Biochemistry, University of CA Irvine, Irvine, CA; The Scripps Research Institute Florida, Jupiter, FL, United States

Ki Hyun Nam
College of General Education, Kookmin University, Seoul, Republic of Korea

Allen M. Orville
Diamond Light Source, Harwell Science & Innovation Campus; Research Complex at Harwell, Rutherford Appleton Laboratory, Didcot, United Kingdom

Fredrik Orädd
Department of Chemistry, Umeå University, Umeå, Sweden

Robin L. Owen
Diamond Light Source Ltd; Research Complex at Harwell, Harwell Science and Innovation Campus, Didcot, Oxfordshire, United Kingdom

Jaehyun Park
Pohang Accelerator Laboratory, Pohang University of Science and Technology, Pohang, Republic of Korea

James M. Parkhurst
Diamond Light Source Ltd; Rosalind Franklin Institute, Harwell Science and Innovation Campus, Didcot, Oxfordshire, United Kingdom

Arwen R. Pearson
Institute for Nanostructure and Solid State Physics, University of Hamburg, HARBOR, Hamburg, Germany

Tomás S. Plivelic
MAX IV Laboratory, Lund University, Lund, Sweden

Yelyzaveta Pulnova
ELIbeamlines, Extreme Light Infrastructure, Dolni Brezany, Czechia

Javier Pérez
Synchrotron SOLEIL, Saint-Aubin - BP, Gif sur Yvette Cedex, France

Wataru Sato
Graduate School of Science, University of Hyogo, Kouto, Kamigori, Ako, Hyogo, Japan

Joachim Schulz
European XFEL, Schenefeld, Germany

Marie Skepö
Department of Chemistry, Division of Computational Chemistry, Lund University, Lund, Sweden

Peter Smyth
European XFEL, Schenefeld, Germany

Henrik Vinther Sørensen
MAX IV Laboratory; Department of Chemistry, Division of Computational Chemistry, Lund University, Lund, Sweden

Ann E. Terry
MAX IV Laboratory, Lund University, Lund, Sweden

Graeme Winter
Diamond Light Source Ltd, Harwell Science and Innovation Campus, Didcot, Oxfordshire, United Kingdom

Daichi Yamada
Graduate School of Science, University of Hyogo, Kouto, Kamigori, Ako, Hyogo, Japan

Briony A. Yorke
School of Chemistry, University of Leeds, Woodhouse Lane, Leeds, United Kingdom

David von Stetten
European Molecular Biology, Laboratory (EMBL), Hamburg, Germany

Preface

The application and development of the methods of structure determination of biological macromolecules and their complexes has not only given us unprecedented insight into biological processes, but it has also had a far-reaching impact on the understanding and treatment of disease (as recognized by many Nobel prizes). However, following the changes that take place over time - such as the turn-over of a substrate by an enzyme, conformational changes upon binding, and changes in redox state - present additional challenges.

The most obvious of these is that most structure determination techniques involve averaging over time and space. Eadweard Muybridge's famous 'Horse in Motion' photographs (where multiple cameras were used to record what would now be considered as cinematography frames) showed how the horse moved with frozen (0.02 second) images over ~0.5 meter gaps. It is tempting to apply the same logic to an enzymatic reaction. By mixing an enzyme and substrates and determining the structures at intervals as the reaction proceeds, we could, in theory, create a "movie" of the process in a manner analogous to continuous flow or pump-probe techniques. However, if we consider that we are not looking at a single horse (or molecule) but the entire field, who may be more or less aligned at the start of the race, but soon spread out, losing any synchronization. Of course, the picture is more complex than this. The "race" is not held on a flat racecourse, there will be different energy levels during the course of the reaction and the population distribution in the different states will be determined not only by time since the start (the pump) but also by their relative energy levels. Deconvolution of the mixtures of different structures is a daunting prospect. If the energy profile of the reaction can be exploited, an intermediate with lower energy levels might be persuaded to accumulate (indeed it may be possible to trap some intermediate states by cryo- or other means). Even when we cannot determine the structures of all the steps in a reaction, the power of the difference Fourier lets us see what has moved over time.

In crystallography, the use of nano- or micro-crystals will reduce the time taken for substances to diffuse in or out of the crystals, the preparation of which is covered in the chapter by Alex McPherson. The extremely bright femtosecond X-ray pulses available from XFELs and the longer, but still useful, short pulses available from the latest synchrotrons, enable fine temporal resolution. In this volume we have included chapters which describe how samples can be delivered to take full advantage of these capabilities (Jaho *et al.*, Kamps *et al.*, Park and Nam, and Dörner *et al.*). Of course, it is essential to be able to monitor and validate the chemical changes taking place, which is also addressed in these chapters and the use of time-resolved IR spectroscopy in microcrystals is further addressed in the chapter by Sato *et al.* Klureza *et al.* discuss the scope and limitations of how planning the use if multiplexing methods can be used to optimize and increase the efficiency of time-resolved experiments. The measurement of many thousands of diffraction images, each from individual microcrystals, requiring indexing and scaling, has presented a challenge which is addressed by Beilsten-Edmands *et al.*

Crystallography is not the only means of time-resolved structural biology, Herranz-Trillo *et al.* describe with examples what can be done to follow conformational changes in solution that are inaccessible in the constrained environment of the crystal by using SAXS/WAXS at their dedicated and flexible MAX IV beam line. Electron Cryo-microscopy (cryoEM) has the advantage that each individual molecule or complex is imaged, allowing different structures to be separated and averaged independently. There is therefore potential for time-resolved studies using single particle cryoEM, by using microfluidic mixing techniques and cryo-trapping on cryoEM grids.

This volume arose from discussions following a meeting, organized by the editors "Time-Resolved Structural Enzymology", held in September 2023 at the University of Leicester, sponsored by the UK's UKRI-BBSRC UK-Japan Partnership, with the help and support of LINXS Institute for Advanced X-ray and Neutron Science and the University of Leicester's Institute for Structural and Chemical Biology. This volume would not have been possible without the help and oversight of the *Methods in Enzymology* staff. We are most grateful for their support. PCEM also wishes to acknowledge the support of the Royal Society Wolfson Fellowship scheme during this time.

PETER MOODY AND HANNA KWON
Leicester UK

CHAPTER ONE

The growth of microcrystals for time resolved serial crystallography

Alexander McPherson[a,b,]*

[a]Department of Molecular Biology and Biochemistry, University of CA Irvine, Irvine, CA, United States
[b]The Scripps Research Institute Florida, Jupiter, FL, United States
*Corresponding author. e-mail address: amcphers@uci.edu

Contents

1. Introduction	1
2. Properties of macromolecular crystals	2
3. Microcrystals for time resolved serial crystallography	4
4. Screening and optimization	5
5. Supersaturation, nucleation and growth	6
6. Matching crystallization methods to the presentation of crystals to the X-ray beam	10
7. Creating a state of supersaturation	12
8. Precipitants	21
9. Visualization	23
References	23

Abstract

The production of enzyme microcrystals for time resolved serial crystallography employing free electron laser or synchrotron radiation is a relatively new variation on traditional macromolecular crystallization for conventional single crystal X-ray analysis. While the fundamentals of macromolecular crystal growth are the same, some modifications and special considerations are in order if the objective is to produce uniform size, microcrystals in very large numbers for serial data collection. Presented here are the basic principles of protein crystal growth with particular attention to the approaches best employed to achieve the goal of microcrystals and some novel techniques, as well as old, that may be useful. Also discussed are the advantages of particular precipitants and certain methods of growing protein crystals that might be advantageous for serial data recording.

1. Introduction

Macromolecular crystallization, which includes the crystallization of proteins, nucleic acids, and larger macromolecular assemblies such as

Methods in Enzymology, Volume 709
ISSN 0076-6879, https://doi.org/10.1016/bs.mie.2024.10.003
Copyright © 2024 Elsevier Inc. All rights are reserved, including those for text and data mining, AI training, and similar technologies.

viruses and ribosomes, is based on a rather diverse set of principles, experiences and intuitions. There is no comprehensive theory to guide efforts. Discovering optimal crystallization conditions is largely empirical and is fundamentally a trial and error process. Crystallization, generally, is a matter of searching, as systematically as possible, the ranges of the individual parameters that influence crystal formation, finding a set, or multiple sets of factors that yield some kind of crystals, and then optimizing the individual variables to obtain the best possible. This is usually achieved by carrying out an extensive series, or establishing a vast matrix of crystallization trials, evaluating the results, and using information obtained to improve conditions in successive rounds of trials. The number of variables is large, and the ranges broad, so experience and insight in designing and evaluating the individual and collective trials becomes important.

2. Properties of macromolecular crystals

Macromolecular crystals (McPherson, 1982, 1989) are composed of approximately 50 % solvent on average (Matthews, 1968), though this may vary from 25–90 % depending on the particular macromolecule (McPherson, 1999). Protein or nucleic acid occupies the remaining volume so that the entire crystal is in many ways an ordered gel permeated by extensive interstitial channels and spaces through which solvent and other small molecules may diffuse. This may be an essential property for time resolved dynamic processes such as enzyme catalysis.

In proportion to molecular mass, the number of bonds (salt bridges, hydrogen bonds, hydrophobic interactions) that a conventional molecule forms with its neighbors in a crystal far exceeds those exhibited by crystalline proteins. Since these contacts provide the lattice interactions essential for crystal maintenance, this largely explains the difference in physical and material properties between crystals of salts, or conventional molecules, and macromolecules. Biological systems are based exclusively on aqueous chemistry and maintained within narrow ranges of temperature and pH. As a consequence, protein and nucleic acid crystals must be grown from aqueous solutions, ones to which they are tolerant, and these solutions are called mother liquors.

Although comparable in their morphologies and appearance, there are important practical differences between crystals of low-molecular-mass compounds and crystals of proteins and nucleic acids. Crystals of conventional molecules are characterized by firm lattice forces, are relatively highly ordered, generally physically hard and brittle, easy to manipulate, usually can be exposed to air, have strong optical properties, and diffract X-rays to high resolution. Macromolecular crystals are by comparison usually more limited in size, are the softest material known (Chernov & Komatsu, 1995) and crush easily, disintegrate if allowed to dehydrate (Bernal & Crowfoot, 1934; Crowfoot, 1935), exhibit weak optical properties and diffract X-rays poorly. Macromolecular crystals are temperature sensitive and undergo extensive damage after prolonged exposure to radiation (De La Mora et al., 2020).

The liquid channels and solvent filled cavities that permeate macromolecular crystals are primarily responsible for the limited resolution of the diffraction patterns. Because of the relatively large spaces between adjacent molecules and the consequent weak lattice forces, all molecules in the crystal may not occupy exactly equivalent orientations and positions but may vary slightly within or between unit cells. Furthermore, because of their structural complexity and their potential for conformational dynamics, protein molecules in a particular crystal may exhibit slight variations in the course of their polypeptide chains or the dispositions of side groups from one to another.

Although the presence of extensive networks of solvent microvolumes are major contributors to the modest diffraction resolution of protein crystals, they are also largely responsible for their value to biochemists. Thanks to the high solvent content, the individual macromolecules in protein crystals are surrounded by layers of water that maintain their structure virtually unchanged from that found in solution. As a consequence, ligand binding, enzymatic, spectroscopic characteristics, and most other biochemical features are essentially the same as for the fully solvated molecule. Conventional chemical compounds, which may be ions, ligands, substrates, coenzymes, inhibitors, drugs, or other effector molecules, may diffuse into and out of the crystals (McPherson and Larson, 2018). Crystalline enzymes, though immobilized, are usually accessible for experimentation simply through alteration of the surrounding mother liquor.

Polymorphism is a common phenomenon with both protein, nucleic acid, and virus crystals. Presumably this is a consequence of their conformational dynamics and the sensitivity of prevailing lattice contacts. Thus, different habits and diverse unit cells may arise from what, by most standards, would be called identical conditions. Multiple crystal forms are, in fact, commonly seen coexisting in the same mother liquor.

There are further differences which complicate the crystallization of macromolecules as compared with conventional, small molecules (Durbin & Feher, 1996; Feher, 1986; Feigelson, 1988; McPherson, 1982; McPherson, 1999; McPherson, Malkin, & Kuznetsov, 1995). Macromolecules may assume several distinctive solid states that include amorphous precipitates, oils, or gels as well as crystals, and most of these alternate states are kinetically favored. Macromolecular crystals nucleate, or initiate development, only at relatively high levels of supersaturation (see below), often two to three orders of magnitude higher than required to sustain growth (Chernov & Komatsu, 1995; McPherson, 1999; Rosenberger, 1986). Finally, the kinetics of macromolecular crystal nucleation and growth are generally two to three orders of magnitude slower than for conventional molecules (Kuznetsov, Malkin, Greenwood, & McPherson, 1995; Malkin, Kuznetsov, & McPherson, 1996; Malkin, Kuznetsov, & McPherson, 1997). This latter difference arises from the considerably larger size, lowered diffusivity, and weaker association tendencies compared with conventional molecules or ions, as well as a lower probability of incorporation of incoming macromolecules into growth steps (Chernov, 2003).

3. Microcrystals for time resolved serial crystallography

Protein crystallographers have, for more than 100 years (Bailey, 1942; Giege, 2013; McPherson, 1991, 1992), spent the better parts of their lives trying to grow large single crystals for X-ray diffraction analysis. Very often their efforts ended in failure, and frequently those failures were manifested by masses of microcrystals of no value for data collection. Technological advances have obviated that problem and given value to those microcrystals for new applications, such as serial time resolved crystallography, and microcrystals have increasingly become an objective (Falkner et al., 2005; Glaeser et al., 2016; Murakawa et al., 2021; Schulz, Yorke, Pearson, & Mehrabi, 2022).

Microcrystals do, in fact, have some important advantages over larger single crystals that are largely unappreciated. The incorporation of impurities (McPherson, Malkin, Kuznetsov, & Koszelak, 1996) and the occurrence of crystal imperfections (Malkin, Kuznetsov Yu, & McPherson, 1996) increase with crystal size, thus microcrystals are generally more perfect. For TRSC, microcrystals have another distinct advantage over large crystals. Small molecule diffuse quickly throughout a microcrystal so that time dependent changes are more readily observable. For large protein crystals the diffusion rates are substantially greater (Kuznetsov Yu & McPherson, 2019; McPherson & Larson, 2018) and preclude most time resolved studies.

Experience has shown that protein microcrystals are more easily grown, particularly in bulk quantities, and are produced by more different crystallization strategies and conditions. Because they are grown at high levels of super-saturation they have little tendency to redissolve, i.e. they are more stable. Microcrystals freeze far more readily than larger crystals and often can be flash frozen from their mother liquor without recourse to cryo-protectants. While it is true that they are subject to Ostwald ripening (Ostwald, 1896, 1897), this is not a significant concern. The mosaicity of microcrystals is less than for larger crystals, and they are more likely to nucleate free of surfaces.

Traditionally, crystallographers have spoken of protein microcrystals as any crystals less than about 100 μm, and for the most part concluded that they are unsuitable for conventional protein crystallography. This attitude is of course not applicable for TRSC. The ideal size of microcrystals for TRSC may not be obvious, and there is indeed more than one opinion. The extent of mosaic blocks in the crystals is a consideration that must be balanced against the need to produce enough photons upon diffraction so that structure amplitudes can be accurately measured. Simulations by James Holton suggest that the ideal crystal size would be about 1 μm (Holton & Frankel, 2010). To obtain a 2.0 Å resolution data set from a lysozyme crystal without unacceptable radiation damage, he estimates a single microcrystal would need to be 1.2 μm in size (tetragonal lysozyme crystals are essentially equidimensional). He further points out that crystals come in many shapes. Therefore, linear dimensions are less important than crystal volumes. Another consideration is that, optimally, there should be a size match between the crystal dimensions and the dimensions of the X-ray beam, though Holton points out that in many studies the micro-crystals are significantly larger. In addition, one would like to have a high degree of uniformity with regard to crystal size (Holton, 2019).

4. Screening and optimization

There are two phases in obtaining protein crystals for conventional X-ray diffraction investigations, and these are (a) the identification of chemical, biochemical, and physical conditions that yield some crystalline material and (b) the systematic alteration of those initial conditions by incremental amounts to obtain optimal samples for diffraction analysis. Whether the objective is large single crystals or micrcrystals, however, it remains essential to identify chemical and physical conditions that will yield crystals at all, hence screening will generally be important.

Crystals, including microcrystals used for TRSC, are for now usually of well characterized enzymes that have been extensively investigated by conventional, single crystal X-ray diffraction analyses. Thus, it is often unnecessary to search for unique crystallization conditions, as it would be for a newly studied macromolecule. As a result the screening considerations presented here, may be of lesser importance. They may be replaced, however, by the necessity of finding modifications to the conditions that will consistantly yield uniform microcrystals of optimal dimensions, or volumes.

There are fundamentally two approaches to screening for crystallization conditions. The first is a systematic variation of what are believed to be the most important variables, precipitant types and their concentrations, pH, temperature, etc. The second is what we might term a shotgun approach, but aimed with accumulated wisdom. While far more thorough in scope and more congenial to the scientific mind, the first method requires a substantially greater amount of protein. In those cases where the quantity of material is limiting, it may simply be impractical. The second technique provides more opportunity for useful conditions to escape discovery, but in general requires less material, and it provides greater convenience and even automation.

There is currently on the commercial market a wide selection of crystallization screening kits. The availability and ease of use of these relatively modestly priced kits, which may be used in conjunction with a variety of crystallization methods, make them the tool of choice in attacking a new crystallization problem. With these kits, nothing more is required than combining a series of potential crystallization solutions against one's protein of interest using a micropipette, sealing, and monitoring the samples.

Once some crystals are observed and shown to be of protein origin then optimization begins. Every component in the solution yielding crystals should be noted and considered (buffer, salt, ions, etc.), along with pH, temperature, and whatever other factors might have an impact on the quality of the results. Each of these parameters is then incremented in additional trial matricies spanning the conditions which gave the initial success. Because the problem is non linear, and one variable may be coupled to another, this process is often more complex and difficult than one might expect (Bergfors, 1999; Ducruix & Giége, 1992; McPherson, 1982; McPherson, 1999).

5. Supersaturation, nucleation and growth

Crystallization of any molecule, including proteins, proceeds in two distinct but coupled stages, nucleation and growth. Nucleation is the most difficult problem to address theoretically and experimentally because it represents a first order phase transition by which molecules pass from a wholly disordered state to an ordered one. Presumably this occurs through the formation of partially ordered or paracrystalline intermediates, in this case protein aggregates having short-range order, and ultimately yields small, completely ordered assemblies which we refer to as critical nuclei.

Critical nuclei must be considered in terms of the molecular dimensions, the supersaturation, and the surface free energy of molecular addition. The critical nuclear size is totally dependent on the supersaturation, which is the driving force for molecular addition. Supersaturation is best characterized by the solubility phase diagram in Fig. 1, and has been extensively discussed elsewhere (Bergfors, 1999; Haas & Drenth, 1999; McPherson, 1999; McPherson & Gavira, 2014). At this time the critical nuclear size has only been described for a few systems, and for several cases, these were only investigated in terms of two-dimensional nuclei developing on the surfaces of already existent crystals (Malkin, Kuznetsov, et al., 1996; Malkin et al., 1997). A theory has also emerged which explains the nucleation phenomenon in terms of statistical fluctuations in solution properties (Haas & Drenth, 1999; Piazza, 1999; Ten Wolde & Frenkel, 1997). This idea holds that a distinctive "liquid protein phase" forms in concentrated protein solutions, and that this "phase" ultimately gives rise to critical nuclei with comprehensive order. This idea has persuasive support from light scattering investigations (Asherie, 2004).

Growth of macromolecular crystals (Malkin, Kuznetsov Yu, Land, DeYoreo, & McPherson, 1995) is a better-characterized process than nucleation, and its mechanisms are reasonably well understood. Protein crystals grow principally by the classical mechanisms of dislocation growth, and growth by two-dimensional nucleation, along with two other less common mechanisms known as normal growth and three dimensional nucleation (Malkin, Kuznetsov, Land, DeYoreo, & McPherson, 1995; McPherson, Malkin, & Kuznetsov, 2000). A common feature of nucleation and growth is that both are critically dependent on what is termed the supersaturation of the mother liquor giving rise to the crystals. Supersaturation is the variable that drives both processes and determines their occurrence, extent, and the kinetics that govern them.

Crystallization requires the creation of a supersaturated state as illustrated by the phase diagram in Fig. 1. This is a non–equilibrium condition in which some quantity of the solute in excess of its solubility limit, under specific chemical and physical conditions is created in solution. Equilibrium, the lowest energy state, is reestablished by formation and

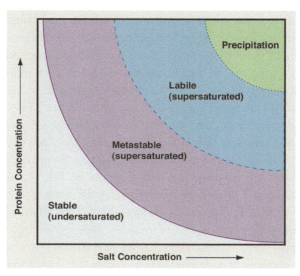

Fig. 1 The phase diagram for the crystallization of macromolecules. The solubility diagram is divided sharply into a region of undersaturation and a region of supersaturation by the line denoting maximum solubility at specific concentrations of a precipitant, which may be salt or a polymer. The line represents the equilibrium between existence of solid phase and free molecule phase. The region of supersaturation is further divided in a more uncertain way into the metastable and labile regions. In the metastable region nuclei will develop into crystals, but no nucleation will occur. In the labile region, both might be expected to occur. The final region, at very high supersaturation is denoted the precipitation region where that result might be most probable. Crystals can only be grown from a supersaturated solution, and creating such a solution supersaturated in the protein of interest is the immediate objective in growing protein crystals. *Illustration courtesy of Hampton Research, Aliso Viejo, CA.*

development of a solid state, which may be crystals, as the saturation limit is reattained. To produce the supersaturated solution, the properties of an undersaturated solution must be modified to reduce the ability of the medium to solubilize the macromolecule (i.e., reduce its chemical activity), or some property of the solute must be altered to reduce their solubility and/or to increase the attraction of one molecule for another.

If no crystals or other solid is present as conditions are changed, then solute will not immediately partition into two phases, and the solution will continue to be supersaturated. The solid state does not develop spontaneously as the saturation limit is exceeded because energy, analogous to the activation energy of a chemical reaction, is required to initiate the second phase, the stable nucleus of a crystal or a precipitate. Thus, a kinetic or energy (or probability) barrier, dependent on the level of supersaturation, characterized for example in Fig. 2, allows conditions to proceed further from equilibrium and further into the zone of supersaturation. Once a stable nucleus appears in a supersaturated solution, however, it will proceed to grow until the system regains equilibrium. So long as non-equilibrium forces prevail and some degree of supersaturation exists to drive events, a crystal will grow or precipitate continue to form.

For the purposes here, large quantities of microcrystals are sought rather than a few large single crystals. Because quantity is dependent on nucleation phenomena, and crystal size on growth, we are less concerned with the latter and more focused on the former. Under conditions of high levels of protein supersaturation nuclei continually appear in great numbers as shown by light scattering experiments (Malkin, Cheung, & McPherson, 1993; Malkin & McPherson, 1993; Malkin & McPherson, 1993a). For the vast majority of the clusters, aggregates, and nascent pre-nuclei, the rate at which molecules leave their surfaces exceeds the rate at which molecules join the ordered aggregates. As a consequence, almost all pre-nuclei dissolve and disappear. If a larger, more improbable nucleus succeeds in exceeding this critical nuclear

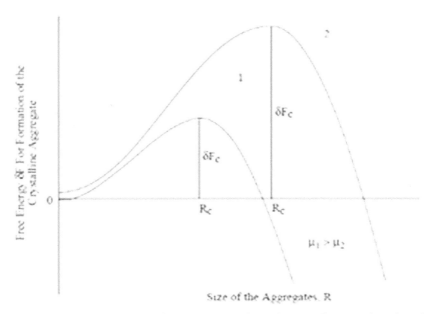

Fig. 2 Dependencies of F, the free energy, on the radius Rc of a critical nucleus for two values of solution supersaturation 1 and 2. Peak 1 corresponds to higher supersaturation and therefore requires a smaller, ordered cluster to create a critical nucleus. Peak 2, at lower supersaturation, demands a larger aggregate size for successful formation of a critical nucleus. *From Malkin and McPherson (1994).*

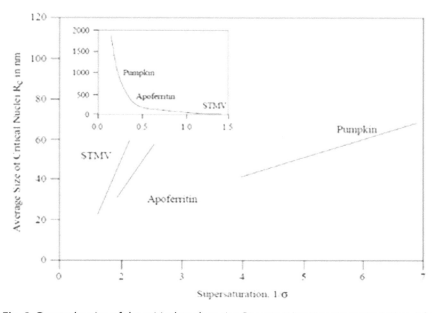

Fig. 3 Dependencies of the critical nuclear size Rc versus inverse supersaturation, 1/σ, for pumpkin globulin, apoferritin and STMV crystallization. The corresponding values of the interfacial energies and molar surface energies are pumpkin globulin, = 6.1109 Jcm^2; apoferritin,= 2.7109 Jcm^2; STMV,= 1.89 Jcm^2. The inset shows the supersaturation dependency of the number of protein molecules or virus particles comprising the critical nucleus. *From Malkin and McPherson (1993).*

barrier, its surface to volume ration lower and the probability of molecule escape increasingly less, then there is a high likelihood that it will become a stable microcrystal. A nucleus that exceeds the probability barrier is termed a critical nucleus (Chernov, 1984; Chernov et al., 1988; Malkin & McPherson, 1993). Its size is dependent on the supersaturation at which it forms (Fig. 3 shows examples), and it is the smallest crystal that can be grown.

6. Matching crystallization methods to the presentation of crystals to the X-ray beam

The particular technique for growing microcrystals has to be matched with the intended means of presenting the crystals to the X-ray beam. Currently there appears to be three general approaches to presentation, (1) projecting or streaming the microcrystals from a reservoir (such as with ink jet technology) into the path of the X-ray beam, (2) displaying the microcrystals on a thin film or other surface, possibly marked with a coordinate grid, and (3) in situ exposure to the X-ray beam (Bingel-Erlenmeyer et al., 2011; McPherson, 2000; Pinker et al., 2013) with, therefore, many possibilities for the crystal growth strategy and mechanics. For TRSC, XFELS has been the principal X-ray source (Schlichting & Miao, 2012; Vartanyants & Yefanov, 2013) and the means of presentation has been projecting or streaming the crystals before the X-ray beam. Synchrotron sources are, however, in use (Hough & Owen, 2004) and for those applications displaying the crystals on surfaces has been a preferred approach.

The idea of disrupting large protein crystals using sonication or hard "seed bead" approaches are unlikely to be productive. Again, they tend to result in gummy masses with few actual if any microcrystals. They are principally useful for creating seeds for the growth of large crystals (Bergfors, 2003). An idea that has seen some application is to invoke some chemical or physical effect that would halt the further growth of crystals after the initiation of crystallization. An example is the infusion of glutaraldehyde, a common protein crosslinking agent (Falkner et al., 2005). Even the slightest exposure to glutaraldehyde, or some similar agent, produces surface crosslinking and the termination of growth. Introduction of PEG or MPD might be considered as well. There are likely other means to accomplish the objective that might be based on pH or temperature shifts.

A further consideration is whether the crystals are to be flash frozen, and data collected at cryogenic temperature, or simply exposed to the X-ray beam at room temperature. For XFELS, where the method of presentation is by streaming, freezing presents some significant technical challenges. If the microcrystals are presented on a two-dimensional grid or support, then the problem may be substantially less. Crystals can be spread, or grown, in a thin film of mother liquor that, if a cryoprotectant is included, can be flash frozen as conventionally done for larger protein crystals. Alternatively, crystals on a surface often can be exposed to X-rays at room temperature. This has the advantage that the experiments can be conducted under more physiological conditions. In choosing the manner of presentation it should be born in mind that, whatever the means, crystal nuclei will always exhibit a strong tendency to form on surfaces, whatever surfaces are offered by the crystallization apparatus or cell. Thus, in situ or two-dimensional arrays may be the best choices for crystal presentation to the X-ray beam. Preferred nucleation, heterogeneous nucleation, on surfaces has firm support on both kinetic and energetic bases (Chernov, 1984; Rosenberger & Meehan, 1988; Rosenberger, Vekilov, Muschol, & Thomas, 1996).

7. Creating a state of supersaturation

Table 1 is a compilation of the approaches from which strategies might be developed for the initial crystallization of a protein. Indeed there may be others, the limit is only a function of the imagination and inginuity of the investigator. The details of these various approaches have been described in detail numerous times elsewhere (Bergfors, 1999; Ducruix & Giége, 1992; McPherson, 1982; McPherson, 1999; McPherson, Cudney, & Patel, 2003). It is probably sufficient to say that if a protein has a propensity to crystallize readily, it can probably be accomplished by variation of precipitant type, precipitant concentration, pH, and to a lesser extent temperature, but with all due consideration of the biochemical properties and eccentricities of the protein under investigation. The purity of the protein and the homogeneity of its physical and chemical state must always be at the forefront (Dale, Oefner, & D'Arcy, 2003; McPherson & Gavira, 2014).

In practice, one begins (with the exception of the batch method, see below) with a solution, a potential mother liquor, which contains some concentration of the protein below its solubility limit, or at its solubility maximum. The objective is then to alter conditions so that the solubility of the protein in the sample is significantly reduced, thereby rendering the solution supersaturated. This may be done through several approaches, (a) altering the protein itself (e.g. by change of pH which alters the ionization state of surface amino acid residues), (b) by altering the chemical activity of the water (e.g. by addition of salt), (c) by altering the degree of attraction of one protein molecule for another (e.g. change of pH, addition of bridging ions),

Table 1 Methods for creating supersaturation.

1. Direct mixing of protein and precipitant solutions to immediately create a supersaturated condition (batch method)
2. Alter the temperature
3. Add salt (increase ionic strength), salting out
4. Remove salt (decrease ionic strength), salting in
5. Alter pH through liquid or vapor phase
6. Add a ligand that changes the solubility of the macromolecule
7. Alteration of the dielectric of the medium (by addition of organic solvents)
8. Evaporation
9. Addition of a polymer that produces volume exclusion
10. Addition of a cross-bridging agent that promotes lattice interactions
11. Concentration of the macromolecule by membrane
12. Removal of a solubilizing agent (chaotropes)

or (d) altering the nature of the interactions between the protein molecules and the solvent (e.g. addition of polymers or ions).

With reference to the phase diagram for crystallization in Fig. 1, it can be surmised that the optimal location in the diagram for the growth of large, single crystals is slightly above the metastable region (which is, unfortunately, always ambiguous). There, whenever a stable nucleus does appear (exceeding the critical nuclear size), its growth reduces the supersaturation and discourages further nucleation (see Figs. 2 and 3), yielding only one, or a few large crystals, rather than many small crystals. It also leads to more measured and stable crystal growth (McPherson, 1999).

Conversely, reason suggests, and experience teaches us that if one's goal is microcrystals, then the crystallization had best take place at the very highest levels of supersaturation possible. That is, in the high reaches of the phase diagram. Creation of a highly supersaturated solution from which to grow microcrystals, however, also creates an important problem, and that is the formation of precipitates, which are disordered aggregates and chains of molecules, rather than ordered crystal nuclei. The competition between precipitate and crystal nuclei often shapes the strategy employed. Microcrystals and precipitate, it has been shown by electron microscopy (Barnes et al., 2016), can, however, coexist. Even a mixture of precipitate and microcrystals could conceivably allow X-ray diffraction data to be obtained.

Table 2 Methods for promoting a solubility minimum.

1. Bulk crystallization
2. Batch method in vials
3. Micro batch under oil
4. Controlled evaporation
5. Bulk dialysis
6. Concentration dialysis
7. Micro dialysis
8. Free-interface diffusion[†]
9. Counter-diffusion in capillaries[†]
10. Liquid bridge[†]
11. Vapor diffusion on plates (sitting drop)
12. Vapor diffusion in hanging drops
13. Sequential extraction
14. pH-induced crystallization
15. Temperature-induced crystallization
16. Crystallization by effector addition

[†]8, 9 and 10 are variations on liquid–liquid diffusion.

For the growth of large, single crystals useful for conventional X-ray crystallography, the approaches listed in Table 2 have, in different cases, proven successful. In Fig. 4a and Fig. 4b are shown, as examples, two widely popular methods among crystallographers based on the principle of vapor diffusion. The first, in Fig. 4a, is known as the sitting drop method, and the second, shown in Fig. 4b, as the hanging drop method. These may not, however, be advisable if microcrystals are the desired outcome. Vapor diffusion approaches are designed to very slowly produce a state of limited supersaturation so to target the growth of large, single crystals. Larger crystals almost invariably nucleate and grow on surfaces of the container or apparatus. They are designed not to intrude on high levels of supersaturation that favor the production microcrystals.

While vapor diffusion is used principally to grow large, single crystals, it has value in identifying conditions that will yield microcrystals as well. Virtually all of the vapor diffusion-based screening kits on the market today provide an array of potential crystallization solutions that are strongly biased

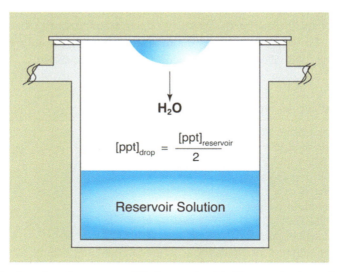

Fig. 4a The hanging drop vapor diffusion method is illustrated schematically here. The components of the drop and reservoir, and the physical equilibration process are the same here as for the sitting drop. The exception is that the protein drop is suspended from a cover slip over the reservoir rather than resting on a surface. Plasticware for carrying out both sitting and hanging drop vapor diffusion are widely, and commercially available in numerous formats. *Illustration courtesy of Hampton Research, Aliso Viejo, CA.*

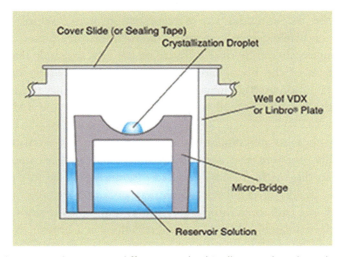

Fig. 4b The sitting drop vapor diffusion method is illustrated in this schematic diagram. The drop on the elevated platform, which is commonly 2 ul to 10 ul, consists of half stock protein solution and half the reservoir solution, which contains some concentration of a salt or polymer precipitant. About 0.5 ml of the reservoir solution is added to the bottom of the cell before sealing. By water equilibration through the vapor phase the drop ultimately approaches the reservoir in osmolarity both raising the concentration of the precipitant in the drop and increasing the protein concentration there. *Illustration courtesy of Hampton Research, Aliso Viejo, CA.*

to high precipitant concentrations, for example 30% PEG 3350. As a consequence, many vapor diffusion experiments are essentially batch experiments. The combination of the protein solution with the precipitant solution in a single trial may immediately produce a high degree of supersaturation and subsequent critical nuclei. Thus, the results of an initial series, or matrix of trials in a screening kit usually yield microcrystals. A vapor diffusion screening kit is therefore of considerable value in directing the investigator to conditions that will be microcrystal productive.

If the objective is to grow microcrystals that are not attached to any surface and are free in solution for projection into the X-ray beam, then other methods, those less used by conventional crystallographers, may be preferred. These include, most prominently the batch method, dialysis, liquid – liquid diffusion, and, potentially, a variation known as counter-diffusion. With all of these methods mother liquors can be created that quickly produce high supersaturation and high rates of nucleation. Batch and dialysis approaches are among the oldest that have been used, and among the most convenient.

With the batch method, a concentrated protein solution is simply combined and mixed rapidly with a salt, or other precipitant containing solution, micro filtered in some applications, and left to stand. The resultant concentration of the precipitant in the mixture is at once sufficient to render the solution highly supersaturated. The most familiar example of the batch method is the crystallization of hen egg lysozyme (Abraham & Robinson, 1937).

Protein chemists are familiar with dialysis as a means to exchange solutes and solvents, but it is also an efficient means to produce supersaturated solutions. In this method a concentrated protein is encased in a permeable membrane, usually of cellulose, which may be a dialysis tube or bag, and submerged in a bath containing a solution at high concentration of salt or other precipitant such as MPD. Although dialysis in a button, as shown in Fig. 5, is usually employed for growing larger crystals for X-ray analysis, it can be used as well to prepare microcrystals. It is found with internal volumes of 10 μl up to 100 μl. Lysozyme, for example, can be obtained as microcrystals by dialysis at 4 °C against 6% NaCl buffered with acetate at pH 4.3. It is surprising how many enzymes can be crystallized, almost always as microcrystals, by this simple technique.

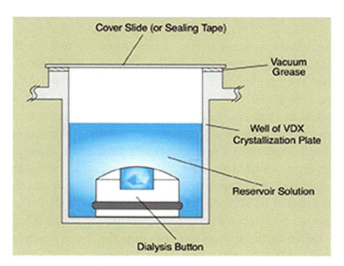

Fig. 5 The use of micro dialysis buttons to dialyze small volumes of protein solution against a precipitating solution is illustrated. The protein solution volumes may be from 10 to 50 ml. The buttons are commercially available. *Illustration courtesy of Hampton Research, Aliso Viejo, CA.*

Even PEG precipitants can be accommodated for dialysis as the molecular weight permeability limit can be chosen by using the proper membrane. The method works best however with salt precipitants and small molecular weight precipitants such as MPD. An advantage of using membranous dialysis bags is that their surfaces do indeed promote nucleation, generally extensively, so that, when successful, the inside of the bag is coated with microcrystals. Slight pulsing of the pliable membrane is generally sufficient to release the microcrystals into the bulk solution inside the bag and without damage to the crystals. Dislodging crystals from surfaces otherwise usually is accompanied by damage to the crystals. Even if the objective is to spread microcrystals over some surface or grid for data collection, it is necessary to have a liquid solution of the microcrystals. Batch and dialysis methods provide those media.

There are proteins, many of them enzymes, that may be easily grown simply by dialysis against distilled water, or a solution at very low ionic strength. The earliest proteins crystallized, as microcrystals, were globulins from squash and Brazil nuts (McPherson, 1999). JB Sumner (Sumner, 1919) obtained microcrystals of concanavalins A and B by water dialysis. Among enzymes, beef liver catalase can be obtained in copious amounts of microcrystals simply by dialysis against cold water, and B. subtilis alpha amylase will do the same.

Approaches that favor the formation of larger crystals can be simply modified to induce the formation of microcrystals. While combination of the NaCl/acetate solution with 50 mg/ml hen egg lysozyme is commonly used to grow large single crystals of lysozyme, if the combined solutions are agitated or vigorously stirred during the crystallization period, microcrystals will result. The same approach can be applied to other proteins that have a propensity to form larger crystals. The author has had much success with other easily crystallized proteins such as rabbit muscle creatine kinase, horse hemoglobin, fungal lipase, and xylanase.

An old technique for growing large single crystals of proteins is the free interface diffusion, or liquid – liquid diffusion approach, which is illustrated schematically in Fig. 6. This can be used with almost any volumes of sample, including large volumes suitable for dialysis. It has been adapted to small volumes in capillaries as well. With this method, a precipitant solution, which may be at a high concentration of salt, PEG, or other polymeric precipitant, is loaded into a tube or capillary, and a concentrated protein solution is carefully layered atop. A distinct, sharp interface is the desired result. By simple diffusion the precipitant intrudes on the protein

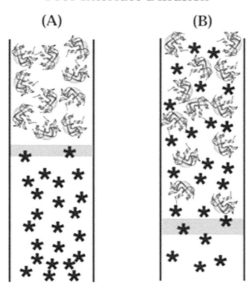

Fig. 6 The process of free interface diffusion to effect crystallization is illustrated here. A protein solution is layered atop a precipitant solution in a narrow bore tube or capillary. Diffusion across the interface, principally of the precipitant, induces nucleation and growth. *Illustration from McPherson (1999).*

phase and forms a gradient. By virtue of its larger size, the protein molecules barely penetrate the precipitant phase. Diffusion normally results in microcrystals at the interface with progressively larger crystal at the upper reaches of the gradient. It might be noted that this technique has seen particular application to the growth of protein crystals in microgravity, and in microfluidic devices.

A relatively novel technique for crystallization has been developed (García-Ruiz, 2003; Ng, Gavira, & Garcia-Ruiz, 2003) that is essentially an extension of the standard liquid – liquid diffusion method (Fig. 7a), and this has been termed "counter – diffusion" (Gavira, Toh, Lopez-Jaramillo, Garcia-Ruiz, & Ng, 2002; Otálora, Gavira, Ng, & García-Ruiz, 2009). With this technique (Fig. 7b) a gel, into which one end of a capillary containing the protein solution is pressed, is then impregnated with a precipitating solution. With time, the precipitant diffuses up the capillary so that a gradient is ultimately established. Thus the protein is exposed to a continuum of precipitant concentration. Because of the interplay of

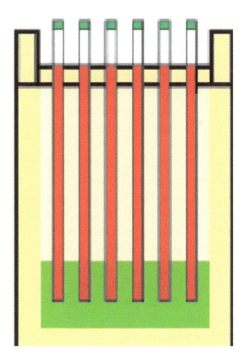

Fig. 7a Diagram illustrating the counter diffusion method for growing protein crystals. Here the protein solution is shown in red, the gel saturated with the precipitant solution is in green. The capillaries are sealed at their distal end but open where they enter the gel. By diffusion of precipitant up the length of the capillary a concentration.

Fig. 7b Photo of the result of crystallizing lysozyme in a counter – diffusion tube. Note the masses of microcrystals appearing near the interface where supersaturation is highest. *Photo courtesy of Hampton Research, Aliso Viejo, CA.*

precipitant diffusion and crystallization, the dynamics of the process in the capillary is more complex than might be thought, but this enhances the probability of nucleation. When successful, microcrystals may be observed where the precipitant concentration is highest, near the surface of the gel, and large crystals near the distal end of the capillary. The method has now been used to obtain crystals for X-ray diffraction for many proteins at both room and cryogenic temperatures (García-Ruiz, 2003; Ng et al., 2003).

Another crystallization method, used in the past for the purification of proteins that appears to have been neglected is known as back – extraction. First described by Jakoby (Jakoby, 1968), the protein is completely precipitated in a test tube by addition of ammonium sulfate (or other salt) and centrifuged to obtain a pellet. At 4 °C, the pellet is then exposed to small volumes of precipitant solution at incrementally decreasing salt concentrations. The solid protein gradually dissolves in these serial extractions. The individual extracted samples are then placed at room temperature. Because most proteins are slightly more soluble at 4 °C than 22 °C, their solubility in the extracts is reduced and the samples become supersaturated in protein. The advantage of this approach for our purposes is that when crystals appear, they are invariably microcrystals, and their sizes vary with the salt concentrations of the extracted samples.

Surfaces offer distinct advantages if they can be effectively utilized because they promote nucleation, and usually lead to smaller crystal size. They reduce the critical nuclear size, and they enhance crystal stability. In addition, as noted previously, they provide a ready means of presenting microcrystals to the X-ray beam. One of the earliest reports on the possibilities of in situ X-ray data collection (McPherson, 2000) took advantage of the growth of protein crystals on wands, or pallets made of thin plastic films. Surfaces, from minerals (McPherson & Schlichta, 1989), synthetic materials (Chayen, Saridakis, El-Bahar, & Nemirovsky, 2001; Chayen, Saridakis, & Sear, 2006), grapho epitaxy (Givargizov et al., 1991), and natural epitaxy have been demonstrated to enhance and promote nucleation and growth.

In addition to crystal size and perfection, uniformity is also a consideration. Little attention has been given this requirement except for those interested in producing microcrystals of pharmaceutically important protein drugs. These include slow-release formulations of injectable insulin (Owens, 2011) and alpha-amylase (Pechenov, Shenoy, Yang, Basu, & Margolin, 2004) and the ever increasing array of crystalline monoclonal antibodies. Data is, for good reason, incomplete and slow in coming, but evidence has mounted that nucleation and growth in microgravity, in bulk solution, supports crystal uniformity (Reichert et al., 2019). The expectation from physical considerations is that this should be true, and the experiments so far seem to bear this out.

8. Precipitants

Table 3 Protein crystals have been grown from solutions containing a variety of precipitants, but the precipitating agents can be roughly divided into five categories. Examples of each appear in Table 3. These are (1) concentrated salt solutions [see the Hofmeister series] (Collins, 2004; Collins & Washabaugh, 1985; Hofmeister, 1888; Hofmeister, 1890) where the salt is most commonly ammonium sulfate, phosphate, or sodium chloride, (2) Polyethyleneglycol (PEG) containing solutions, generally at concentrations ranging from 2 % to 30 %, and where the PEG molecules have a wide range of molecular weights from 400 Da to 20,000 Da (Ingham, 1990; McPherson, 1976), (3) solutions containing organic solvents (generally at cold temperatures), such as ethanol or butanol, (4) curious small molecules such as 2,5 pentane diol (MPD) (Anand, Pal, & Hilgenfeld, 2002) or very low molecular weight PEGs, and (5) distilled water or solutions at very low ionic strength (Harris, Skaletsky, & McPherson, 1995; Osborne, 1892; Thomas B Osborne, 1895; Ritthausen, 1880, 1881; Sumner, 1919). No attempt will be made here to present details of the various chemical and physical mechanisms by which these precipitants exert their function, as these have been extensively described

Table 3 Precipitants used in macromolecular crystallization.

Salts

Ammonium sulfate Ammonium phosphate Lithium sulfate Lithium chloride
Sodium citrate Ammonium citrate Sodium phosphate Sodium chloride
Potassium chloride Sodium acetate Ammonium acetate Magnesium sulfate
Magnesium chloride Calcium chloride Sodium formate Sodium tartrate
Cadmium sulfate Sodium succinate Sodium malonate

Volatile organic solvents

Ethanol
Propanol Isopropanol Dioxane
Acetone Isobutanol n-Butanol tert-Butanol Acetonitrile Dimethyl sulfoxide
1,3-Butyrolactone

Polymers

Polyethylene glycol 1000, 3350, 6000, 8000, 20,000 Jeffamine T, Jeffamine M
Polyethylene glycol monomethyl ester Polyethylene glycol monostearate

Nonvolatile organic solvents

2-Methyl-2,4-pentanediol (MPD) 2,5-Hexanediol
1,3-Propanediol
Polyethylene glycol 400 Jeffamine 400

Distilled water or very low ionic strength

and discussed elsewhere (Bergfors, 1999; Cohn & Edsall, 1943; Cohn & Ferry, 1943; Cohn et al., 1947; Ducruix & Giége, 1992; McPherson, 1982; McPherson, 1999; McPherson & Gavira, 2014). This does not, however, preclude a discussion of the advantages and manifestations of the precipitants with regard the problem at hand.

The first issue is that the protein molecule of interest should crystallize reproducibly. Generally, the protein will crystallize from only a few very specific precipitant solutions, over a narrow range of pH and temperature, and frequently only in the presence of certain ions or small molecules, such as inhibitors or cofactors. Thus, the crystallographers do not usually have the luxury of choosing the medium in which they have to work, the protein does it for him/her.

A conclusion that appears to be widely shared is that protein crystals of any size grown from PEG solutions are significantly more stable than crystals grown from salt solutions. Crystals grown from salts tend to dissolve with small decreases in the salt concentration or from minor changes in pH or temperature. Generally, they are more fragile and more subject to physical damage. Crystals grown from organic solvents are notoriously difficult to work with, and the volatility of many organic solvents inspire instability. Crystals grown from PEG, on the other hand, are often, in fact, difficult to dissolve even when placed in pure water or solutions containing no PEG. Crystals from PEG or other polymeric precipitants (Patel, Cudney, & McPherson, 1995) should, therefore, be the first choice. Similar conclusions might equally be drawn for crystals grown from MPD and like precipitants.

A further advantage of PEG is that, generally, crystals grown from salts or organic solvent mother liquors can be exchanged into PEG solutions. Often MPD will act to stabilize protein crystals as well, and both PEGs and MPD act in concert with or compliment cryoprotectant solutions used in cryo-crystallography. Exchange must be done in steps and with care, but it is usually successful and bestows upon crystals otherwise grown the stability advantages of PEG or MPD. Because microcrystals flash freeze far more readily, and without damage, than do large crystals, PEG growth or exchange may remove some technical problems that arise in handling and presentation for data collection.

As noted above, growing crystals by simple, extended dialysis against distilled water, particularly at cold temperature is one of the earliest practices of protein biochemists to purify proteins by crystallization. Fortunately, for the purposes here, the procedure usually produces microcrystals of unusual uniformity. Though little used currently, production of microcrystals by dialysis against water or low ionic strength solution offers the opportunity of exchange into PEG or MPD solutions which confer enhanced stability.

9. Visualization

There are a number of ways that microcrystals can be detected or visualized. The oldest and surest method is to simply agitate or swirl the putative crystalline solution in front of a light and, if microcrystals are present, observe a silky sheen produced by the Tyndal effect (a light scattering phenomenon) (Tyndall, 1896). Using more sophisticated instrumentation, elastic and inelastic light scattering (Asherie, 2004; Malkin & McPherson, 1993; Malkin & McPherson, 1993; Malkin & McPherson, 1994) can be used to detect, not only stable crystal nuclei, but even transient nuclei less than critical nuclear size. For observing two-dimensional nucleation on growing crystal surfaces (which give estimates of what critical nuclear sizes might be for a specific protein crystal) atomic force microscopy is the technique of choice (Malkin, Kuznetsov, et al., 1995; Malkin, Land, Kuznetsov, McPherson, & DeYoreo, 1995; McPherson et al., 2000; McPherson, Malkin, Kuznetsov, & Plomp, 2001). More recently it has been shown that conventional electron microscopy can be effective in visualizing microcrystals even in the presence of other solid states such as precipitate (Barnes et al., 2016).

References

Abraham, E. P., & Robinson, R. (1937). Crystallization of lysozyme. *Nature, 140,* 24–28.
Anand, K., Pal, D., & Hilgenfeld, R. (2002). An overview on 2-methyl-2,4-pentanediol in crystallization and in crystals of biological macromolecules. *Acta Crystallographica. Section D, Biological Crystallography, 58*(Pt 10 Pt 1), 1722–1728. https://doi.org/10.1107/s0907444902014610.
Asherie, N. (2004). Protein crystallization and phase diagrams. *Methods (San Diego, Calif.), 34*(3), 266–272. https://doi.org/10.1016/j.ymeth.2004.03.028.
Bailey, K. (1942). The growth of large protein crystals. *Transactions of the Faraday Society, 38,* 186.
Barnes, C. O., Kovaleva, E. G., Fu, X., Stevenson, H. P., Brewster, A. S., DePonte, D. P., ... Calero, G. (2016). Assessment of microcrystal quality by transmission electron microscopy for efficient serial femtosecond crystallography. *Archives of Biochemistry and Biophysics, 602,* 61–68. https://doi.org/10.1016/j.abb.2016.02.011.
Bergfors, T. (2003). Seeds to crystals. *Journal of Structural Biology, 142*(1), 66–76.
Bergfors, T. M. (1999). *Protein crystallization: Techniques, strategies and tips. Protein crystallization: Techniques, strategies and tips.* La Jolla, CA: International University Line.
Bernal, J. D., & Crowfoot, D. (1934). X-ray photographs of crystalline pepsin. *Nature, 133,* 794.

Bingel-Erlenmeyer, R., Olieric, V., Grimshaw, J., Gabadinho, J., Wang, X., Ebner, S. G., ... Schulze-Briese, C. (2011). SLS crystallization platform at beamline X06DA-A fully automated pipeline enabling in situ X-ray diffraction screening. *Crystal Growth and Design, 11*(4), 916–923. https://doi.org/10.1021/cg101375j.

Chayen, N. E., Saridakis, E., El-Bahar, R., & Nemirovsky, Y. (2001). Porous silicon: An effective nucleation-inducing material for protein crystallization. *Journal of Molecular Biology, 312*(4), 591–595. https://doi.org/10.1006/jmbi.2001.4995.

Chayen, N. E., Saridakis, E., & Sear, R. P. (2006). Experiment and theory for heterogeneous nucleation of protein crystals in a porous medium. *Proceedings of the National Academy of Sciences of the United States of America, 103*(3), 597–601.

Chernov, A. A. (1984). *Modern crystallography: Crystal growth. Modern crystallography: Crystal growth Vol. III.* Berlin: Springer-Verlag.

Chernov, A. A. (2003). Protein crystals and their growth. *Journal of Structural Biology, 142*, 3–21.

Chernov, A. A., & Komatsu, H. (1995). Principles of crystal growth in protein crystallization. In J. P. V. E. A. O. S. L. Bruinsma (Ed.). *Science and technology of crystal growth* (pp. 67). Dordrecht, The Netherlands: Kluwer.

Chernov, A. A., Rashkovich, L. N., Smolískii, I. L., Kuznetsov, Y. G., Mkrtchyan, A. A., & Malkin, A. I. (1988). Growth of crystals. In E. I. Givargizov & S. A. Grinberg (Eds.), (Vol. 15, pp. 43–91). New York.

Cohn, E. J., & Edsall, J. T. (Eds.). (1943). *Proteins, amino acids and peptides as ions and dipolar ions.* Princeton, NJ: Van Nostrand-Reinhold.

Cohn, E. J., & Ferry, J. D. (Eds.). (1943). *The interactions of proteins with ions and dipolar ions.* New Jersey: Van Nostrand-Rheinhold.

Cohn, E. J., Hughes, W. L., Jr., & Weare, J. H. (1947). Preparation and properties of serum and plasma proteins; crystallization of serum albumins from ethanol water mixtures. *Journal of the American Chemical Society, 69*(7), 1753–1761.

Collins, K. D. (2004). Ions from the Hofmeister series and osmolytes: Effects on proteins in solution and in the crystallization process. *Methods (San Diego, Calif.), 34*(3), 300–311. https://doi.org/10.1016/j.ymeth.2004.03.021.

Collins, K. D., & Washabaugh, M. W. (1985). The Hofmeister effect and the behaviour of water at interfaces. *Quarterly Reviews of Biophysics, 18*(4), 323–422.

Crowfoot, D. (1935). X-ray single crystal photographs of insulin. *Nature, 135*, 591–592.

Dale, G. E., Oefner, C., & D'Arcy, A. (2003). The protein as a variable in protein crystallization. *Journal of Structural Biology, 142*(1), 88–97.

De La Mora, E., Coquelle, N., Bury, C., Rosenthal, M., Holton, J., Carmichael, I., ... Weik, M. (2020). Radiation damage and dose limits in serial synchrotron crystallography at cryo- and room temperatures. *Proceedings of the National Academy of Sciences of the United States of America, 117*(8), 4142–4151.

Ducruix, A., & Giége, R. (1992). *Crystallization of nucleic acids and proteins, a practical approach. Crystallization of nucleic acids and proteins, a practical approach.* Oxford: IRL Press.

Durbin, S. D., & Feher, G. (1996). Protein crystallization. *Annual Review of Physical Chemistry, 47*, 171–204.

Falkner, J. C., Al-Somali, A. M., Jamison, J. A., Zhang, J., Adrianse, S. L., Simpson, R. L., ... Colvin, V. L. (2005). Generation of size-controlled, submicrometer protein crystals. *Chemistry of Materials: A Publication of the American Chemical Society, 2005*(17), 10.

Feher, G. (1986). Mechanisms of nucleation and growth of protein crystals. *Journal of Crystal Growth, 76*, 545–546.

Feigelson, R. S. (1988). The relevance of small molecule crystal growth theories and techniques to the growth of biological macromolecules. *Journal of Crystal Growth, 90*, 1–13.

García-Ruiz, J. M. (2003). Counterdiffusion methods for macromolecular crystallization. In *Methods in enzymology* (Vol. 368, pp. 130–154). Academic Press.

Gavira, J. A., Toh, D., Lopez-Jaramillo, J., Garcia-Ruiz, J.-M., & Ng, J. D. (2002). Ab initio crystallographic structure determination of insulin from protein to electron density without crystal handling. *Acta Crystallographica. Section D, Biological Crystallography, 58*(7), 1147–1154. https://doi.org/10.1107/S0907444902006959.

Giege, R. (2013). A historical perspective on protein crystallization from 1840 to the present day. *The FEBS Journal, 280*(24), 6456–6497. https://doi.org/10.1111/febs.12580.

Givargizov, E. I., Kliya, M. O., Melik-Adamyan, V. R., Grebenko, A. I., DeMattei, R. C., & Feigelson, R. S. (1991). Artificial epitaxy (graphoepitaxy) of proteins. *Journal of Crystal Growth, 112*(4), 758–772.

Glaeser, R., Facciotti, M., Walian, P., Shahab, R., Holton, J. M., MacDowell, A., & Padmore, H. (2016). Characterization of conditions required for X-ray diffraction experiments with protein microcrystals. *Biochemical and Biophysical, 602*, 61–68.

Haas, C., & Drenth, J. (1999). Understanding protein crystallization on the basis of the phase diagram. *Journal of Crystal Growth, 196*(2-4), 388–394.

Harris, L. J., Skaletsky, E., & McPherson, A. (1995). Crystallization of intact monoclonal antibodies. *Proteins, 23*(2), 285–289.

Hofmeister, F. (1888). Zur Lehre von der Wirkung der Saltz. *Naunyn-Schmiedebergs Archiv für Experimentelle Pathologie und Pharmakologie, 24*, 247.

Hofmeister, T. (1890). *Hoppe-Seyler's Zeitschrift für Physiologische Chemie, 14*, 165.

Holton, J. M. (2019). Challenge data set for macromolecular multi-microcrystallography. *Acta Crystallographica Section D Structural Biology, 75*(Pt 2), 113–122. https://doi.org/10. 1107/S2059798319001426.

Holton, J. M., & Frankel, K. A. (2010). The minimum crystal size needed for a complete diffraction data set. *Acta Crystallographica. Section D, Biological Crystallography, 66*(Pt 4), 393–408. https://doi.org/10.1107/S0907444910007262.

Hough, M. A., & Owen, R. L. (2004). Serial synchrotron and XFEL, crystallography for studies of metalloprotein catalysis. *Methods (San Diego, Calif.), 34*, 266–272.

Ingham, K. C. (1990). Precipitation of proteins with polyethylene glycol. *Methods in Enzymology, 182*, 301–306.

Jakoby, W. B. (1968). A technique for the crystallization of proteins. *Analytical Biochemistry, 26*(2), 295–298. https://doi.org/10.1016/0003-2697(68)90340-0.

Kuznetsov, Y. G., Malkin, A. J., Greenwood, A., & McPherson, A. (1995). Interferometric studies of growth kinetics and surface morphology in macromolecular crystal growth. Canavalin, thaumatin and turnip yellow mosaic virus. *Journal of Structural Biology, 114*(3), 184–196.

Kuznetsov Yu, G., & McPherson, A. (2019). Penetration of dyes into protein crystals. *Acta Crystallographica Section F, 75*(2), 132–140. https://doi.org/10.1107/S2053230X18018241.

Malkin, A. J., Cheung, J., & McPherson, A. (1993). Crystallization of satellite tobacco mosaic virus I. Nucleation Phenomena. *Journal of Crystal Growth, 126*, 544–554.

Malkin, A. J., Kuznetsov, Y. G., Land, T. A., DeYoreo, J. J., & McPherson, A. (1995). Mechanisms of growth for protein and virus crystals. *Nature Structural & Biology, 2*(11), 956–959.

Malkin, A. J., Kuznetsov, Y. G., & McPherson, A. (1996). Defect structure of macromolecular crystals. *Journal of Structural Biology, 117*, 124–137.

Malkin, A. J., Kuznetsov, Y. G., & McPherson, A. (1997). An in situ investigation of catalase crystallization. *Surface Science, 393*, 95–107.

Malkin, A. J., Kuznetsov Yu, G., Land, T. A., DeYoreo, J. J., & McPherson, A. (1995). Mechanisms of growth for protein and virus crystals. *Nature Structural Biology, 2*(11), 956–959.

Malkin, A. J., Kuznetsov Yu, G., & McPherson, A. (1996). *Journal of Structural Biology, 117*, 124–137.

Malkin, A. J., Land, T. A., Kuznetsov, Y. G., McPherson, A., & DeYoreo, J. J. (1995). Investigation of virus crystal growth mechanisms by in situ atomic force microscopy. *Physical Review Letters, 75*(14), 2778–2781.

Malkin, A. J., & McPherson, A. (1993). Crystallization of pumpkin seed globulin: Growth and dissolution kinetics. *Journal of Crystal Growth, 133,* 29–37.

Malkin, A. J., & McPherson, A. (1993). Light scattering investigations of protein and virus crystal growth: Ferritin, apoferritin and satellite tobacco mosaic virus. *Journal of Crystal Growth, 128,* 1232–1235.

Malkin, A. J., & McPherson, A. (1993a). Crystallization of satellite tobacco mosaic virus II. Postnucleation events. *Journal of Crystal Growth, 126,* 555–564.

Malkin, A. J., & McPherson, A. (1994). Light scattering investigations of nucleation processes and kinetics of crystallization in macromolecular systems. *Acta Crystallographica Section D, 50,* 385–395.

Matthews, B. W. (1968). Solvent content of protein crystals. *Journal of Molecular Biology, 33*(2), 491–497.

McPherson, A. (1976). Crystallization of proteins from polyethylene glycol. *The Journal of Biological Chemistry, 251,* 3600–6303.

McPherson, A. (1982). *The preparation and analysis of protein crystals.* New York: John Wiley and Sons.

McPherson, A. (1989). Macromolecular crystals. *Scientific American, 260,* 62.

McPherson, A. (1991). A brief history of protein crystal growth. *Journal of Crystal Growth, 110,* 1–10.

McPherson, A. (1992). Two approaches to the rapid screening of crystallization conditions. *Journal of Crystal Growth, 122,* 161–167.

McPherson, A. (1999). *Crystallization of biological macromolecules. Crystallization of biological macromolecules.* Cold Spring Harbor, NY: Cold Spring Harbor Laboratory Press.

McPherson, A. (2000). *In situ* X-ray crystallography. *Journal of Applied Crystallography, 33,* 397–400.

McPherson, A., Cudney, R., & Patel, S. (2003). The crystallization of proteins, nucleic acids, and viruses for X-ray diffraction analysis. In S. R. Fahnestock, & A. Steinbuchel (Vol. Eds.), *Biopolymers: Vol. 18,* (pp. 427–468). (16).

McPherson, A., & Gavira, J. A. (2014). Introduction to protein crystallization. *Acta Crystallographica Section F: Structural Biology Communications, 70*(Pt 1), 2–20. https://doi.org/10.1107/S2053230X13033141.

McPherson, A., & Larson, S. B. (2018). Investigation into the binding of dyes within protein crystals. *Acta Crystallographica Section F: Structural Biology Communications, 74*(Pt 9), 593–602. https://doi.org/10.1107/S2053230X18010300.

McPherson, A., Malkin, A., Kuznetsov, Y. G., & Koszelak, S. (1996). Incorporation of impurities into macromolecular crystals. *Journal of Crystal Growth, 168,* 74–92.

McPherson, A., Malkin, A. J., & Kuznetsov, Y. G. (1995). The science of macromolecular crystallization. *Structure (London, England: 1993), 3*(8), 759–768.

McPherson, A., Malkin, A. J., & Kuznetsov, Y. G. (2000). Atomic force microscopy in the study of macromolecular crystal growth. *Annual Review of Biophysics and Biomolecular Structure, 29,* 361–410.

McPherson, A., Malkin, A. J., Kuznetsov, Y. G., & Plomp, M. (2001). Atomic force microscopy applications in macromolecular crystallography. *Acta Crystallographica. Section D, Biological Crystallography, 57*(Pt 8), 1053–1060.

McPherson, A., & Schlichta, P. (1989). The use of heterogeneous and epitaxial nucleants to promote the growth of protein crystals. *Journal of Crystal Growth, 90,* 40–46.

Murakawa, T., Suzuki, M., Arima, T., Sugahara, M., Tanaka, T., Tanaka, R., ... Okajima, T. (2021). Microcrystal preparation for serial femtosecond X-ray crystallography of bacterial copper amine oxidase. *Acta Crystallographica Section F: Structural Biology Communications, 77*(Pt 10), 356–363. https://doi.org/10.1107/S2053230X21008967.

Ng, J. D., Gavira, J. A., & Garcia-Ruiz, J. M. (2003). Protein crystallization by capillary counterdiffusion for applied crystallographic structure determination. *Journal of Structural Biology, 142*(1), 218–231.

Osborne, T. B. (1892). Crystallised vegetable proteids. *American Chemical Journal, 14,* 662–689.

Osborne, T. B. (1895). The proteids of barley. *Journal of the American Chemical Society, 17*(7), 539–567.

Ostwald, W. (1896). *Lehrbuch der Allgemeinen Chemie, 2*(1).

Ostwald, W. (1897). Studien über die Bildung und Umwandlung fester Körper [Studies on the formation and transformation of solid bodies]. *Zeitschrift für Physikalische Chemie, 22,* 289–330.

Otálora, F., Gavira, J. A., Ng, J. D., & García-Ruiz, J. M. (2009). Counterdiffusion methods applied to protein crystallization. *Progress in Biophysics and Molecular Biology, 101*(1-3), 26–37.

Owens, D. R. (2011). Insulin preparations with prolonged effect. *Diabetes Technology & Therapeutics, 13*(Suppl 1), S5–S14. https://doi.org/10.1089/dia.2011.0068.

Patel, S., Cudney, R., & McPherson, A. (1995). Polymeric precipitants for the crystallization of macromolecules. *Biochemistry and Biophysics Research Communications, 207,* 819–828.

Pechenov, S., Shenoy, B., Yang, M. X., Basu, S. K., & Margolin, A. L. (2004). Injectable controlled release formulations incorporating protein crystals. *Journal of Controlled Release: Official Journal of the Controlled Release Society, 96*(1), 149–158. https://doi.org/10.1016/j.jconrel.2004.01.019.

Piazza, R. (1999). Interactions in protein solutions near crystallization: A colloid physics approach. *Journal of Crystal Growth, 196,* 415–423.

Pinker, F., Mathieu, B., Morin, P., Deman, A. L., Chateaux, J. F., Olieric, V., ... Sauter, C. (2013). ChipX: A novel microfluidic chip for counter-diffusion crystallization of biomolecules and in situ crystal analysis at room temperature. *Crystal Growth & Design, 13*(8), 3333–3340. https://doi.org/10.1021/cg301757g.

Reichert, P., Prosise, W., Fischmann, T. O., Scapin, G., Narasimhan, C., Spinale, A., ... Strickland, C. (2019). Pembrolizumab microgravity crystallization experimentation. *NPJ Microgravity, 5,* 28. https://doi.org/10.1038/s41526-019-0090-3.

Ritthausen, H. (1880). Ueber die Eiweisskörper verschiedener Oelsamen. *Flügers Arch.*

Ritthausen, H. (1881). Krystallinische Eiweisskörper aus verschiedenen Oelsamen. *Journal für Praktische Chemie/Chemiker-Zeitung.*

Rosenberger, F. (1986). Inorganic and protein crystal growth—Similarities and differences. *Journal of Crystal Growth, 76,* 618.

Rosenberger, F., & Meehan, E. J. (1988). Control of nucleation and growth in protein crystal growth. *Journal of Crystal Growth, 90,* 74–78.

Rosenberger, F., Vekilov, P. G., Muschol, M., & Thomas, B. R. (1996). Nucleation and crystallization of globular proteins—What we know and what is missing. *Journal of Crystal Growth, 168,* 1–27.

Schlichting, I., & Miao, J. (2012). Emerging opportunities in structural biology with X-ray free electron lasers. *Current Opinion in Structural Biology, 22*(5), 613–626.

Schulz, E. C., Yorke, B. A., Pearson, A. R., & Mehrabi, P. (2022). Best practices for time-resolved serial synchrotron crystallography. *Acta Crystallographica Section D Structural Biology, 78*(Pt 1), 14–29. https://doi.org/10.1107/S2059798321011621.

Sumner, J. B. (1919). The globulins of the jack bean, Canavalia ensiformis. *Journal of Biological Chemistry, 37*(1), 137–142.

Ten Wolde, R., & Frenkel, D. (1997). Enhancement of protein crystal nucleation by critical density fluctuations. *Science (New York, N. Y.), 277,* 1975–1978.

Tyndall, J. (1896). *The glaciers of the alps.* London, UK: Longman, Green and Co.

Vartanyants, I. A., & Yefanov, O. M. (2013). Coherent X-ray diffraction imaging of nanostructures. *arXiv preprint arXiv:*1304.5335.

CHAPTER TWO

Use of fixed targets for serial crystallography

Sofia Jaho[a,b,*], Danny Axford[a,b], Do-Heon Gu[a,b], Michael A. Hough[a,b], and Robin L. Owen[a,b,*]

[a]Diamond Light Source Ltd, Harwell Science and Innovation Campus, Didcot, Oxfordshire, United Kingdom
[b]Research Complex at Harwell, Harwell Science and Innovation Campus, Didcot, Oxfordshire, United Kingdom
*Corresponding authors. e-mail address: sofia.jaho@diamond.ac.uk; robin.owen@diamond.ac.uk

Contents

1. Introduction	30
2. Common aspects of SSX and SFX: optimising the experiment	32
2.1 Preparing for the experiment	32
2.2 Moving from standard MX to serial experiments	33
2.3 Sample preparation	34
2.4 Anaerobic fixed target SSX and SFX	37
2.5 Fixed target SSX beamline hardware: I24	38
2.6 Fixed target SFX beamline hardware: SACLA	40
3. Data collection	40
3.1 Setup and data collection strategies	40
3.2 Fast-feedback from data acquisition	41
3.3 Dose	42
3.4 Choice of X-ray energy	44
4. Record keeping, log and beamtime logistics – best practice	46
4.1 Iterative optimisation of experiment design	48
4.2 Unit cell as a metric for reaction progression: contrasting examples of crystallin and myoglobin	48
4.3 Exploring radiation damage and X-ray driven reactivity using fixed-target SSX	51
5. Summary	53
Acknowledgements	53
References	53

Abstract

In serial crystallography, large numbers of microcrystals are sequentially delivered to an X-ray beam and a diffraction pattern is obtained from each crystal. This serial approach was developed primarily for X-ray Free Electron Lasers (XFELs) where crystals are destroyed by the beam but is increasingly used in synchrotron experiments. The combination of XFEL and synchrotron-based serial crystallography enables time-resolved experiments over an extremely wide range of time domains - from femtoseconds to seconds - and allows intact or pristine structures free of the effects of radiation damage to be obtained. Several approaches have been developed for sample delivery with varying levels of sample efficiency and ease of use. In the fixed target approach, microcrystals are loaded onto a solid support which is then rastered through the X-ray beam. The key advantages of fixed targets are that every crystal loaded can be used for data collection, and that precise control of when crystals are moved into the beam allows for time-resolved experiments over a very wide range of time domains as well as multi-shot experiments characterising the effects of the X-ray beam on the sample. We describe the application of fixed targets for serial crystallography as implemented at beamline I24 at Diamond Light Source and at the SACLA XFEL. We discuss methodologies for time-resolved serial crystallography in fixed targets and describe best practices for obtaining high-quality structures covering sample preparation, data collection strategies and data analysis pipelines.

1. Introduction

Serial crystallography conducted at XFEL (Serial Femtosecond Crystallography [SFX]) or synchrotron (Serial Synchrotron Crystallography [SSX]) beamlines has revolutionised structural biology enabling room temperature structure determination from macromolecules with minimal to no radiation damage. In the serial approach, a single diffraction pattern is obtained from each crystal with the data subsequently merged from many crystals to obtain complete, good quality datasets. When coupled with appropriate reaction initiation approaches, serial crystallography also enables dynamic, time-resolved structural biology helping to illustrate enzymatic mechanisms, ultrafast conformational changes in response to light or signalling mechanisms (Barends et al., 2022; Hough and Owen, 2021; Orville, 2020; Pearson and Mehrabi, 2020). Serial approaches require large numbers of microcrystals to be delivered efficiently to the X-ray beam. This requirement has led to the development of various sample delivery methods including gas dynamic virtual nozzles injectors (GDVN), high viscosity extruders, fixed targets, microfluidic delivery, droplet on demand tape drive systems amongst others. Each sample delivery method is associated with different sample consumption and hit rates, different accessible timescales for time–resolved approaches and vary in their applicability to different reaction triggering methods. A full discussion of sample delivery approaches is beyond the scope of this chapter but readers are directed to relevant recent reviews (Cheng, 2020; Grünbein and Nass Kovacs, 2019; Orville, 2020; Zhao et al., 2019).

Here we describe the application of silicon fixed targets for serial data collection at Diamond Light Source and XFEL beamlines. These two types of X-ray facilities each have different strengths and weaknesses and can be considered complementary sources. XFEL experiments enable diffraction data to be recorded in tens of femtoseconds, allowing for the characterisation of ultrafast time-resolved processes and effectively outrun manifestations of X-ray induced radiation damage. XFEL experiments are, however, infrequent and often require a dedicated experimental setup to be constructed by the users or beamline staff at the start of their experiment. Additionally, XFEL data analysis can be challenging due to variability in the intensity and wavelength of the pulses and the poorly defined experimental geometry of the end station. In contrast, synchrotron experiments can be regularly conducted allowing for fast iteration of sample preparation, reaction initiation and data collection parameters. The experiment geometry of synchrotron beamlines is typically well-characterised and can often be experimentally verified and refined using conventional rotation data collection if required. SSX experiments also allow for the characterisation of dose-dependent processes and are very well-suited for probing longer timescales in time-resolved experiments, typically in the millisecond to second range. However, while X-ray doses in SSX are very low compared to those in conventional macromolecular crystallography (because each crystal is exposed only once), radiation damage from the beam cannot be entirely avoided.

The fixed target approach is compatible with both light-activated and substrate mixing- reaction initiation approaches and is highly sample efficient (Horrell et al., 2021; Mehrabi et al., 2020, 2019; Owen et al., 2017; Schulz et al., 2018; Sherrell et al., 2015). Moreover, the precise positioning of microcrystals on the solid support opens new possibilities for measuring multiple diffraction patterns from the same microcrystal, which is not possible using the majority of other sample delivery methods. Of crucial importance is the ability to compare data structures from XFEL and synchrotron experiments. In the approaches we describe here, sample delivery and reaction initiation for SFX and SSX experiments are near identical, allowing different aspects of synchrotron and XFEL data collection to be exploited. This consistency ensures that structures from both sources can be seamlessly compared minimising concerns that observed changes could be artefacts arising from differing sample preparation, delivery methods or environments.

2. Common aspects of SSX and SFX: optimising the experiment

2.1 Preparing for the experiment

Producing crystals of high diffraction quality is considered one of the main barriers in structural biology. Crystal harvesting, manual handling and cryo-cooling in conventional single-crystal macromolecular X-ray crystallography (MX) under cryogenic conditions can damage the fragile protein crystals and sometimes can hinder conformational changes (Fraser et al., 2011). Collecting diffraction data at room temperature is considered the best strategy for studying dynamics and performing time-resolved experiments. *In situ* experiments can be performed either in SBS standard 96-well plates at dedicated beamlines (Sandy et al., 2024) or using microfluidic devices (Gu et al., 2023). However, moving from single-crystal or multi-crystal data collection with rotation to serial experiments poses an additional hurdle in sample preparation which is the necessity of moving from the most commonly used crystallisation method of vapour diffusion (hanging or sitting drop) to batch crystallisation. Fixed targets are well suited for crystals with maximum dimensions in the range 10–50 μm. While crystallisation methods to produce microcrystals are outside the scope of this article and have been reviewed elsewhere (Beale et al., 2019; Shoeman et al., 2023), we will comment on some particularly relevant aspects here.

Fixed targets, in particular, offer the opportunity for the majority of produced crystals to be used thereby achieving very high sample efficiency. They are well suited for projects where crystals of a suitable size have been grown within vapour diffusion sitting or hanging drops. These crystals may be pooled from several crystallisation drops, typically with volumes ranging from 200 nL to 10 μL, and then transferred into a suitable plastic tube for subsequent loading onto the fixed target.

A key advantage of fixed target approaches over other sample delivery methods is that homogeneity of crystal size is not strictly required. The chevron opening of fixed targets allows crystals of different sizes to be caught at various depths of the apertures. The narrowest opening of the funnel-like apertures varies from 5–20 μm, the depth is 50 μm and the widest opening (on the surface where the sample is loaded) is 98 μm (Ebrahim et al., 2019b). While it is important for pump-probe or dose series experiments to work with crystals with a narrow size distribution, the ability to work with crystals of very different sizes is particularly advantageous in the early stages of work on a protein target. For example, large crystals obtained by batch or vapour diffusion method can be fragmented using vortexing, or with a glass rod, to produce a wide range of crystal fragments suitable for data collection on a fixed target. Similarly, preliminary and unoptimised crystallisation batches can simply be loaded into the fixed target. Crystals that are too small will pass through the apertures, while crystals that are too large will lie on the surface of the fixed target. This can be very helpful in obtaining initial information and serial structures for further optimisation.

An example of this is provided by the copper radical oxidase enzyme GlxA for which conversion of conventional crystallisation conditions to produce microcrystals in batch had not been optimised by the date of XFEL beamtime. Therefore large crystals grown in drops were merged and then fragmented by vortexing with a seed bead in a 1.5 mL plastic tube, followed by further manual disruption using a pipette tip. This step resulted in sufficient microcrystal-sized fragments to load onto chips and produce a good quality dataset to 2.8 Å resolution (Fig. 1) despite being in a monoclinic space group (P2$_1$).

2.2 Moving from standard MX to serial experiments

While using crystals grown in vapour diffusion droplets for loading sample onto the fixed targets can be a way of transitioning to serial experiments, the task itself can be arduous and might require tens of crystallisation drops pulled together for loading each chip. This approach can be useful for obtaining an initial structure or information about the unit cell at room temperature and comparing to the respective data from the cryogenic MX experiment. However, we strongly recommend continuous efforts and experimentation to transition to batch crystallisation as the volumes required for continuing serial experiments particularly for time-resolved or dose series studies will be larger.

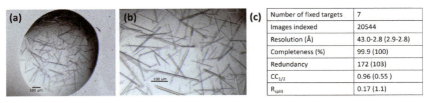

Fig. 1 Droplets of long needle-shaped crystals of the copper enzyme GlxA (a, b). These crystals were fragmented by vortexing with a seed bead before loading onto fixed targets providing the data summarised in (c).

Another approach that we recommend, and routinely use, for bridging the gap between cryogenic MX and serial data collection, is the use of thin films. As shown in Fig. 2, this approach can be applied using either a tweezer-pin which simulates a cryo-pin and can be mounted on standard goniometer bases (Axford et al., 2016) or dedicated SSX hardware (Doak et al., 2018) making it easy to fit into either type of experimental session. For the pin-based approach, merely 1 μL of the crystal slurry is loaded on the aperture of a double-sided adhesive spacer sandwiched between two small pieces of 6 μm thick Mylar (SPEX SamplePrep) sheets (Fig. 2a). This way, protein crystals can be scanned with the X-ray beam using low dose and wedges of data can be consecutively collected from multiple crystals and merged into a full dataset. Similarly, multiple wells of a double-sided adhesive spacer can be attached to a "chipless chip" format (Fig. 2b). In this case, the use of adhesive spacers allows multiple samples or variants of the same sample (*e.g.* ligand soaks or different crystallisation conditions) to be loaded in the same chipless chip for screening prior to a full data collection on the fixed targets.

2.3 Sample preparation

Crystal morphology and symmetry can also play key roles in sample loading and the acquisition of a complete dataset. As a general principal, high symmetry space groups require fewer diffraction patterns in order to obtain

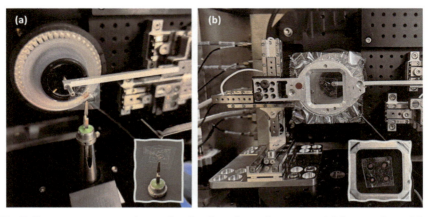

Fig. 2 Room temperature data collection from horse heart myoglobin crystals on (a) a thin-film sandwich mounted on a tweezer-pin and (b) a "chipless" chip mounted on the XYZ stages of the SSX hardware at beamline I24.

a high-quality dataset. Where feasible, it is advisable to pursue high symmetry space groups when optimising micro-crystallisation, although there are notable examples of successful data collection on fixed-target SFX from crystals of monoclinic symmetry (Bolton et al., 2024). Crystals' external morphology also significantly influences the orientation of crystals that are held within the apertures of the fixed targets. Needle-shaped crystals with only one long dimension are prone to passing through the fixed targets without being captured within the apertures, while thin plates may be more prone to preferential orientations leading to a low-completeness dataset or sticking on the chip surface.

Apart from morphology, the density of the micro-crystalline slurry (measured usually as the number of crystals per mL of batch) is very important to optimise before loading fixed targets to ensure efficient use of allocated beamtime. Fig. 3 shows images of various densities of thaumatin crystals obtained with a high-resolution optical microscope featuring motorised triple zoom lenses (Hirox HR-5000E). Briefly, thaumatin from *Thaumatococus danielii* (Sigma-Aldrich) was dissolved in double-distilled water to a final concentration of 45 mg mL^{-1}. The thaumatin solution was then mixed in a 1:1 ratio with the precipitant solution (40 % w/v sodium potassium tartrate, 0.1 M bis-Tris propane pH 6.5) (Martiel et al., 2021), briefly vortexed and left overnight on a rotator at 293 K. The average

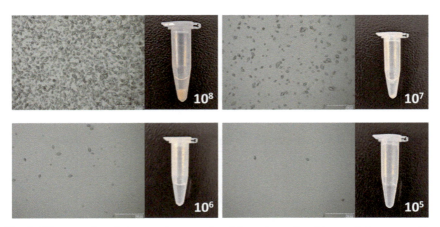

Fig. 3 Micrographs of thaumatin crystals with various densities and the respective Eppendorf tubes with the microcrystalline solution. Precipitation of a visible crystal pellet can be an indication of high density (10^7–10^8 crystals mL^{-1}) batches, as opposed to lower densities (10^5–10^6 crystals mL^{-1}) where the solution appears homogenised/almost clear. The scale bar represents 100 μm.

crystal size was approximately $12\,\mu m$ (ranging from 5–$20\,\mu m$) and the density was 2×10^8 crystals mL^{-1}. The other tubes shown in Fig. 3 were produced by simply diluting the original high density batch.

To date, we have typically advised fixed-target users to load 100–$200\,\mu L$ of crystal slurry onto the fixed targets as a good 'default' or starting point (Ebrahim et al., 2019b; Horrell et al., 2021). Over time however it has become apparent that loading volume and sample density are strongly correlated and affect both the hit-rate and the total number of indexed patterns per collected chip. Using thaumatin crystals as a model sample, we evaluated the effect of loading volume and crystal density on the hit-rate as shown by per-image-analysis plots (or 'live hit-rate') reported by the DIALS software package (Winter et al., 2018), as well as the final numbers of total hits and indexed images from the auto-processing pipeline for serial experiments at I24 (Table 1). The live hit-rate differs from the final number of hits as the focus in obtaining and reporting the live estimate to the user is speed rather than accuracy. Commenting first on the impact of the sample density on the hit-rate when loading volume is similar $(50\,\mu L)$, it is apparent from Fig. 7 that lower hit-rates are observed for lower density samples suggesting either collection from multiple chips or higher concentration of the sample is required for obtaining sufficient hits and a complete dataset.

Concerning the effect of the loading volume on the hit-rate and the final number of indexed images (Table 1), the results suggest a counter-intuitive trend. For sample density of 2×10^8 crystals mL^{-1}, the hit-rate reported by the auto-processing pipeline and the total number of hits increase with higher loading volumes, with significant differences observed

Table 1 Sample (thaumatin) loading volumes and crystal densities affecting the hit rate and the total indexed images for a serial experiment on fixed targets.

Loading volume (µL)	Density (crystals mL^{-1})	Live hit-rate (%)	Number of hits	Indexed images
50	2×10^8	71.2	23,152	3,816
90	2×10^8	70.7	22,839	3,441
130	2×10^8	86.0	24,231	2,768
50	2×10^7	17.5	16,308	3,274

Note. Two data collections were performed per loading volume value and per crystal density and the outcome hits and number of indexed images were averaged.

when moving from 50 to 130 μL. However, the respective number of total indexed images decreases significantly for higher volumes. This can be attributed to multiple lattices and/or less stringent criteria for a "hit" contributing to higher live hit-rates. Even though live hit-rates can be an adequate metric for near to real-time results facilitating fast decision-making and progression during serial beamtime, especially when time is limited, slower feedback from other processing runs should be also considered to ensure sufficient data collection for a complete dataset prior to the most meticulous post-beamtime processing. It is important to note that the guidance provided here on sample preparation and loading should be considered as best practice guidelines rather than absolute scenarios as these parameters are highly sample dependent. Loading sub-areas of fixed targets using different loading volumes and concentrations of the sample can be a time- and sample-efficient way of optimising these for a new sample. Data collection from multiple chips (usually 2–4) might be necessary for protein crystals in low symmetry space groups (*e.g.* P1) even in cases of high-density samples with relatively high hit-rates.

Manual loading on the fixed targets can present variations which can be attributed to expected experimental error. Therefore, even when loading two or more different chips with the same volume of a sample from the same batch, hit-rates will vary depending on how efficiently the crystalline sample was spread on the chip, the number of crystals captured in the apertures or how the chip is manipulated when sealed. For the examples detailed in Table 1, two chips were collected per loading volume value (50, 90 or 130 μL) and per crystal density (10^8 or 10^7 crystals mL^{-1}) and the resulting hits and number of indexed images were averaged.

In our experience, samples with a crystal density below 10^7 crystals mL^{-1} are not suitable for serial experiments with fixed targets, as many chips must be loaded and collected for a full dataset. If concentrations of this order are not achievable, or if sample availability is a pressing concern, acoustic drop ejection is a more automated approach for loading fixed targets and can significantly reduce sample consumption (Davy et al., 2019).

2.4 Anaerobic fixed target SSX and SFX

Structural studies of oxygen sensitive enzymes have traditionally relied on X-ray crystallography under cryogenic conditions, exploiting the formation of vitreous ice at low temperatures for extremely low oxygen diffusion rates into the protein crystal. However, an adaptation of the "chipless" chip format has been implemented for anaerobic and low–dose data collection at room temperature at beamline I24 (Rabe et al., 2020). Sample loading was performed in a glovebox and the localised environment of the sample was maintained anaerobic over the duration of transfer, alignment and data collection (approximately 10–15 min) at the beamline. Low gas permeability was secured by using vacuum grease and a 50 µm adhesive inter-seal in between two Mylar thin foils on each side of the fixed-target sample holders. Another approach for anaerobic data collection using sandwiched silicon nitride membranes has recently been developed at MAX IV Laboratory (Bjelčić et al., 2023), where microcrystals of haemoglobin retained their deoxy state for at least 30 min as shown by complementary XANES (X-ray Absorption Near-Edge Structure) and UV-Vis spectroscopic data. We have pursued the development of novel sample environment for anaerobic data collection on fixed targets, replacing Mylar with other thin films made of materials with high oxygen barrier, frequently used in other advanced and daily applications (*e.g.* food packaging where lipid oxidation during long-term storage must be minimised). Polymers of ethylene vinyl alcohol (EVOH) and polyvinylidene chloride (PVDC) have both been extensively tested for oxygen diffusion over time and have proven to be excellent choices for anaerobic experiments. Oxygen-sensitive targets such as myoglobin have remained anaerobic in their ferrous unbound form for the duration of a data collection with the fixed targets at I24 and SACLA (results not shown here).

2.5 Fixed target SSX beamline hardware: I24

Used as an exemplar here, beamline I24 at Diamond Light Source has constructed a flexible and modular sample environment to accommodate a range of fixed-target and serial crystallography experiments, in addition to more traditional single-crystal cryo-crystallography (Fig. 4a). Magnetic kinematic fixtures atop the vertical goniometer table enable quick installation of hardware and seamless changes in experiment mode. Experiments are facilitated by an accessible and relatively spacious sample environment with breadboard mountings (Fig. 4c) and a second motion table, allowing for the precise integration of complementary hardware such as liquid substrate addition (Fig. 4b). For light sensitive samples, the local environment around the sample position can be made dark with covers and shielding, and fixed target alignment can be performed with either filtered visible light or infra-red illumination (Fig. 4d).

At the synchrotron, X-rays are delivered in a pseudo-continuous manner and the sample needs to be held in the beam for typically ten or

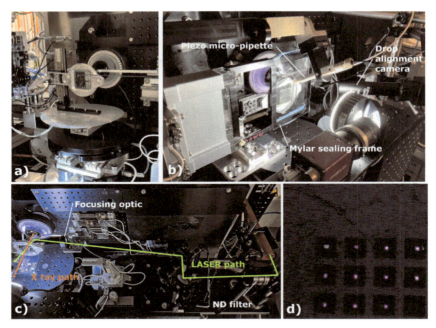

Fig. 4 SSX experimental setups at beamline I24. (a) Fixed target XYZ stages installed on vertical goniometer table using magnetic kinematic mounts. (b) Drop-on-chip hardware including piezoelectric microcapillary to add liquid to the chip aperture by aperture. The frame with mylar film has a small hole to allow micro-droplets of substrate to pass through directly to individual crystal-containing wells. The film presses against the chip holder allowing chip motion but providing a seal to limit dehydration of the sample. (c) Optical path to deliver focused laser light to each aperture. (d) Replacing the standard backlight with an infra-red version allows alignment of a chip in complete darkness with the apertures clearly visible in this on-axis camera view of a mounted chip.

more milliseconds. For standard SSX experiments the minimum 'in position' time is dictated by the frame-rate of the detector in use. The X-ray shutter remains open during data collection (with the exception of some pump-probe experiments when the fast shutter opens and closes at each position), and the progression of the experiment is controlled by a GeoBrick (Sherrell et al., 2015). In this regime, the fixed-target data acquisition rate is dictated by the motion stages, with an 'in position' flag initiating an experiment specific, time-controlled sequence of events, such as the triggering of a laser pulse, adding a droplet of substrate, opening/closing the X-ray shutter, and reading-out of the detector.

2.6 Fixed target SFX beamline hardware: SACLA

A strength of the fixed-target instrumentation summarised above is that the hardware and software for sample motion and signal triggering are largely self-contained. The setup is thus readily portable (Sherrell et al., 2015) to other light sources providing regular users of SSX with a familiar, repeatable experiment, consistent look-and-feel and identical sample preparation requirements. The Diamond hardware described in Fig. 4 is shown mounted at SACLA XFEL in Fig. 5; while different interfacing brackets are used, key components are the same.

At an XFEL, X-ray pulses of a few tens of femtoseconds are provided at a regular rate and the motion of fixed targets must be synchronised to deliver these pulses to the sample position. The rate and timing of data acquisition is thus dictated by the source. In this case, the XFEL serves as the master clock of the experiment with a machine provided TTL pulse starting the sequence of events at each position on a fixed target from motion to laser illumination, insertion of an attenuator and detector read-out.

3. Data collection
3.1 Setup and data collection strategies

Fixed target data collection using this hardware follows a similar procedure, irrespective of location. First a chip is aligned (a process detailed in Horrell et al., 2021) using the main user interface (Fig. 6a). The portion

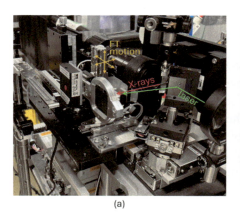

(a) (b)

Fig. 5 Fixed target SFX at SACLA. (a) Hardware layout with sample stage motion, X-ray beam direction and laser beam position highlighted. (b) An upstream air gap allows an oscillating flipper that can attenuate selected X-ray pulses to be added to the beam path.

Use of fixed targets for serial crystallography 41

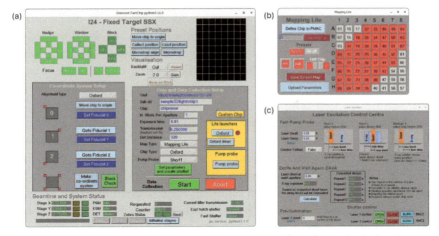

Fig. 6 User interfaces for SSX at I24. (a) Main user interface from which fixed targets are aligned to the X-ray beam and data collection parameters selected. (b) Mapping lite interface for selecting the sections of the fixed target data should be collected from. (c) Laser excitation control centre where the timing of triggers for dynamic experiments are defined.

of the fixed target for data collection is then selected (Fig. 6b). At I24, the main interface also allows user to define exposure time, the number of X-ray images recorded at each position, attenuation and detector distance. The duration and delay associated with any pump process are defined through a second sub-window of the user interface (Fig. 6c). At Diamond, to facilitate automated data processing, it is important that part of the file path (*e.g.* text 'sampleID' in Fig. 6a) matches the filename of a yml file containing the space group and unit cell in the data collection directory. Data collection from other fixed target types, or thin films, is also possible through the interface, provided a standard descriptor is provided (Owen et al., 2023).

At I24, pressing 'Start' moves beamline hardware (*i.e.* attenuators, hutch shutter and detector) into the correct position and passes relevant timing information to the GeoBrick controller which then starts fixed target motion and triggering. At SACLA and PAL, starting data collection from the user interface leaves fixed target hardware in a 'ready' state, awaiting the triggers required for data collection.

3.2 Fast-feedback from data acquisition

A crucial tool for an effective serial fixed-target experiment is providing near real-time results feedback. Automated software pipelines have been built to provide analysis, such as diffraction spot counts, as soon as data from the detector becomes available (Fig. 7). This information can then be used as a diagnostic for sample preparation (effectiveness of sample loading onto a chip), indicate rates of progress towards sufficient data for structure solution, and identifies desired or undesired changes to the sample in dynamics investigations. When time and amount of sample are typically limiting factors, efficient data acquisition and timely decisions regarding experiment variables are essential. Fig. 7 shows a per-image-analysis plot that updates as diffraction data are collected offering the experimenter a quick visual indicator of diffraction quality. Clicking on an individual marker on the plot opens a diffraction image viewer, allowing close inspection of the raw data, e.g. spot shapes and resolution. Rapid feedback is further enhanced by live indexing which can be performed by automatic routines when sample unit cell and space group information is provided in an appropriate format, as described in section "Setup and data collection strategies".

Fast feedback during the experiment provides additional information on sample quality and loading strategies for the fixed targets, allowing flexibility on decision-making for the optimal output of data collection. Once the sample preparation protocol is optimised (as discussed in previous sections) and a relatively dense batch can be consistently obtained, the experimental parameter that most significantly affects successful data collection is the loading volume.

3.3 Dose

An advantage of serial experiments is that they facilitate room temperature (RT) data collection. It is well known that cryo-cooling of crystals greatly

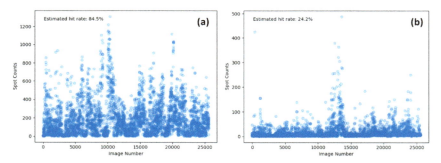

Fig. 7 Automated/real-time per-image-analysis plots of diffraction spot counts found on each detector frame, produced by the DIALS software package during SSX experiments at beamline I24. 50 μL of thaumatin crystals with a density of (a) 10^8 and (b) 10^7 crystals mL^{-1} were loaded onto the fixed target.

reduces the rate of radiation damage and consequently a drawback of RT crystallography is a lower dose-threshold for radiation damage (Garman and Weik, 2023). Radiation damage is observed both in terms of global effects and site-specific changes. Global changes include a reduction in diffracting power and change in unit cell volume, while site-specific changes refer to structural and chemical rearrangements within the asymmetric unit (Shelley and Garman, 2024).

Serial crystallography minimises the manifestation of radiation damage by, typically, collecting only a single image from each crystal. In XFEL-based serial crystallography, the short duration of the pulse combined with its intensity results in crystal destruction (Neutze et al., 2000). The short duration of the pulse means, however, that diffraction data are recorded before site-specific damage affects the obtained structure providing extreme beam parameters are not used as observed in PSII at SACLA (Ibrahim et al., 2020). Observations of site-specific damage at XFEL sources have been made but, to our knowledge, they usually relate to experiments which utilise extremely long or intense pulses, or exploit multiple pulses (Bolton et al., 2024; Nass et al., 2020). This absence of beam-induced changes places a premium on XFEL structures, particularly for metalloenzymes.

In synchrotron serial crystallography, beam induced damage cannot be completely avoided and a strategy to minimise it should be followed. X-ray induced damage is directly related to the absorbed dose which can be calculated using *RADDOSE* (Bury et al., 2018). At synchrotrons, site-specific damage has been observed to occur at doses of only tens of kGy. Recently reported examples of site-specific damage include disulphide damage at $\sim$ 10 kGy (de la Mora et al., 2020), side chain rearrangements in an iron-containing protein at $\sim$22 kGy (Bolton et al., 2024), changes in heme geometry at $\sim$30 kGy (Ebrahim et al., 2019b) and reduction of heme iron at $\sim$40 kGy (Pfanzagl et al., 2020).

Typical doses for I24 SSX experiments are shown in Table 2 for the iron-containing protein FutA, the copper-containing enzyme AcNiR and thaumatin. The doses estimated for I24 are the dose that can be delivered in a single 10 ms image and it is striking that these doses are significantly greater than the doses reviewed above at which site-specific damage has been observed. It is easy to over-expose, negating the gain made by distributing the dose required for structure solution over many crystals. Dose is still, therefore, a major experimental concern in SSX and should be reduced as much as practically possible. This is especially important for metalloproteins but it holds true for all samples. The benefits of energy choice are discussed below.

Table 2 Doses in kGy for typical serial experiments at I24 and SACLA.

	FutA	AcNiR	Thaumatin
I24 DWD dose 12.4 keV (in 10 ms)	71	79	79
I24 DWD 20 keV(in 10 ms)	19	21	19
SACLA RD3D	132	155	153
SACLA ADER	15	13	16

Note. Dose calculations for I24 assume an incident flux of 5×10^{12} ph s^{-1} and a beam size of $8 \times 8\,\mu m^2$ while SACLA dose calculations assume a pulse duration of 10 fs and pulse energy of 100 µJ. To aid comparison, a photon energy of 12.4 keV is used for XFEL dose calculations and a crystal size of $10 \times 10 \times 10\,\mu m^3$ is used throughout. For XFEL dose calculations, both the RADDOSE-3D style average dose whole crystal (RD3D) and RADDOSE-XFEL average absorbed doses in the exposed region (ADER) are given. I24 dose calculations performed using RADDOSE-3D, SACLA dose calculations used RADDOSE-XFEL.

While doses can also be calculated for XFEL experiments using *RADDOSE-XFEL* (Dickerson et al., 2020), it is difficult to directly compare the doses obtained with those realised in synchrotron-based serial crystallography. Similar sites (*e.g.* metals or disulphide bonds) may be damage hot-spots at both synchrotron and XFEL for site-specific damage, but very different damages processes are occurring (*i.e.* primarily reduction at the synchrotron and oxidation at XFEL) on very different timescales. In summary, less concern is required in seeking to minimise the absorbed dose, but the user should be aware of pulse duration and pulse intensity.

3.4 Choice of X-ray energy

A wide range of X-ray wavelengths/energies can be employed for X-ray diffraction, allowing for data collection below or above elemental absorption edges or at high energies to minimise the contribution of air scatter. The majority of 'standard' SFX and SSX experiments are carried out at the default energy of the beamline without optimisation however. With the recent advent of new cadmium telluride-based sensor technologies (Zambon et al., 2018), the opportunity has arisen to exploit high-energy ($\geq$20 keV) X-rays for crystallography experiments utilising synchrotron radiation. In contrast to lower energies, high-energy X-rays offer the advantage of recording stronger spot intensities and an enhanced signal-to-noise ratio for a given absorbed X-ray dose (Storm et al., 2021).

Compared to cryo-cooled or room temperature rotation data collection, SSX typically utilises a much shorter total X-ray exposure time per crystal with the dose required for structure solution divided over many hundreds of crystals. As outlined above however, significant structural rearrangements can occur even at dose scales of tens of kGy. Attenuating the X-ray beam provides an obvious gain but results in a concomitant decrease in spot intensity and signal-to-noise ratio (SNR), posing a challenge for data indexing and accurate measurement of Bragg peaks. High-energy usage in fixed target serial crystallography can further mitigate the adverse effects of X-rays by providing improved signal-to-noise without compromising data quality.

The comparative graphs shown in Fig. 8 indicate the impact of varying the incident X-ray energy on fixed-target serial data collected at I24 using lysozyme crystals. For accurate comparison, data collection was conducted three times at each energy, averaging results with crystals from the same batch. The average crystal size was $\sim 20 \times 10 \times 10\,\mu m^3$ and the sample density 2×10^8 crystals mL^{-1}. We ensured the incident photon flux and resolution of data recorded were consistent across all data with the inscribed circle on detector set at 1.7 Å resolution.

Serial data were collected at the two 'default' energies of 12.4 and 20 keV at beamline I24 to illustrate the possible advantages of high energy data collection. To allow a direct comparison, the incident flux at 12.4 and 20 keV was chosen such that the same absorbed dose was delivered to the sample in each 10 ms image at each energy. The signal-to-noise ratio (SNR, $I/\sigma(I)$) and metrics such as R_{split} and $CC_{1/2}$, all show a significant improvement as a function of energy. At the higher energy the SNR increases, the overall R_{split} decreases while $CC_{1/2}$ increases with the most significant improvement in the higher resolution shells (Table 3, Fig. 8). For both datasets, an equal number of images (5000) was used in the comparison.

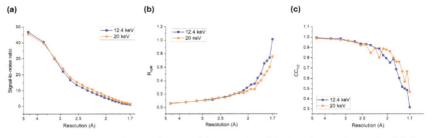

Fig. 8 Comparison graphs evaluating (a) intensity, (b) signal-to-noise ratio, (c) R_{split} and (d) $CC_{1/2}$ from data collected at different energies. The blue and orange line graph correspond to 12.4 keV and 20 keV respectively in all panels.

Table 3 Comparison of scaling statistics from high and low energy serial data.

	12.4 keV	20 keV
Incident flux ($\times 10^{12}$ ph s^{-1})	0.11	0.32
Absorbed dose (kGy)	1.82	1.84
SNR	12.4 (1.0)	13.4 (1.5)
R$_{split}$	0.112 (1.013)	0.113 (0.468)
CC$_{1/2}$	0.991 (0.316)	0.989 (0.758)

Note. Data range 50–1.70 Å, outer shell 17.3–1.70 Å. Dose calculations for each energy was conducted with a Gaussian beam size of $8 \times 8\,\mu m^2$ using RADDOSE-3D.

Using high-energy crystallography in fixed-target serial experiments presents a solution to help mitigate some of the challenges associated with lower energies, specifically room temperature radiation damage. Despite higher fluxes typically being available at lower energies from synchrotron undulators – benefiting 'flux hungry' experiments, high-energy data collection can deliver enhanced signal-to-noise and improved scaling statistics for a given absorbed dose indicating a more dose-efficient route to structural data acquisition.

4. Record keeping, log and beamtime logistics – best practice

While record keeping and storage of metadata have been largely automated in conventional, modern macromolecular crystallography experiments at synchrotron sources *via*, for example, ISPyB/SyncWeb (Fisher et al., 2015), meticulous keeping of an experimental log is vital to the success of serial crystallography experiments. While serial data collections at I24 are recorded, and can be viewed, in SynchWeb, metadata for dynamic experiments can be incomplete with information such as laser power or droplet size often absent. Limitations in metadata are particularly prevalent in XFEL experiments where frequently minimal, or no, metadata is recorded. Data files containing the recorded diffraction data may contain little more than a best guess for the experiment geometry and may be simply numbered rather than having a user-defined name which can be interpreted later.

When operating in shifts to maximise extended, through-the-night beamtime, a comprehensive log is essential for ensuring smooth information transfer regardless of who is conducting a particular role at any given moment. Typically, this log should be in two parts: a narrative log and a spreadsheet listing all data collections. The narrative log is typically verbose, recording the ongoing status of, and changes to the experiment. The spreadsheet should detail all parameters relevant to data collection such as sample information, X-ray beam parameters and details of any laser or droplet induced reaction initiation. It is often useful to pair the spreadsheet with a data processing spreadsheet allowing progress of the experiment to be easily tracked.

In our experience, for long duration experiments, it is optimal to subdivide the experimental team into groups based on discipline – *i.e.* sample preparation and loading, beamline operation and data collection, and data analysis. Each group can swap in and out of beamtime as desired, ensuring continuity within the beamtime without wholesale personnel changes. Paramount to this approach is clear communication between the different teams. This includes communicating priorities for data analysis, providing feedback from the data analysis team regarding the completeness of experiments and indicating which samples are being prepared or are required.

Another parameter that falls into the necessity to keep clear and updated logistics during beamtimes is the data collection time for different experimental modes. 'Standard' fixed-target data collection runs complete in a relatively short period (8 min at I24 or 14 min at SACLA for 25600 aperture silicon chips) while other types of data collection, such as dose series, laser pump-probe or X-ray pump/X-ray probe (XRPP) experiments, can last up to 40 min per chip. In all cases, it is important to have a supply of loaded chips ready to ensure the best use of beamtime. The simplest experiments are straightforward static data collections with a single X-ray exposure per aperture on the fixed targets, without special requirements for sample preparation (*e.g.* anaerobic or dark room conditions). Such chips can be prepared in advance and used to fill any gaps in the data collection schedule, for example in between more challenging samples that require extra preparation time, or when a fixed target containing another sample fails to get good results and is terminated early. It is important to factor in that, in addition to variation in the duration of data collection, the time required for sample preparation and loading into fixed targets can be highly variable. Some samples are relatively easy to prepare, such as static or stable ligand soaks, while others can be time-sensitive (medium stability soaks or semi-stable intermediates) or necessitate challenging preparation (*e.g.*, loading within an anaerobic chamber or dark room or both).

Different samples have different requirements for in-beamtime feedback to guide subsequent experiments within the beamtime and the data processing group should be made aware of the priorities. For example, assessing whether photo-activation by laser has been effective or whether a ligand is present may require inspection of electron density maps. Therefore, these experiments should be conducted first to allow sufficient time for data processing and iteration of the experiment if required. In any case, at the start of beamtime, a small volume of exemplar data should be collected - usually from well-diffracting crystals of model proteins. This data should be processed and refined to establish a robust experimental model and identify if any changes to the experimental setup are required. An example of the positive impact this can have is detailed below.

4.1 Iterative optimisation of experiment design

Small changes to the experimental setup can have a large effect on the quality of data collected. This example demonstrates the impact of optimising the beamstop for one aspect of the experiment (laser illumination) on another (X-ray diffraction). Initially the position of the laser beamstop was optimised for the setup and operation of the visible laser, allowing relatively straightforward alignment. While the shadow observed on diffraction images appeared large (Fig. 9a), the initial diffraction data appeared to be of reasonable quality. However, subsequent processing revealed difficulties with refinement and artefacts in electron density maps. As the position of the beamstop in the beam direction was fixed, adjustment was non-trivial. The interaction point and other beamline hardware were moved 10 mm upstream while the beamstop was left in the same position. Data collected with this new experiment geometry is shown in Fig. 9b and contrasting plots of multiplicity *vs.* resolution in reciprocal space shown in Fig. 9c and d. Subsequent datasets proved to be of high quality, with stable refinement of models against them, illustrating the impact of in-beamtime data processing and using the results to optimise the experiment.

4.2 Unit cell as a metric for reaction progression: contrasting examples of crystallin and myoglobin

Fast feedback is an important aspect of dynamic serial experiments. In addition to hit-rates as outlined above, real-time feedback during beamtime

Use of fixed targets for serial crystallography

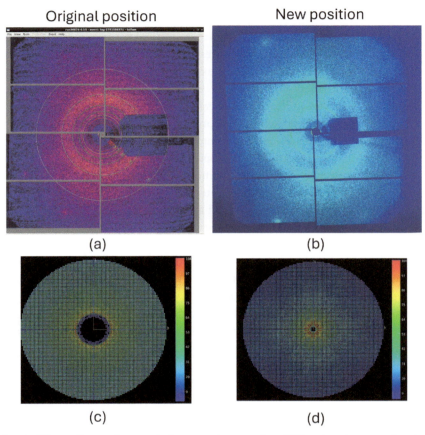

Fig. 9 Effect of experiment design on data quality. Diffraction images showing shadow of X-ray and laser beamstops in original (a) and modified (b) positions. The lower panels show multiplicity *vs.* resolution in reciprocal space for datasets comprised of the same number of images. With the initial experiment geometry, a lack of data and then a low multiplicity low resolution ring at 8 Å can be seen (c), contrasting with the optimised setup where a lower resolution cutoff and normal resolution dependence of the multiplicity is observed (d).

can provide indications of whether the desired action has been triggered in the crystal and if the activity *in crystallo* is progressing. Reaction progression and conformational changes are often linked to small changes in unit cell dimensions. These changes can be detected rapidly, long before the capture of sufficient reflection observations, and their integration, scaling and merging to a dataset capable of generating high quality electron density maps.

Fig. 10 shows the live hit-rate and indexing from crystallin crystals (Hill et al., 2024) observed during a 'dose series' type experiment. In this setup, each crystal has five X-ray exposures recorded with laser illumination occurring during the second X-ray exposure. In contrast to an identical experiment performed without laser excitation, a split in unit cell distribution is observed indicating successful reaction initiation.

Care should be taken when interpreting live feedback however as, by necessity, the emphasis is on providing information as quickly as possible rather than obtaining a carefully and optimally processed dataset. When the experiment geometry is not well defined, the unit cell can act as a compensating variable for an incorrect detector distance or position. This can result in the unit cell initially obtained appearing shifted or even split. This is illustrated here for a fixed-target experiment performed at SACLA where data collection was performed on the reduced state of horse heart myoglobin under anaerobic conditions without any external trigger such as photoexcitation or exposure to oxygen. During the beamtime, data processing with CrystFEL (White et al., 2012) showed an apparent split in the unit cell (Fig. 11a). This anomaly couldn't be attributed to any rational experimental parameter, such as crystals from different batches or different morphologies. Reprocessing the data after the beamtime and carefully optimising the detector geometry led to a uniform unit cell distribution as shown in Fig. 11b.

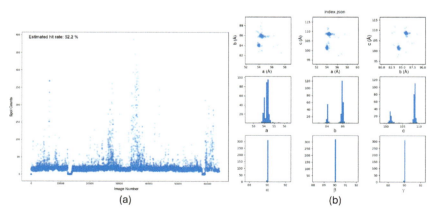

Fig. 10 Automated feedback on a dynamic fixed target experiment at I24 showing (a) the live hit-rate and (b) a clear laser light induced expansion in crystallin unit cell dimensions.

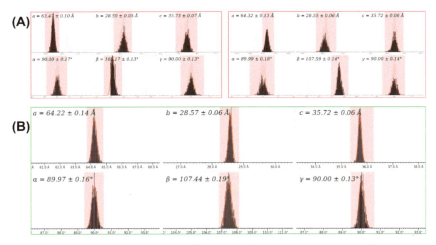

Fig. 11 Screenshots of cell_explorer from CrystFEL showing the unit cell parameters distribution (a) during the beamtime suggesting the presence of two different unit cells as shown on the left and right panel and (b) single cell resulting after reprocessing and refinement of the detector geometry file.

4.3 Exploring radiation damage and X-ray driven reactivity using fixed-target SSX

As described above, a key advantage of fixed-target SSX is the precise control over the position of crystals and the ability to expose them to the beam on demand. This, in turn, allows the possibility of measuring multiple diffraction patterns from a single fixed-target aperture with each successive diffraction pattern corresponding to an iteratively higher dose. As well as characterising classical radiation damage progression at room temperature, for example the loss of diffracting power, crystalline order and unit cell changes, this approach can also enable the use of X-ray generated photoelectrons to drive reactions in proteins containing redox active centres. This approach has been termed Multiple Serial Structures (MSS) (Ebrahim et al., 2019a).

In practical application, a series of images are obtained from each aperture during data collection and subsequently binned into dose points. Typical data collection times are 10–20 ms per exposure corresponding to e.g. 100–200 ms dwell time at each aperture for a series of 10 dose points. While data collection over the whole fixed-target may take several tens of minutes, full data collection for each crystal is complete within a very short space of time, thus limiting the progression of radiation damage, that occurs in a time-dependent as well as dose-dependent manner.

Using this approach, a fixed target containing resting state microcrystals is prepared using the standard methods described above. Within the GUI, set the parameter 'number of shots per aperture' to the desired number of dose points. It is useful to consider the likely maximum dose that can be tolerated by the crystals before losing diffracting power and resolution such that the site-specific changes being studied can no longer be reliably followed. In some cases, the higher dose points of an MSS series may be discarded due to excessive radiation damage.

Two brief examples are described here with full details available in the citations to the original publications. In the first example (Ebrahim et al., 2019a), a series of dose points were obtained from microcrystals of copper nitrite reductase that exhibited unit cell polymorphism with crystals divided into small cell and large cell populations with a difference in unit cell parameter in a cubic space group of > 2 Å. In response to increasing X-ray dose, unit cell volume for each population progressively increased slightly (a common response to radiation damage) but unexpectedly crystals with the small unit cell flipped into the large unit cell producing a gradual shift of population. At high X-ray doses only the large unit cell population was observed.

In the second example, microcrystals of a heme peroxidase enzyme were studied by MSS (Ebrahim et al., 2019b). In this enzyme, a water molecule is bound to the Fe(III) resting state of the heme group and is lost upon reduction to Fe(II). The first dose points in the MSS series showed a small increase in the iron–water bond length compared to an XFEL structure. With increasing dose, the iron–water bond length increased linearly with the water eventually becoming completely disassociated from the iron atom. By fitting this bond length to a straight line, a zero-dose bond length was estimated which was identical within experimental error to that obtained in the damage-free XFEL structure.

The equivalent experiment to MSS at an XFEL beamline is the X-ray pump/X-ray probe (XRPP) approach. This again takes advantage of the ability to hold the fixed target stationary for more than one exposure. Alternate XFEL pulses are used for the pump and probe after which the fixed target is translated to the next aperture. Given the high level of damage caused by XFEL pulses to protein crystals, an attenuator should be used to reduce the dose of pump pulse to well below the experimentally determined dose limit for the system under study. A more typical XFEL pulse energy is then used for the probe pulse to produce a dataset which can be compared to a conventional single pulse dataset in order to characterise processes initiated by the radiation. In a recent example (Bolton et al., 2024), a sapphire wafer fitted to a fast flipper was used to attenuate the pump pulse before being flipped out of the beam to allow the pump pulse to pass (Fig. 5b). In this case, the repetition rate of SACLA resulted in a time delay between pump and probe of 33 ms.

5. Summary

Fixed targets offer an efficient way of conducting serial crystallography at both synchrotron and XFEL sources. They not only provide dynamic structural data across a broad range of timescales, but also facilitate a transition from traditional rotation crystallography which typically uses one or a few tens of crystals to serial crystallography using thousands of crystals. Suggestions for optimising sample preparation, experiment design and the best use of beamtime during data collection sessions are presented here.

Acknowledgements

We would like to thank Jonathan Worrall for useful discussions and providing images of GlxA crystals, SACLA staff especially Kensuke Tono and Shigeki Owada for much assistance before and during beamtime, and Cicely Tam for thaumatin crystallisation.

References

Axford, D., Aller, P., Sanchez-Weatherby, J., & Sandy, J. (2016). Applications of thin-film sandwich crystallization platforms. *Acta Crystallographica Section F, 72*, 313–319.

Barends, T. R. M., Stauch, B., Cherezov, V., & Schlichting, I. (2022). Serial femtosecond crystallography. *Nature Reviews Methods Primers, 2*, 1–24.

Beale, J. H., Bolton, R., Marshall, S. A., Beale, E. V., Carr, S. B., Ebrahim, A., ... Owen, R. L. (2019). Successful sample preparation for serial crystallography experiments. *Journal of Applied Crystallography, 52*, 1385–1396.

Bjelčić, M., Sigfridsson Clauss, K. G. V., Aurelius, O., Milas, M., Nan, J., & Ursby, T. (2023). Anaerobic fixed-target serial crystallography using sandwiched silicon nitride membranes. *Acta Crystallographica Section D, 79*, 1018–1025.

Bolton, R., Machelett, M. M., Stubbs, J., Axford, D., Caramello, N., Catapano, L., ... Tews, I. (2024). A redox switch allows binding of Fe(II) and Fe(III) ions in the cyanobacterial iron-binding protein FutA from Prochlorococcus. *Proceedings of the National Academy of Sciences, 121*, e2308478121.

Bury, C. S., Brooks-Bartlett, J. C., Walsh, S. P., & Garman, E. F. (2018). Estimate your dose: RADDOSE-3D. *Protein Science, 27*, 217–228.

Cheng, R. K. (2020). Towards an optimal sample delivery method for serial crystallography at XFEL. *Crystals, 10*, 215.

Davy, B., Axford, D., Beale, J. H., Butryn, A., Docker, P., Ebrahim, A., ... Aller, P. (2019). Reducing sample consumption for serial crystallography using acoustic drop ejection. *Journal of Synchrotron Radiation, 26*, 1820–1825.

de la Mora, E., Coquelle, N., Bury, C. S., Rosenthal, M., Holton, J. M., Carmichael, I., ... Weik, M. (2020). Radiation damage and dose limits in serial synchrotron crystallography at cryo- and room temperatures. *Proceedings of the National Academy of Sciences, 117*, 4142–4151.

Dickerson, J. L., McCubbin, P. T. N., & Garman, E. F. (2020). RADDOSE-XFEL: Femtosecond time-resolved dose estimates for macromolecular X-ray free-electron laser experiments. *Journal of Applied Crystallography, 53*, 549–560.

Doak, R. B., Nass Kovacs, G., Gorel, A., Foucar, L., Barends, T. R. M., Grünbein, M. L., ... Schlichting, I. (2018). Crystallography on a chip—Without the chip: Sheet-on-sheet sandwich. *Acta Crystallographica Section D, 74*, 1000–1007.

Ebrahim, A., Appleby, M. V., Axford, D., Beale, J., Moreno-Chicano, T., Sherrell, D. A., ... Owen, R. L. (2019a). Resolving polymorphs and radiation-driven effects in microcrystals using fixed-target serial synchrotron crystallography. *Acta Crystallographica Section D, 75*, 151–159.

Ebrahim, A., Moreno-Chicano, T., Appleby, M. V., Chaplin, A. K., Beale, J. H., Sherrell, D. A., ... Hough, M. A. (2019b). Dose-resolved serial synchrotron and XFEL structures of radiation-sensitive metalloproteins. *IUCrJ, 6*, 543–551.

Fisher, S. J., Levik, K. E., Williams, M. A., Ashton, A. W., & McAuley, K. E. (2015). SynchWeb: A modern interface for ISPyB. *Journal of Applied Crystallography, 48*, 927–932.

Fraser, J. S., van den Bedem, H., Samelson, A. J., Lang, P. T., Holton, J. M., Echols, N., & Alber, T. (2011). Accessing protein conformational ensembles using room-temperature X-ray crystallography. *Proceedings of the National Academy of Sciences, 108*, 16247–16252.

Garman, E. F., & Weik, M. (2023). Radiation damage to biological macromolecules*. *Current Opinion in Structural Biology, 82*, 102662.

Grünbein, M. L., & Nass Kovacs, G. (2019). Sample delivery for serial crystallography at free-electron lasers and synchrotrons. *Acta Crystallographica Section D, 75*, 178–191.

Gu, K. K., Liu, Z., Narayanasamy, S. R., Shelby, M. L., Chan, N., Coleman, M. A., ... Kuhl, T. L. (2023). All polymer microfluidic chips—A fixed target sample delivery workhorse for serial crystallography. *Biomicrofluidics, 17*, 051302.

Hill, J. A., Nyathi, Y., Horrell, S., von Stetten, D., Axford, D., Owen, R. L., ... Yorke, B. A. (2024). An ultraviolet-driven rescue pathway for oxidative stress to eye lens protein human gamma-D crystallin. *Communications Chemistry, 7*, 1–8.

Horrell, S., Axford, D., Devenish, N. E., Ebrahim, A., Hough, M. A., Sherrell, D. A., ... Owen, R. L. (2021). Fixed target serial data collection at diamond light source. *JoVE (Journal of Visualized Experiments)*, e62200.

Hough, M. A., & Owen, R. L. (2021). Serial synchrotron and XFEL crystallography for studies of metalloprotein catalysis. *Current Opinion in Structural Biology, 71*, 232.

Ibrahim, M., Fransson, T., Chatterjee, R., Cheah, M. H., Hussein, R., Lassalle, L., ... Yano, J. (2020). Untangling the sequence of events during the S2 → S3 transition in photosystem II and implications for the water oxidation mechanism. *Proceedings of the National Academy of Sciences, 117*, 12624–12635.

Martiel, I., Beale, J. H., Karpik, A., Huang, C.-Y., Vera, L., Olieric, N., ... Padeste, C. (2021). Versatile microporous polymer-based supports for serial macromolecular crystallography. *Acta Crystallographica Section D, 77*, 1153–1167.

Mehrabi, P., Müller-Werkmeister, H. M., Leimkohl, J.-P., Schikora, H., Ninkovic, J., Krivokuca, S., ... Miller, R. J. D. (2020). The HARE chip for efficient time-resolved serial synchrotron crystallography. *Journal of Synchrotron Radiation, 27*, 360–370.

Mehrabi, P., Schulz, E. C., Agthe, M., Horrell, S., Bourenkov, G., von Stetten, D., ... Miller, R. J. D. (2019). Liquid application method for time-resolved analyses by serial synchrotron crystallography. *Nature Methods, 16*, 979–982.

Nass, K., Gorel, A., Abdullah, M. M., V. Martin, A., Kloos, M., Marinelli, A., ... Schlichting, I. (2020). Structural dynamics in proteins induced by and probed with X-ray free-electron laser pulses. *Nature Communications, 11*, 1814.

Neutze, R., Wouts, R., van der Spoel, D., Weckert, E., & Hajdu, J. (2000). Potential for biomolecular imaging with femtosecond X-ray pulses. *Nature, 406*, 752–757.

Orville, A. M. (2020). Recent results in time resolved serial femtosecond crystallography at XFELs. *Current Opinion in Structural Biology, Catalysis and Regulation • Protein Nucleic Acid Interaction, 65*, 193–208.

Owen, R. L., Axford, D., Sherrell, D. A., Kuo, A., Ernst, O. P., Schulz, E. C., ... Mueller-Werkmeister, H. M. (2017). Low-dose fixed-target serial synchrotron crystallography. *Acta Crystallographica Section D, 73*, 373–378.

Owen, R. L., de Sanctis, D., Pearson, A. R., & Beale, J. H. (2023). A standard descriptor for fixed-target serial crystallography. *Acta Crystallographica Section D, 79*, 668–672.

Pearson, A. R., & Mehrabi, P. (2020). Serial synchrotron crystallography for time-resolved structural biology. *Current Opinion in Structural Biology, Catalysis and Regulation • Protein Nucleic Acid Interaction, 65*, 168–174.

Pfanzagl, V., Beale, J. H., Michlits, H., Schmidt, D., Gabler, T., Obinger, C., ... Hofbauer, S. (2020). X-ray–induced photoreduction of heme metal centers rapidly induces active-site perturbations in a protein-independent manner. *Journal of Biological Chemistry, 295*, 13488–13501.

Rabe, P., Beale, J. H., Butryn, A., Aller, P., Dirr, A., Lang, P. A., ... Owen, R. L. (2020). Anaerobic fixed-target serial crystallography. *IUCrJ, 7*, 901–912.

Sandy, J., Mikolajek, H., Thompson, A. J., Sanchez-Weatherby, J., & Hough, M. A. (2024). Advancing protein structure analysis for drug development (video) | JoVE. *JoVE (Journal of Visualized Experiments)*, e65964.

Schulz, E. C., Mehrabi, P., Müller-Werkmeister, H. M., Tellkamp, F., Jha, A., Stuart, W., ... Miller, R. J. D. (2018). The hit-and-return system enables efficient time-resolved serial synchrotron crystallography. *Nature Methods, 15*, 901–904.

Shelley, K. L., & Garman, E. F. (2024). Identifying and avoiding radiation damage in macromolecular crystallography. *Acta Crystallographica Section D, 80*, 314–327.

Sherrell, D. A., Foster, A. J., Hudson, L., Nutter, B., O'Hea, J., Nelson, S., ... Owen, R. L. (2015). A modular and compact portable mini-endstation for high-precision, high-speed fixed target serial crystallography at FEL and synchrotron sources. *Journal of Synchrotron Radiation, 22*, 1372–1378.

Shoeman, R. L., Hartmann, E., & Schlichting, I. (2023). Growing and making nano- and microcrystals. *Nature Protocols, 18*, 854–882.

Storm, S. L. S., Axford, D., & Owen, R. L. (2021). Experimental evidence for the benefits of higher X-ray energies for macromolecular crystallography. *IUCrJ, 8*, 896–904.

White, T. A., Kirian, R. A., Martin, A. V., Aquila, A., Nass, K., Barty, A., & Chapman, H. N. (2012). CrystFEL: A software suite for snapshot serial crystallography. *Journal of Applied Crystallography, 45*, 335–341.

Winter, G., Waterman, D. G., Parkhurst, J. M., Brewster, A. S., Gildea, R. J., Gerstel, M., ... Evans, G. (2018). DIALS: Implementation and evaluation of a new integration package. *Acta Crystallographica Section D, 74*, 85–97.

Zambon, P., Radicci, V., Trueb, P., Disch, C., Rissi, M., Sakhelashvili, T., ... Broennimann, C. (2018). Spectral response characterization of CdTe sensors of different pixel size with the IBEX ASIC. *Nuclear Instruments and Methods in Physics Research Section A: Accelerators, Spectrometers, Detectors and Associated Equipment, 892*, 106–113.

Zhao, F.-Z., Zhang, B., Yan, E.-K., Sun, B., Wang, Z.-J., He, J.-H., & Yin, D.-C. (2019). A guide to sample delivery systems for serial crystallography. *The FEBS Journal, 286*, 4402–4417.

CHAPTER THREE

Sample efficient approaches in time-resolved X-ray serial crystallography and complementary X-ray emission spectroscopy using drop-on-demand tape-drive systems[☆]

Jos J.A.G. Kamps[a,b], Robert Bosman[a,c], Allen M. Orville[a,b], and Pierre Aller[a,b,*]

[a]Diamond Light Source, Harwell Science & Innovation Campus, Didcot, United Kingdom
[b]Research Complex at Harwell, Rutherford Appleton Laboratory, Didcot, United Kingdom
[c]University Medical Center Hamburg-Eppendorf (UKE), Hamburg, Germany
[*]Corresponding author. e-mail address: pierre.aller@diamond.ac.uk

Contents

1. Introduction	58
1.1 Complementary spectroscopy	62
1.2 X-ray emission spectroscopy	65
1.3 XES setup components	67
2. Materials	69
2.1 Sample delivery	69
2.2 XES, detector and analyzer crystals	71
3. Methods	74
3.1 Preparing microcrystal slurries	74
3.2 Reaction initiation scheme	78
3.3 Data collection/beamtime	84
3.4 Future developments	93
References	96

Abstract

Dynamic structural biology enables studying biological events at the atomic scale from 10's of femtoseconds to a few seconds duration. With the advent of X-ray Free Electron Lasers (XFELs) and 4th generation synchrotrons, serial crystallography is becoming a major player for time-resolved experiments in structural biology. Despite significant

[☆] The Application of tape-drive and in-situ techniques for time-resolved crystallography

Methods in Enzymology, Volume 709
ISSN 0076-6879, https://doi.org/10.1016/bs.mie.2024.10.008
Copyright © 2024 Elsevier Inc. All rights are reserved, including those for text and data mining, AI training, and similar technologies.

progress, challenges such as obtaining sufficient amounts of protein to produce homogeneous microcrystal slurry, remain. Given this, it has been paramount to develop instrumentation that reduces the amount of microcrystal slurry required for experiments. Tape-drive systems use a conveyor belt made of X-ray transparent material as a motorized solid-support to steer deposited microcrystals into the beam. For efficient sample consumption on-demand ejectors can be synchronized with the X-ray pulses to expose crystals contained in droplets deposited on the tape. Reactions in the crystals can be triggered via various strategies, including pump-probe, substrate/ligand mixing, or gas incubation in the space between droplet ejection and X-ray illumination. Another challenge in time-resolved serial crystallography is interpreting the resulting electron density maps. This is especially difficult for metalloproteins where the active site metal is intimately involved in catalysis and often proceeds through multiple oxidation states during enzymatic catalysis. The unrestricted space around tape-drive systems can be used to accommodate complementary spectroscopic equipment. Here, we highlight tape-drive sample delivery systems for complementary and simultaneous X-ray diffraction (XRD) and X-ray emission spectroscopy (XES) measurements. We describe how the combination of both XRD and XES is a powerful tool for time-resolved experiments at XFELs and synchrotrons.

1. Introduction

Serial crystallography methods merge diffraction data from several thousand microcrystals delivered sequentially at room temperature into the X-ray beam. Due to the ambient conditions used during data acquisition, the sample delivery systems can be coupled with reaction initiation schemes making serial methods well suited to time-resolved studies. These can be conducted at synchrotron or X-ray Free Electron Laser (XFEL) sources with microsecond or femtosecond long X-ray exposures, respectively. However, interpreting electron density maps of putative intermediates across a time series is often complicated by the presence of mixtures of species. Such ambiguity can be reduced significantly by collecting additional complementary data that often comes from parallel experiments with a commensurate increase in sample consumption and potential discrepancies derived from similar but not identical reaction conditions. The ideal situation is for complementary data to be collected simultaneously from the same sample and X-ray pulse, thereby providing higher sample efficiency and greater confidence in functional insights. Thus, the primary goals for our on–demand sample delivery system for serial crystallography are i) to conserve sample by delivering low volume crystal slurry droplets at the same frequency as the X-ray pulse arrival at a pulsed X-ray source (i.e. XFEL) and/or detector readout rate, ii) to enable a variety of reaction initiation strategies, and iii) to enable simultaneous collection of complementary spectroscopic data such as X-ray Emission Spectroscopy (XES).

One of the first sample delivery systems developed for XFEL sources was the Gas-focused dynamic virtual nozzles (GVDNs or liquid jets) (DePonte et al., 2008). GVDNs or improved iterations have been widely adopted and are still used for serial data collection at most XFELs (Boutet et al., 2012; Wiedorn et al., 2018). However, a characteristic of XFEL sources is that the X-rays are delivered at a pulsed frequency. As such, a major drawback of GVDNs is their sample consumption rate, due to the continuously dispensing operational mode, wasting precious sample between each X-ray pulse (Zhao et al., 2019). Sample consumption is a major concern for serial methods and limits the number of viable enzyme systems that can be studied. On-demand methods allow for the conservation of precious protein sample by synchronizing the X-ray pulse (10–100 fs duration) with the sample position at the X-ray interaction region.

The current iterations of tape-drive applications typically use on-demand Acoustic Droplet Ejection (ADE) or piezoelectric devices to dispense discrete nanoliter to picoliter droplets of the microcrystal slurry onto the moving tape (Fig. 1). Distinct advantages of this geometry are to offset the reaction time from the X-ray interaction region (decoupling data collection throughput from the reaction timepoint) and provide space for a variety of reaction initiation strategies. The ADE approach consists of a transducer immersed in the sample which can produce an acoustic pressure wave. This is then spherically focused such that acoustic energy concentrates at the liquid surface, providing sufficient energy to overcome surface tension thereby ejecting a droplet. The droplet volume is inversely related to the acoustic pulse frequency, while velocity is proportional to the amplitude. In an ideal case, a frequency ranging from 5 to 300 MHz corresponds to a droplet volume from 14 nL down to 0.065 pL (Elrod et al., 1989). For tape-drive applications, 10–15 MHz transducers are most frequently used to yield $\sim$ 1–3 nL droplets. By contrast, piezoelectric microdroplet dispensers typically use a fluid-filled capillary passed through a piezoelectric material "sleeve" that induces acoustic pressure waves to eject a microdroplet from the open end of the capillary. The droplet diameter is usually equal to the inner capillary diameter with current droplet volumes ranging from approximately 5–150 pL and at rates up to approximately 50,000 drops per second. In either case, acoustic droplet dispensing is unique in combining rapid dispensing rates and precise timing of nanoliter to picoliter volumes to maximize sample efficiency.

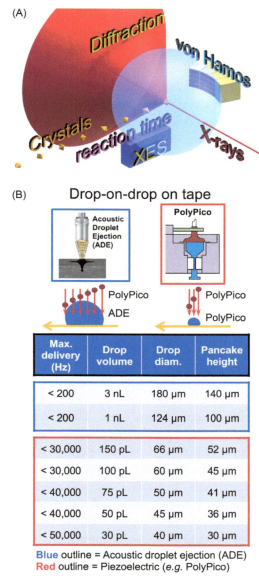

Fig. 1 Concepts and characteristics of on-demand microdroplet sample delivery strategies and drop-on-drop on tape reaction initiation methods for time-resolved crystallography. (A) Microcrystals in random orientations within droplets (not shown) are delivered into the X-ray interaction region from left to right on transport belt (not shown). X-rays intersecting crystals yield diffraction data along the X-ray path and X-ray emission photons emerging as a sphere from the same sample and X-ray pulse, especially if the sample contains first-row transition elements (either in solution or in

The first experiments using ADE to mount microcrystals of insulin and lysozyme (10 to 20 μm) on micro-mesh were directed by Orville, Allaire, and Soares at Brookhaven National Laboratory in collaboration with Labcyte Inc. (Sunnyvale CA) (Soares et al., 2011). The authors used the Echo liquid handling device at Labcyte, exploiting a 10 MHz transducer to transfer a 2.5 nL droplet containing microcrystal slurry on a micro-mesh. This experiment demonstrated that the acoustic energy could be used to eject microcrystals and was not harmful to the sample. Subsequently, these results inspired Roessler et al. (Roessler et al., 2013) to use ADE in combination with a Kapton conveyor belt to present the sample to the X-ray beam at a synchrotron facility. The authors ejected a nanoliter droplet containing thermolysin crystals (20 to 200 μm) onto a moving Kapton tape before cryo-cooling the droplet using a cryostream. Subsequently, a full dataset was obtained by merging 30° sweeps from 22 crystals at 100 K. Several additional applications followed using the ADE technique to transfer droplets containing ligand onto crystals or alternatively eject crystallization buffer and protein solution (Cole et al., 2014; Cuttitta et al., 2015; Soares et al., 2014; Yin et al., 2014). In 2013, ADE was used for the first time to our knowledge as an on-demand sample delivery system at the XPP instrument at Linac Coherent Light Source (LCLS) (proposal L748, Orville et al. (2013)). The microcrystal containing droplets were ejected through free space and synchronized with the X-ray pulse at a frequency up to 60 Hz. Complete serial datasets of lysozyme, thermolysin or stachydrine demethylase were collected to 2.13 Å, 2.52 Å and 2.20 Å resolution respectively (Roessler et al., 2016).

Synchronizing droplets moving in free space with arrival of the XFEL pulse at the interaction region can be challenging. A more robust solution combines ADE droplet deposition onto a conveyor belt that transports the sample into the interaction region as demonstrated by earlier experiments at the National Synchrotron Light Source (Roessler et al., 2013). A new tape-drive using ADE was designed in collaboration between Brookhaven National Laboratory and Lawrence Berkeley National Laboratory (LBNL) (Fuller et al., 2017). The current incarnation of this sample delivery system

---

crystals). The von Hamos geometry analyzer crystals fill four, independent, 4 × 1 arrays that focus the energy dispersive X-ray spectrum of the emitted photons on a 2D detector (labeled XES) located under the sample and X-ray interaction region. (B) ADE (assuming 10–15 MHz transducer frequency) and piezoelectric (assuming an inner diameter capillary of ∼ 30–150 μm) droplet characteristics used for tr-SSX or tr-SFX and tr-XES studies.

is versatile allowing a large variety of time-resolved experiments such as pump-probe (Bhowmick, Hussein, et al., 2023; Bhowmick, Simon, et al., 2023; Burgie et al., 2020; Hussein et al., 2021; Ibrahim et al., 2020; Kern et al., 2018; Makita et al., 2023; Simon et al., 2023; Young et al., 2016) anaerobic and/or O_2 gas mixing (Lebrette et al., 2023; Miller et al., 2020; Ohmer et al., 2022; Rabe et al., 2021; Srinivas et al., 2020), or active drop-on-drop mixing experiments in combination with conveyor belt droplet delivery (Butryn et al., 2021; Nguyen et al., 2023) (Fig. 2). One of the advantages of this sample delivery approach is the possibility to combine it with complementary spectroscopic methods.

1.1 Complementary spectroscopy

Understanding the relationship between structure and function is of fundamental importance for assigning biochemical functions to macromolecules including proteins and enzymes. Structural biology employs techniques such as macromolecular X-ray crystallography to determine atomic resolution structures of enzymes. The complex relationship between electronic and atomic structure underlies the interplay between structure and function, particularly during enzymatic catalysis. Recent advances in room temperature, time resolved crystallography have enabled the observation of transient and metastable intermediates, which are often found as convoluted mixtures of different chemical states (Hough & Owen, 2021). While macromolecular crystallography (MX) provides exceptional atomic resolution structural information, it lacks direct electronic observations which is important to characterize different intermediate states. In addition, due to the high intensity X-rays used for diffraction studies, some chemical moieties, such as metal ions and disulfide bridges, are particularly prone to undergo radiation induced alterations (often referred to as site-specific radiation "damage") during data acquisition (Beitlich et al., 2007; Chreifi et al., 2016; Hajdu et al., 2000). In order to reduce these radiation-induced changes, serial synchrotron MX permits low-dose structural determination (Ebrahim et al., 2019), while serial femtosecond X-ray MX provides "damage" free structural determination – typically referred to as SSX and SFX, respectively.

Monitoring of X-ray radiation-induced alterations can be done using spectroscopic signals and is important to validate the refined atomic model(s). Moreover, spectroscopic studies provide information about electronic states of various intermediates, adding valuable information about observed chemical species. Consequently, significant efforts have been made to complement

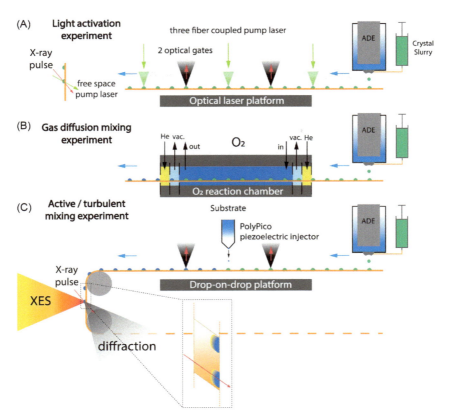

Fig. 2 Schematic of the time-resolved experiments offered by the LBNL tape-drive associated with XES. The slurry of microcrystals is stored in a syringe refilling the well for ADE ejection. The microcrystal slurry droplets pass through either the optical laser platform (A), the O_2 reaction chamber (B), or the drop-on-drop platform (C) before reaching the interaction region with X-ray. The X-ray beam grazes the Kapton belt and allows XRD in the forward path and XES data in ~90° direction. (A) The optical laser platform consists of 3 evenly separated fiber-coupled pump laser (green arrows) and 1 free space pump laser. The free space pump laser is positioned just before the interaction region with X-ray allowing shorter time-points. (B) The gas reaction chamber is filled with 100 % O_2 and covers about 60 mm of the tape. The system is enclosed in a chamber filled with He with a concentration of < 350 ppm O_2. (C) The drop-on-drop capability includes a piezoelectric injector depositing at high frequency picoliter volume ligand droplets on the ADE crystal slurry drops. Only the optical laser and the drop-on-drop platforms can accommodate the IR gates. The inclusion panel shows that the drops are sitting at the front edge of the Kapton tape while the X-ray beam is grazing the tape.

crystallographic observations with various spectroscopic techniques (Bachega et al., 2015; Cohen et al., 2016; Daughtry et al., 2012; Engilberge et al., 2024; Fitzpatrick et al., 2007; Héroux et al., 2009; Huo et al., 2015; Li et al., 2015; Major et al., 2009; Orville et al., 2011; Pompidor et al., 2013; Rose et al., 2024; Sakaguchi et al., 2016; Stoner-Ma et al., 2011; Von Stetten et al., 2015; Yi et al., 2010). These spectroscopies cover the electromagnetic spectrum from UV to near-infrared, and encompass techniques such as electronic absorption, fluorescence, and Raman spectroscopies. It is vital to focus the optical photon spot size typically to $\sim 20\,\mu m$ and to interrogate the same sample volume with X-ray and spectroscopic photons. The electronic absorption spectrum from single crystals is almost always anisotropic and varies with crystal rotation (De La Mora-Rey & Wilmot, 2007; Hutchison et al., 2022; Kendrew, 1950; Ronda et al., 2015). The collection of spectroscopic and X-ray diffraction data is often interleaved rather than simultaneous. In some experimental geometries the optical and X-ray paths are orthogonal to each other. However, since the (often cryocooled) sample and the two paths intersect at the eucentric sample rotation point, the same volumes are interrogated without a need to move any of the optical components for electronic absorption spectroscopy in a $180°$ transmission path. Some spectral acquisition geometries deploy the microphotospectrometer axis with the X-ray beam, that is coaxial, which helps ensure the ideal overlap between the optical and X-ray pathways, but also requires motions of the collection objective since it will block the diffraction pattern.

There are particular challenges associated with data collection of optical signals in microcrystals (Dworkowski et al., 2015). Collection of electronic absorption spectroscopic data requires the presence of a chromophore that undergoes changes during enzymatic activity. In contrast, Raman spectroscopy does not rely on the presence of a chromophore for data acquisition in a backscatter geometry and can theoretically investigate specific protein and substrate bonds directly (Carey, 2006). However, any molecule with N atoms can elicit $3N - 6$ normal modes of vibration; therefore, a typical enzyme-substrate complex has many Raman vibration bands, only some of which are of interest to catalysis and/or are shifted outside those of the enzyme. Thus, the number and similarities in their chemical nature make Raman spectroscopy assignment challenging, often requiring isotopic labeling for validation. Fortunately, resonance Raman (RR) allows for selective excitation of specific bond vibrations, reducing the complexity of the observed spectrum. Metal-ligand bonds often undergo electronic transitions at significantly lower energies than the protein backbone and are thus usually chosen for selective enhancement by RR. Nevertheless, the higher energy absorption associated with laser irradiation may lead to radiation induced alteration of chemical bonds.

Additionally, optical spectroscopy on solid materials such as microcrystals containing high extinction coefficient chromophores, requires thin enough material to allow the probing light to pass through (Grunbein et al., 2020). The scattering nature of microcrystals, which are often presented in random orientations to the probing source, and their surrounding liquid, or the dissolved compounds (e.g. chromophores) can contribute to the deterioration of signal to noise for complementary spectroscopy.

Various X-ray spectroscopy techniques, including X-ray emission spectroscopy (XES), can similarly provide valuable insights into enzymatic activity and are not affected by optical scattering and crystal size. Non-resonant XES is particularly well-suited for investigating first-row transition metals commonly found in metalloenzymes (Makita et al., 2023). The photon energy required for excitation above 1 s core holes, a principle behind the observation of XES, is often used for X-ray diffraction studies and does not require the photon energy to be precisely tuned or monochromatic. Observation of XES is limited to metalloenzymes which represent about 40 % of the total enzymes (Feehan et al., 2021); however, the detailed element-specific information obtained is invaluable due to the crucial role metals typically play in enzymatic catalysis. Additionally, the spectroscopic signal could, in principle, be obtained from the exact same molecules that provide the X-ray diffraction (XRD) data, ensuring a direct correlation between structural and electronic information.

1.2 X-ray emission spectroscopy

XES is an inner-shell spectroscopic technique that probes the electronic structure of metal complexes and is sensitive to their chemical environment, spin states and geometries (De Groot, 2001; Glatzel & Bergmann, 2005; Zimmermann et al., 2020). In K-shell XES, X-ray radiation causes photo-ionization, which excites a 1 s electron into the continuum, creating a core hole. This 1 s vacancy is subsequently filled by an electron from a higher orbital, accompanied by the emission of a fluorescent photon (Fig. 3A). The energy of the emitted photon corresponds to the difference in energy level between the different electronic orbitals. For 3d transition metals commonly found in enzymes these transitions lie in the hard X-ray regime. Selection rules dictate fluorescence transitions and their relative intensity. The most pronounced fluorescence signal is observed from the 2p-1s transitions, i.e. $K\alpha$ lines.

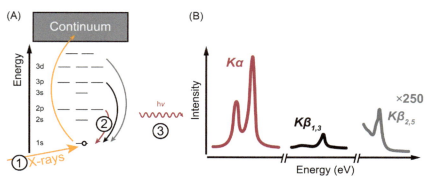

Fig. 3 Schematic representation of X-ray emission spectroscopy (XES) phenomena. (A) Diagram showing relative energy levels of different orbitals in a first row transition metal; 1) X-rays excite an electron into the continuum, 2) 1s core hole is filled by an outer shell electron, 3) a photon is emitted because of the change in orbital energy state. (B) Schematic representation of XES data, showing relative intensity of the different signals as observed for first row transition metals. Note that the $K\beta_{2,5}$ signal is enhanced × 250 fold for illustration. Typically, analyzer crystal arrays (see Table 1) are used to collect $K\alpha_{1,2}$ lines from one 4x1 bank, whereas the weaker $K\beta_{1,3}$ and $K\beta_{2,5}$ signals require a larger solid angle and may use up to four 4x1 banks with up to 16 analyzer crystals that each reflect photons across the energy range of both lines.

Transitions from higher orbitals such as 3p-1s and 3d-1s transitions, i.e. $K\beta_{1,3}$ and $K\beta_{2,5}$ lines respectively, are observed with lower intensity (Fig. 3B). These core-to-core transitions do not probe the valence shell directly and instead rely on indirect sensitivity. The greater overlap of 3p/3d orbitals compared to 2p/3d does, however, result in more pronounced spectral differences in $K\beta$ versus $K\alpha$ lines (Lafuerza et al., 2020). To compensate for the lower signal intensity observed for $K\beta$ in comparison to $K\alpha$, longer acquisition times or collection over a larger solid angle are necessary. Depending on experimental constraints, a trade-off must be made. Currently, even though capturing $K\beta$ spectra with a larger solid angle (which comes at a significant upfront cost for analyzer crystals) is possible, most of the data collection focuses on $K\alpha$ spectroscopy due to the more efficient data collection process. It is important to note that when simultaneously collecting XRD and XES data, acquiring a sufficiently strong XES signal often requires more time than required to collect enough high-resolution lattices to support refinement of atomic resolution models. Only when XRD data collection is particularly challenging (e.g. low symmetry space group, and/or high spatial resolution) does the XRD process become the time-limiting factor.

1.3 XES setup components

The core component of an emission spectrometer are Bragg reflection analyzer crystals (Table 1), which spectrally analyze the emitted fluorescent signal from the sample and reflects it onto a photon detector. Various experimental setups for the collection of XES data have been conceived over time, all relying on a change in reflective angle to scan an energy range. To improve sensitivity and collection efficiency, attempts at capturing a larger solid angle of the emitted signal have produced multi-Bragg crystal analyzer setups. These setups can frequently be found at spectroscopy beamlines of various synchrotron radiation facilities, relying on point-to-point scanning spectrometers (Fig. 4A). However, the deployment of the drop-on-demand system at XFELs, where shot-by-shot variation in beam position, intensity and energy, in combination with the slight variation in droplet positioning, produces considerable challenges for point-to-point scanning spectrometers. These considerations have led to the decision to collect an entire spectrum on a per-shot basis, deploying a multi-crystal von Hamos geometry setup (Fig. 4B), where emitted photons are reflected by cylindrically bent analyzer crystals. These disperse the energy in one dimension (vertical) while focusing in the other (horizontal), creating a line on a 2D position-sensitive detector (PSD). This approach has been demonstrated successfully at XFELs and is compatible with synchrotron facilities (Alonso-Mori, Kern, Gildea, et al., 2012; Alonso-Mori, Kern, Sokaras, et al., 2012; Fransson et al., 2018; Fuller et al., 2017; Makita et al., 2023; Sahle et al., 2023).

This book chapter describes the various aspects of a drop-on-demand tape-drive experiment that combines X-ray diffraction with XES studies, including the experimental setup, requirements for the crystalline sample, planning and preparations for the experiment, reaction initiation sequences, and data collection of both diffraction images and XES spectra. The success of the LBNL tape-drive described in this chapter, has inspired research and development projects at other facilities to build tape-drive systems for sample delivery. At Diamond Light Source, we are currently spearheading projects for a robust and reliable drop-on-demand tape-drive system that is simple to operate for regular users and allows for simultaneous XES and XRD for time-resolved experiments at both synchrotrons and XFELs. We will briefly describe our work at the end of this chapter.

Table 1 Examples of von Hamos analyser crystals for XES.

Element (redox states)	Spectral line	Suggested #, crystal cut	Energy (keV)	Bragg angle	Reflectivity[a]	ΔE[b] (meV)
Mn (II, III, IV)	$K\alpha_{1,2}$	4x Ge(3,3,3)	5.899 5.888	74.8° 75.3°	72.9 74.9	119.1 118.9
	$K\beta_{1,3}$	16x Si(4,4,0)	6.490	84.2°	102.5	61.3
Fe (II, III, IV)	$K\alpha_{1,2}$	4x LiNbO$_3$(2,3,−4)	6.404 6.391	84.3° 85.6°	97.0 103.4	101.4 101.2
	$K\alpha_{1,2}$	4x Ge(4,4,0)	6.404 6.391	75.5° 75.9°	90.4 93.4	140 139.8
	$K\beta_{1,3}$	16x Ge(6,2,0)	7.058	79.1°	84.4	106.8
Ni (I, II)	$K\alpha_{1,2}$	4x LiNbO$_3$(4,2,2)	7.478 7.461	82.4° 83.4°	78.8 91.4	77.3 77.2
	$K\beta_{1,3}$	16x Si(5,5,1)	8.265	80.5°	20.1	25.8
Cu (I, II)	$K\alpha_{1,2}$	4x Si(4,4,4)	8.048 8.028	79.3° 80.1°	29.1 31.5	39.4 39.3
	$K\beta_{1,3}$	16x Si(5,5,3)	8.905	79.9°	14.6	21.4

www.aps.anl.gov/Analyzer-Atlas/Analyzer-Atlas

[a]Integrated Reflectivity, a measure of the reflection strength the dynamical reflectivity is numerically integrated over the whole rocking curve.

[b]Intrinsic Energy Resolution, ΔE, energy resolution of the analyzer reflection due to its intrinsic (Darwin-) width.

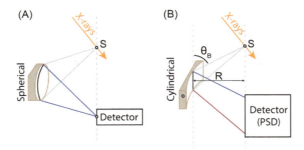

Fig. 4 Schematic representation of X-ray emission spectroscopy (XES) setup. (A) Point-to-point XES setup utilizing spherically bend analyzer crystals to refocus signal (one λ at a time from a scanning monochromator) onto a photon counting detector; (B) Point-to-line XES setup utilizing spherically bend analyzer crystals in a von Hamos geometry for energy dispersion spectrum in a shot-by-shot mode with photon X-ray energy higher than the absorption edge for the element of interest. On the image are indicated: S, sample location; R, focusing radius under Bragg reflection condition; θ$_B$, angle of incidence. PSD: two-dimensional, position sensitive area detector.

2. Materials
2.1 Sample delivery

Sample delivery is an important aspect for room temperature serial crystallography experiments; consequently, various approaches have been explored and most excel with one or more use cases (Henkel & Oberthur, 2024). Although some strategies, e.g. GVDNs and viscous extruders, have been adopted at various facilities, no single sample delivery method has been fully standardized for serial crystallography at synchrotron or XFEL facilities. Tape-drive sample delivery systems have been used by different research groups for serial crystallography. They all include two main components i) a tape or motorized conveyor belt, typically made of Kapton (polyimide X-ray transparent material) that transports the sample to the interaction region ii) a sample injector that deposits the sample onto the tape. The sample injector can be on demand, like ADE or piezoelectric dispensers, or continuous like microfluidic devices. Access to a tape-drive sample delivery system depends upon circumstances and facility; in some cases, access may include formalized collaborations with groups outside the particular facility. While several tape-drive systems are available (Table 2), the LBNL tape-drive approach provides the user with the widest range of reaction initiation strategies and enables simultaneous collection of spectroscopic data in the form of XES. Because XRD and XES data are often

Table 2 Tape drive sample delivery systems available (or soon to be) at XFELs and synchrotrons.

	Time-resolved experiments			Facilities		Complementary techniques			
Tape Drive *Current Capability*	Pump probe	Liquid-liquid mixing	Gas diffusion	Synchrotrons	XFELs	XES	UV–VIS/Raman/XAS	X-ray path	Tape drive system
LBNL	Free-space ~ ns − ms 3 fibers 100 s ms	10 s ms–10 s ~60 s	100 s ms–10 s	(ALS) (SSRL)	LCLS, SACLA, SwissFEL, PAL-XFEL	Yes (Fe, Cu, Mn, Ni)	In progress	Parallel to tape surface	Reuse tape after cleaning/drying
CFEL TapeDrive2.0	No	Few s – ~50 ms	No	Petra III, ESRF-EBS	No	No	No	Through tape	Reel to Reel
Inject-and-transfer system	No	No	No	No	PAL-XFEL	No	No	Through tape	Reel to Reel
Near Future									
Diamond tape drive	Yes	10 s ms–20 s	100 s ms–20 s	Diamond	SwissFEL	Yes (Fe, Cu, Ni)	In development	~ Parallel to tape surface	Reuse tape after cleaning/drying
SACLA	Yes	1.3 to 19 s	No	SRring−8	SACLA	In development	In development	Through tape	Reel to Reel
LCLS/SSRL (in collaboration with LBNL)	Yes	Yes	Yes	SSRL	LCLS	Yes (Fe, Cu, Mn)	In development	Parallel to tape surface	Reuse tape after cleaning/drying

measured simultaneously with the LBNL tape-drive, the X-ray beam passes approximately parallel to and above the belt. Consequently, the stronger XRD data is collected along the beam path with a commensurate small background effect from the belt, and the XES analyzer array located about 90° from the flat Kapton surface reflects the weaker XES signal to a second detector approximately in the plane of the sample.

Oberthuer and collaborators developed a simple and robust tape-drive for time-resolved experiments (Beyerlein et al., 2017; Henkel et al., 2023; Zielinski et al., 2022). Rather than relying on droplet ejection the CFEL TapeDrive 2.0 uses microfluidics to dispense a continuous stream of sample on the tape before being exposed to X-ray. Reaction initiation for time-resolved experiments is achieved by diffusion mixing in the nozzle before dispensing onto the tape with the desired reaction time point depending on the tape speed and the nozzle distance to the X-ray interaction region. Experiments with the TapeDrive 2.0 are possible at the P11 and ID29 instruments from the Petra III and ESRF-EBS synchrotrons, respectively. In this case the X-ray beam is perpendicular to and passes through the tape before interacting with the sample, consequently the diffraction pattern is convoluted with a relatively smooth background. The CFEL tape-drive is simpler in its composition compared to the LBNL tape-drive. It currently only supports mixing type time-resolved experiments, and the conveyor belt system uses a "reel to reel" strategy, which eliminates the need for a tape washing procedure in belt tape-drives. Like the CFEL TapeDrive 2.0, a similar system has been developed at PAL-XFEL with capillary dispensing sample on a polyimide film (Lee et al., 2022). Other sample delivery systems based on tape-drive strategy are in development for XFELs at LCLS and SACLA, but at the time of writing no further information is publicly available. A tape-drive sample delivery system in combination with XES at Diamond Light Source is under commissioning and we anticipate accepting user experiments in 2025. The Diamond tape-drive will also be available for beamtime at the Cristallina endstation at SwissFEL.

2.2 XES, detector and analyzer crystals

The essential component to the von Hamos geometry XES setup is the analyzer crystal array. Design decisions, such as working distance, the number of and material type for the analyzer crystals, and the orientation of the analyzer array are all constrained by experimental objectives and end-station design. The XES signal is emitted spherically from the sample as a point source from the irradiated sample interaction region. The angle of incidence (θ_B, Fig. 5A and B) defines the Bragg reflection and dispersion of X-ray energies of the element under investigation. Von Hamos style analyzer crystals typically capture a fraction of the total required photons to produce a clear X-ray emission spectrum on a shot-by-shot basis. Due to the specific energy range that each analyzer crystal can reflect, different sets of crystals, each cut for a specific energy range of interest for a given element, are often required for analyzing different elements and emission lines (Table 1). It is, however, theoretically possible to deploy different analyzer crystals in a separate bank, which, when adjusted for height to match the right geometry, can reflect XES signal from a different element onto the same or different detector provided its active area has sufficient width to measure the signals without overlap. In order to capture a larger solid angle of the emitted photons, multiple analyzer crystals are deployed, increasing the signal-to-noise in the resulting data. The optimal signal-to-noise ratio is achieved at a 90° angle of the incidence X-rays, where energy dispersion and elastic scattering noise is lowest (Alonso-Mori, Kern, Sokaras, et al., 2012), this also improves energy resolution.

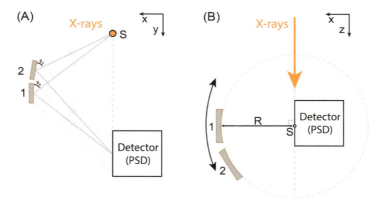

Fig. 5 Schematic representation of the X-ray emission spectroscopy (XES) setup highlighting how von Hamos geometry affects recorded data. (A) Front view that demonstrates how only a single analyzer crystal per bank (crystal 1) can be optimally positioned geometrically to provide a well-focused spectrum on the detector, while flanking analyzer crystal(s) (crystal 2) do not match the geometry exactly, causing some level of aberration; (B) Top view of the setup showing the Rowland circle on which the crystals are positioned to refocus the signal onto the position sensitive detector (PSD). Note that analyzer crystal 2 in b), is not positioned at 90°, which causes some degradation of signal-to-noise due to an increase of elastic scattering signal captured. S: sample interaction point; R: bending radius; θ_B: angle of incidence.

When refocusing the XES signal onto the position sensitive detector (PSD), the signals from the various analyzer crystals can be overlapped or projected separately. Due to the geometrical requirements for this experiment, only a single analyzer crystal can be positioned at the ideal position within a given vertical bank (Fig. 5A). The other crystals within the same bank will adopt a slightly different geometry, leading to a less well focused spectrum onto the detector. Whether to overlap or separate these spectra is a trade-off between signal to noise and resolution. Due to the photon starved nature of the experiment, reducing background noise can significantly enhance the observed signal. Replacing the air/atmosphere between the sample, the analyzer crystals and the detector with a $He_{(g)}$ environment, reduces air scatter, improving signal to noise. Depending on the available space and access to $He_{(g)}$, this can either be a specific cone, between the sample position and the analyzer crystals, or a larger section of the whole drop-on-demand tape-drive system, being put under $He_{(g)}$ environment. Note that the use of $He_{(g)}$ can come at significant costs and that some level of recycling is advisable.

The von Hamos style spectrometer is operated in a back-scattering geometry, aiming for angles of incidence close to $90°$ in order to achieve high photon collection efficiency and resolution. The emitted photons at the sample position are reflected under Bragg conditions onto the PSD. Apart from the angle, the particular cut of material and the lattice spacing govern the energy range reflected. In the hard X-ray regime (>5 keV), the most common materials for analyzer crystals for XES include Si, Ge, and $LiNbO_3$ (Table 1) (Gog et al., 2013). Silicon is commonly used as it provides good mechanical and thermal properties, and well-understood manufacturing techniques. Due to the higher Z, Ge provides better performance for certain X-ray wavelengths. The advantage of $LiNbO_3$ is that it offers a finer variability in d spacing. This allows for the selection of a material cut that is close to the backscattering angle, therefore often improving resolution and photon collection efficiency.

The overall energy resolution (ΔE_{total}) of a given setup consists of a combination of geometrical factors such as beam size and pixel pitch (ΔE_{geom}), the Darwin width of the analyzer crystal (ΔE_{Darwin}), and imperfections in the crystal lattice, often referred to as crystal strain (ΔE_{imper}):

$$\Delta E_{total} = \sqrt{E_{geom}^2 + E_{Darwin}^2 + E_{imper}^2} \tag{1}$$

Increasing the working distance of the analyzer crystals (R in Fig. 5B), would spread the reflected energies back across more pixels on the detector, effectively improving resolution. However, achieving an equivalent solid angle in such a configuration is often more challenging and comes at increased cost. Conversely, reducing the pixel pitch will theoretically improve resolution. At a certain level, the contributions of crystal strain will become the limiting factor for resolution. Additionally, charge sharing effects will be more common at smaller pixel pitches, which complicates the data (Pellegrini et al., 2007). Correction algorithms are commonly applied to raw data to mitigate this issue (Jirsa et al., 2024).

3. Methods
3.1 Preparing microcrystal slurries
3.1.1 Sample requirements
In order to perform serial crystallography experiments using a drop-on-demand sample delivery system such as the LBNL tape-drive, access to sufficient quantities of a microcrystal slurry is important. There are excellent protocols available online on how to approach the optimization for microcrystal slurries (Barends et al., 2022; Beale et al., 2019; Dods et al., 2017; Shoeman et al., 2023). Here we will focus on aspects that are important specifically for drop-on-demand sample delivery. Prior to the experiment, it is advisable to prepare a plan outlining the number of time points and consider any contingency measures. This will help estimate the necessary amount of crystalline material. Operating at a sample consumption rate ranging from 1 to 10 µL/minute, the drop-on-demand system is more efficient than most GVDNs, though it does require more protein sample compared to traditional cryo-crystallography. There are no universally accepted definitions of what dimensions constitute a microcrystal or even a microcrystal slurry for that matter. However, the design of the drop-on-demand system provides a rough framework for defining a microcrystal slurry suitable for efficient data collection. The acoustic droplet dispenser operates continuously by being fed through capillaries (typically 200 µm), which sets an upper limit to the crystal size in a given sample. Although it is possible to remove a limited number of large crystals by filtering through a capillary, ideally, the average size of crystals is such that the longest dimension does not exceed half the capillary diameter (i.e. <100 µm). Crystals are three dimensional objects, however, their size is

often described as a one-dimensional unit, defined by the longest dimension (Fig. 6). What defines the longest dimension, depends on the crystal morphology, e.g. for needle like crystals the longitudinal dimension, for rhombohedron shaped crystals the diagonal. Due to the different morphologies, there is inherently larger variation in average crystal size, which may make it more difficult for some morphologies to match the size requirement. Moreover, short dimensions along a crystal can facilitate rapid diffusion of ligand or substrate through a crystal lattice, which benefits the homogenous initiation of reactions throughout the crystal lattice.

The volume of microcrystal slurry required per dataset for drop-on-demand methods depends on the crystal concentration. Experimental data showed that slurries with a density of 10^6–10^7 crystals/mL, which should result in a so called "hit rate" of 20–50% (it's a ratio of images with indexable Bragg spots over the total number of images collected) are well suited for this approach. Lower density slurries tend to require extended acquisition times due to the fewer indexed lattices per unit time, while more dense crystal slurries do not necessarily lead to a proportionately higher number of indexed patterns because of problems related to indexing/integrating multiple lattices (see also data processing in Chapter 6). Furthermore, to homogenously initiate changes in crystals inside droplets, a sufficiently high concentration of (co-)substrate/reactant/ligand must be obtained relative to the protein concentration, which depending on the solubility of the substrate/reactant/ligand can be challenging for high crystal density slurries. When these conditions are met, depending on space group symmetry and desired resolution, data collection can be completed in anywhere from 15–45 min per time point (~75 to ~225 μL assuming ~3–6 nL per droplet delivered at ~ 30 Hz) with the LBNL tape-drive. Although there is no strict

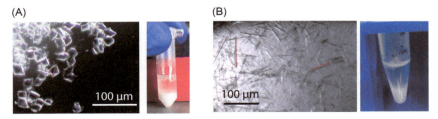

Fig. 6 Comparison of typical crystal morphologies for time-resolved MX studies. (A) Microcrystal slurry of Papain-like protease from SARS-CoV-2, both microscopic and macroscopic view of the slurry; (B) Microcrystal slurry of oxacillinase-10, both microscopic and macroscopic view. Note, the red arrow indicates the size of the a few crystals in one-dimension.

definition of what constitutes a microcrystal or microcrystal slurry, the technical aspects discussed above specific to collecting data using a drop-on-demand system establish some guidelines and boundary conditions for what is considered a microcrystal slurry.

3.1.2 Sample preparation

In theory it is possible to directly probe microcrystals obtained from screening conditions; however, assessing diffraction quality from one or more macrocrystals is typically done to evaluate characteristics (i.e. morphology, crystal lattice packing, etc.), before optimization for serial crystallography strategy exploiting microcrystals. As with other crystallization optimization procedures, it is important to identify parameters that affect the quality of the sample, e.g. number of crystals formed, size, and diffraction quality. When possible, it can be beneficial to eliminate or minimize the amount and number of components in the precipitation solution as well as consider overall viscosity of the final conditions. The drop-on-demand system requires the formation of droplets, which is more challenging for exceptionally viscous liquids. Proteins that crystallize in lipid cubic phase (LCP) for example, are unsuited and require a different sample delivery approach (Weierstall et al., 2014). Nearly all of the most common crystallization conditions work for ADE and PEI droplet ejection. Apart from changing physical properties of the solution, it is also important to be aware of any additives in the crystallization solution that could compete for binding at the active site or elsewhere of interest in a protein. Even when these components have a low affinity for the binding site, due to their relatively high concentration, they can still occupy areas of interest and make interpretation of electron density maps more challenging. In addition, when simultaneously collecting XES data, it is important to keep in mind that the signal arises from all the relevant metal atoms in X-ray interaction region, both in solution and bound to the protein crystal. As such, it is highly advisable to avoid the addition of excess metal to the solution, and to verify the proportion of bound and unbound metal in the mixture.

Initial microcrystallization optimization attempts are usually completed using vapor diffusion methods (sitting or hanging drop), which are conservative on sample consumption. It is possible to obtain enough microcrystal slurry through vapor diffusion-based methods to conduct a limited number of serial crystallography experiments. However, the sample volumes required for more comprehensive studies involving multiple perturbed states, favor a batch crystallization approach that scales readily, yields consistently, and reduces the burden of setting up each experiment. Thus, over the past decade, many new technologies have been developed for the preparation of microcrystal slurries in batch conditions.

Formation of microcrystals is often achieved by inducing secondary nucleation points within a crystallization solution, through microseeding. The average size of crystals in the final slurry is strongly dependent on the density of seeds, i.e. a higher density of seed stock, results in more nucleation and thus smaller crystals. Batch crystallization can be performed using Eppendorf tubes, but also 96 well plates.

When microcrystallization conditions are established, variations in seed batches will affect both size and by extension density of the crystals in the slurry. Due to the strict upper size limit set by mechanical constraints for drop-on-demand systems, it is important to avoid inhomogeneity of seed stocks. Thus, it can be beneficial to filter seed stocks to avoid larger macroseeds (i.e. larger, visibly observable crystals) from growing in the crystal slurry. Others (Dods et al., 2017; Shoeman et al., 2023) have reported higher diffraction quality when seeding with filtered seeds. Similarly, it is also possible to obtain a microcrystal slurry from crushing larger crystals. Alternatively, microfluidic technologies have demonstrated the potential avenue to control the size of microcrystals (Stubbs et al., 2024). To avoid clogging of capillaries, it is advisable to filter a crystal slurry by passing the slurry through a short capillary line attached to a syringe.

One challenge for obtaining a good microcrystal slurry, is achieving a sufficiently high density. The density of a crystal slurry can be increased through settling of crystals by gravity or centrifugation and removing excess mother liquor. Care must be taken when applying centrifugal forces to a slurry since they might damage the crystals, resulting in poor diffraction quality.

Homogenization of the crystal size and morphology can be challenging. Some reports have demonstrated that continued agitation during crystallization can lead to more homogenous slurries, but not always (Mahon et al., 2016). We and others have shown that microfluidics can yield good quantities of homogenous sized crystals suitable for serial MX studies (Abdallah et al., 2015; Echelmeier et al., 2019; Heymann et al., 2014; Stubbs et al., 2024).

To ensure consistent quality of a microcrystal slurry, it is advisable to test a combination of protein batches, precipitant solution, and seed stock, prior to planned experiments. It is important that the precipitant solution and protein solution are free of debris which can lead to formation of non-diffracting particles. It may be necessary to centrifuge and/or filter these solutions prior to crystallization.

3.1.3 Sample characterization

Prior to diffraction experiments, a crystal slurry can be visually evaluated using a microscope by checking the i) crystal size, ii) crystal density, iii) settled pellet size (Fig. 6). These parameters combined can give an indication of the sample performance, with exception of diffraction quality. When estimating the size of the crystals it is important to perform multiple measurements to obtain a reliable crystal size distribution. At present, there is no robust, accurate, and reliable automated method for determining the crystal size distribution, with most reports relying on manual counts from microscope images. Estimating the density of a crystal slurry can be done using a Neubauer counting chamber. The settled pellet is the volumetric ratio between the total slurry volume and the volume occupied by the crystals after settling. Testing for diffraction with X-rays is the ultimate validation of the crystal slurry.

3.2 Reaction initiation scheme

The LBNL tape-drive sample delivery system can accommodate three different types of experiments in combination with complementary XES (Table 2, Fig. 2). The reaction initiation strategies are i) light activation, ii) gas diffusion, iii) liquid droplet-on-droplet addition. Scientists typically choose the appropriate strategy for their studies; however, switching between different strategies during beamtime is not advised because of the time and effort required to alter the setup. A given reaction time point (Δt) is defined by two factors: i) the conveyor belt speed, and ii) the distance between the X-ray interaction point, and the position where the reaction initiation occurs (Table 3). The tape-drive has three locations to mount equipment for reaction initiation, predefined upstream from the X-ray interaction region at 180, 120, and 60 mm, respectively. Varying both the initiation positions and the conveyor belt speed, one can achieve reaction times (Δt) ranging from 0.02 to 10 s.

3.2.1 Light activation experiment

For time-resolved serial crystallography one way to create an excited state or initiate a reaction in a microcrystal slurry, is to expose (pump) the samples to a light source of a given wavelength (λ) at a given point in time (pump at $t = 0$). Subsequently, the sample is probed with X-rays after a variable time delay (Δt) in order to catch intermediate states (Colletier et al., 2018). Usually, the light is provided by a pump laser, but it can also be generated by a LED. To be reactive to light the sample must contain a chromophore; studied examples include bacteriorhodopsin, Photosystem II

Table 3 Possible reaction times according to the tape speed of the LBNL tape drive.

| Tape speed (mm/s) | Reaction time (Δt) in s | | |
	Position 1 (180 mm)	Position 2 (120 mm)	Position 3 (60 mm)
30	6	4	2
50	3.6	2.4	1.2
100	1.8	1.2	0.6
200	0.9	0.6	0.3
300	0.6	0.4	0.2
500	0.36	0.24	0.12
600	0.3	0.2	0.1
3000	0.06	0.04	0.02

or phytochrome. Alternatively, one can use a photosensitive-caged compound mixed into the microcrystal slurry, upon exposure to a light pulse the compound (substrate, ligand or inhibitor) will be released and promote changes within the crystals (Monteiro et al., 2021).

Applying a high-power laser on biological sample is not without consequences. One should adjust the laser power to provide about 1 photon per chromophore to limit the absorption of multiple photons. Recent work showed that this multi-photon absorption could produce non-physiological response from the sample (Barends et al., 2024). It is strongly advised to perform a pump laser power titration to establish the linear photo-excitation regime.

The most notable pump-probe experiment with the LBNL tape-drive is Photosystem II (PSII). The optical laser platform (Fig. 2A) is specifically designed to accommodate the reaction scheme of PSII within the Kok's S-state cycle. The optical laser platform and the free space laser drive the transition between S-states of the oxo-bridged tetra manganese calcium (Mn_4CaO_5) cluster in the oxygen evolving center. Depending on the tape speed, the laser position (Table 3) used from the optical laser platform or the free space laser (near the interaction region), one can achieve time point ranging from 100 s ms to 10 s μs (Bhowmick, Hussein, et al., 2023; Bhowmick, Simon, et al., 2023; Hussein et al., 2021; Ibrahim et al., 2020; Kern et al., 2018; Simon et al., 2023).

Another example of the use of pump-probe experiment with the LBNL tape-drive is the study of the phytochrome Agp2 to reveal intermediates after photoisomerization of the chromophore with time point ranging from 33 ms to 6 s ("A phytochrome at work – serial-femtosecond crystallography reveals how a photoreceptor couples chromophore and protein structural changes.", Sauthof et al., under peer-review).

In those examples the fastest time point is in the order of 10 s μs but the use of a free space femtosecond laser one can achieve light initiation within ~100 fs before probing with X-ray pulse (Simon et al., 2023). The tape-drive therefore provides considerable flexibility for the light initiation sequences, with multiple or single exposures from a few seconds to ~100 fs before X-ray pulse interaction, but a strong attention must be applied to the laser power to avoid multi-photon effects.

3.2.2 Gas diffusion experiment

A different way to initiate a reaction in microcrystal slurries, is through gas diffusion into the microdroplets. The sample delivery setup must be contained in an enclosed environment deprived of the molecular gas of interest. After dispensing the microcrystal slurry onto the Kapton tape, the droplets are passed through a gas incubation chamber, where the gaseous (co-)substrate can be added. Among alternative sample delivery systems for time-resolved serial crystallography it is a unique way to initiate the reaction *in-crystallo* only currently possible with the LBNL tape-drive.

This set-up has been successfully used for mixing O_2 with microcrystal slurry (Fig. 2B). The tape-drive is enclosed in an environment chamber flushed with He, initially giving a concentration of O_2 lower than 0.1 % (Fuller et al., 2017). Improvements on the enclosure design and the installment of an O_2 removing catalyst mean the standard operation for this system is now below 350 ppm. The time spent by the droplet containing microcrystals between the ejection onto the tape and the entry in the reaction chamber is less than a second. Once the droplet enters the reaction chamber, which covers 60 mm of the tape, the crystals are exposed to 100 % O_2 atmosphere for a varying amount of time depending on the tape speed (~100 ms – 10 s). After incubation with O_2, the sample is transported to the X-ray interaction region. Due to the experimental setup relying on diffusion of gaseous O_2 into aqueous droplets and passively diffusing through the crystal lattice, it is difficult to deduce an exact time (Δt) for when the reaction is triggered before X-ray exposure. For these experiments the reported reaction time is the incubation time in the reaction chamber and the time to reach the X-ray interaction region after incubation.

Several examples of enzymes requiring O_2 for their enzymatic reactivity have used the drop-on-demand tape-drive experimental setup for diffraction studies in combination with XES at XFEL facilities. Isopenicillin N synthase (IPNS) is a non-heme Fe(II) containing enzyme that catalyzes the formation of isopenicillin (IPN), a precursor for many of the clinically relevant β-lactam based antibiotics, from L-δ-(α-aminoadipoyl)-L-cysteiny-D-valine (ACV) in presence of O_2. Time resolved crystallography experiments were conducted at the MFX instrument (LCLS) by exposing microcrystals of IPNS in complex with Fe(II) and ACV to O_2 with reaction time ranging from 400 ms to 3.0 s (Rabe et al., 2021). Similarly, a di-iron containing enzyme, soluble methane monooxygenase (sMMO) which can catalyze the oxidation of methane in presence of O_2 into methanol, was studied using the same setup. The total reaction time for sMMO was 10 s: 4 s incubation time with O_2 and additional 6 s to reach the interaction region with X-ray (Srinivas et al., 2020). For both samples it was possible to confirm the Fe oxidation-reduction state after exposure to O_2 by using XES.

3.2.3 Drop-on-drop, active or turbulent mixing experiment

Mixing experiments can be done by either passive diffusion (as discussed above with gas diffusion) or by active turbulent mixing using the drop-on-drop method (Fig. 2C). The latter allows time-resolved SSX and SFX experiments with any crystalline sample that can accommodate ligand or substrate. Alternatively, premixing the crystal slurry with a photocaged compound, followed by laser illumination to release a ligand or substrate, removes diffusion as a consideration for reaction initiation (Mehrabi et al., 2019). However, for each specific experiment caged compounds require either commercial availability or potentially significant synthetic effort (see above).

The drop-on-drop mixing methodology has been jointly developed by LBNL and Diamond Light Source (Butryn et al., 2021). Two injectors are used sequentially, ADE to eject droplet containing microcrystal slurry onto the Kapton tape and a piezoelectric injector (PEI, from PolyPico, https://www.polypico.com/product/miniature-dispensing-head/) to eject a burst of ligand / substrate droplets onto the ADE drop (Fig. 7). The ADE can eject ~2–5 nL droplets containing microcrystals at varying frequency (10 to 120 Hz) depending on the X-ray detector or the pulse frequency at the

XFEL facility. The PEI was positioned at 60 mm before the X-ray interaction region where it can eject a burst (up to ~50 kHz) of 30–150 pL droplets at frequencies matching the ADE. One can adjust the ligand droplet intra-burst frequency to effectively titrate ligand concentration for each time point (Table 3, position 3). In practice, before an experiment we choose a time point (Δt) which will dictate the tape speed of the sample delivery. We then decide how many droplets of ligand should merge with the ADE drop to give the desired final ligand concentration in the microcrystal slurry containing drop. We use the equations below to determine the intra-burst frequency of the ligand droplet ejection to achieve the required number of droplets merging with the ADE drop. During the experiment we use a fast camera looking at the PEI to check the number of ligand droplet intersecting with each ADE drop.

We use the Eq. 2 to calculate the number of ligand droplets (n_L) intersecting with the ADE drop in order to achieve the desired final concentration of ligand (Lig_f).

$$n_L = \frac{Lig_f \cdot V_{ADE}}{V_L \left(C_L - Lig_f \right)} \tag{2}$$

Where V_{ADE} is the volume of the ADE drop, C_L the initial concentration of the ligand in the PEI cartridge and V_L the volume of individual ligand droplet ejected by the PEI. The precise control of sample consumption by the syringe pumps, combined with the ADE dispensing frequency, allow for accurate determination of the average V_{ADE}.

Due to the hydrophilic coating of the Kapton tape means the droplets spread out adopting an ovoid shape (Fig. 7). Therefore, the droplet diameter ($\varnothing_{ADE}$) can be best estimated by considering a droplet with twice the volume (e.g. 5 nL droplet eject has a diameter of a 10 nL droplet).

$$\varnothing_{ADE} = 2 \sqrt[3]{\frac{3 \cdot 2 V_{ADE}}{4\pi}} \tag{3}$$

By using Eqs. 2 and 3, we can now calculate with Eq. 4 the intra-burst frequency (f_{ib}) for the PEI delivering the ligand droplets.

$$f_{ib} = \frac{V_{tape} \cdot (n_L - 1)}{\varnothing_{ADE}} \tag{4}$$

Where V_{tape} is the tape velocity used for data collection (see Table 3).

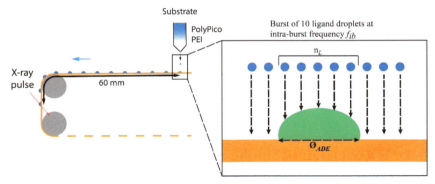

Fig. 7 Schematic of a drop-on-drop, time resolved serial experiment. Because of the velocity of the droplet ejection and the hydrophobic treatment of the Kapton tape, the 2–3 nL ADE drop adopts a flattened shape. In this example, the PEI ejects a burst of 10 droplets each of 10 s–100 s pL volume, and 5 of them merge with the ADE drop. After mixing with ligand, the ADE drop travels 60 mm to reach the X-ray interaction region. The reaction time is defined by the Kapton belt velocity and distance traveled with overall timing accuracy limited by the time required to add multiple PEI drops of ligand and mixing throughout the ADE drop. This principle applies to microcrystal slurries for tr-SSX and tr-SFX datasets that can be correlated with simultaneous tr-XES data; however, tr-XES can also be collected from samples in solution.

Thanks to the equations above, we can keep the same number of droplets intersecting with the ADE drop, while varying the tape speed or Δt, by changing the intra-burst frequency of the ejection of the ligand droplets.

The main advantage of this mixing technique is the turbulent mixing created by the ligand droplets traveling at 1 to 2 m/s when colliding with the ADE drop. We have shown that the mixing in the ADE drop is more efficient than model using diffusion only (Butryn et al., 2021).

In our original work we shown that Hen Egg White Lysozyme (HEWL) bound N-acetyl-D-glucosamine within 0.6 s *in-crystallo*. For the β-lactamase CTX-M-15 we could identify the antibiotic ertapenem at the active site within 2 s. Further work, utilized drop-on-drop mixing to observe the formation of compound 0 after 200 ms incubation with peracetic to the P450 enzyme CYP121 (Nguyen et al., 2023).

An important factor to consider is the diffusion of the substrate within the crystal through the solvent channels. Unfortunately, some crystal systems are not compatible with ligand diffusion to initiate reactions because lattice contacts may occlude the active site. Simple examination of the atomic model and lattice packing of the resting structure in the appropriate space group will quickly reveals these unfortunate situations. Several tools are available for analyzing the solvent channels in protein crystals (Juers & Ruffin, 2014; Pletzer-Zelgert et al., 2023) and can be used to estimate access to the active site via the solvent channel. Ligand diffusion inside the crystal is the bottleneck for most mixing techniques for time-resolved experiments, however the diffusion time inside the crystal can be estimated (Schmidt, 2013) giving an approximation for the shortest accessible time point.

3.3 Data collection/beamtime

Usually, time-resolved experiments involve several groups of people specialized in different areas, such as installing and running hardware, XES alignment, data processing and analysis, and sample preparation. In this section, we will describe few steps - from the hardware installation to the data analysis - in which particular team members may not be directly involved, but should nevertheless be aware of the entire process.

3.3.1 Beamline and tape-drive preparations

3.3.1.1 Hardware installation

Prior to the beamtime, the sample delivery and XES hardware must be installed and tested. For instance, at the LCLS the LBNL team sets up their tape-drive, or at PETRA III the CFEL TapeDrive 2.0 is set up with team members at CFEL and DESY. In the case of the LBNL tape-drive, due to its complexity, more time and access to the MFX endstation are required to install the tape-drive and the XES spectrometer.

3.3.1.2 XES alignment

Once the X-ray interaction point is defined, the analyzer crystal array and 2D array detector can be positioned. After fitting the analyzer crystals, rough alignment onto the detector can be achieved using a class 2 laser scattering from adhesive tape at the X-ray interaction point in a low light environment. This creates significant diffuse scattering of visible light that can be refocused via reflection through the analyzers, mimicking Bragg diffraction. The main goal is to move the crystals such that a well-focused line is visible on the face of the detector. Importantly, this alignment does not require X-rays, meaning that the approximate focus point for each crystal is pre-positioned onto the detector prior to the beginning of beamtime. This ensures that once X-rays are available, signal from each analyzer is visible for optimization. For, multi-analyzer setups this greatly reduces the amount of beamtime used for aligning the XES setup.

Once X-rays are available, final alignment can start. It is good practice to use high concentration solutions of metals as references in the oxidation states anticipated for the studies, when available. At this stage, final focus and location of the spectra on the detector must be decided. Ideally, the region of interest away from the edges of the active area or spanning between panels on the detector. In this multi-crystal analyzer setup, at most, only one crystal per column can meet the perfect alignment geometry, while others will be slightly tilted to ensure refocusing on to the detector (as explained above, Fig. 5A). Due to the asymmetric nature of this alignment, small imperfections in the focus of the lines on the detector should be expected. Careful evaluation of the reference metal spectra during calibration, can guide the decision to overlay spectra from different crystals. This usually comes down to a trade-off between spectral resolution and signal-to-noise.

It is crucial that the geometry of the setup remains identical as any drift in position can lead to signal loss and misinterpretation of signal changes. When a beamtime is divided and spread out over a longer period, it is advisable to realign and recollect reference spectra to ensure the best and most consistent data are obtained.

3.3.2 Experimental set-up and monitoring

The tape-drive includes several individual parts such as: sample ejection with ADE and/or PEI strategies, tape-drive motion, IR gates, experimental platform (either optical platform, O_2 reaction chamber or drop-on-drop mixing), the von Hamos spectrometer, the environment chamber, sample storage (syringe pusher and rocker) and the clean and dry unit for the tape. Most of the parts are fixed on frame mounted with three axis motorized stages that allow fine motion control, including the overall apparatus, the ADE and PEI transducer(s). The XES spectroscopy and X-ray diffraction detectors are always provided by the facility given that they must be integrated into data storage servers and DAQ (Data acquisition) systems. However, it is critical that all these instruments work together and therefore correct alignment and synchronization needs to be validated prior to the experiment. We rely on several monitoring tools for checking the status of the system and debugging errors that arise. We will describe here the tools and methods we regard as essential.

Before starting data collection, the tape-drive operators and instrument scientists must align and confirm the X-ray beam position at the sample position. For this task a YAG or Gallium Arsenide (GaAs) crystal is positioned at the sample position. Fluorescence indicates the X-ray beam position that is recorded and marked on the viewing system, and ultimately to which tape-drive and/or reference tools are subsequently aligned. To help generate a precise geometric description of the experiment for serial crystallographic data processing (see Chapter 6) the X-ray diffraction detector distance to the sample position is experimentally determined. This is usually done by exposing a crystal powder standard like Silver Behenate (AgBe) or Cerium(IV) oxide (CeO_2) and fitting the powder diffraction rings to determine the detector distance. The von Hamos spectrometer will also need to be aligned using the metal foils and other standards as discussed above.

Cameras are essential and give visual confirmation that everything is in order. One camera is used to monitor the sample ejected by ADE onto the tape. We can then assess the size of the drop and that the tape motion is working correctly. If air bubbles prevent the ejection, it is immediately obvious on the camera. An oscilloscope is also used to track the ADE echo signal in the sample well (Fuller et al., 2017). The position of the echo gives important feedback on the level of sample in the ejection well or the presence of air bubbles.

An overview camera will show where the ADE droplet is positioned on the tape. Ideally to reduce the Kapton tape absorption the ADE drop should be sitting at the front edge of the Kapton tape (inclusion panel, Fig. 2C). If the ejected drops are misplaced, the tape-drive operator can move the ADE transducer accordingly.

A separate camera monitors the X-ray interaction region, due to the intensity of the XFEL pulses, droplets that are well aligned and synchronized with X-rays should produce droplet explosions (Stan et al., 2016). If no explosion is observed, the operator can adjust the synchronization of the pulses and drop ejection or move the tape closer to the beam position.

During time-resolved experiments using either light-triggered or drop-on-drop platforms, there is space for two IR gates. When the ADE drops pass an IR gate, the IR light transmitted to a photodiode is reduced which can be monitored on an oscilloscope. The two IR gates are at known positions on the tape and are facilitate synchronization with ADE drop.

For the drop-on-drop experiment, we often rely on a fast camera (Photron Fastcam mini Ax 200, Japan) looking at the PEI ejection onto the ADE drop using a motorized, long working distance telescopic lens (ULWZ-500M Optosigma, Japan). Analysis of the calibrated, highspeed video reveals the timing, ligand droplet size and volume, and the exact number of droplets merging with ADE drop. Depending on the feedback, one can adjust the frequency of ejection of the ligand droplets.

3.3.3 Before and during data collection

In this section we will review good practices to ensure smooth data collection, especially at XFEL facilities where the beamtime is precious. In our experience, a limited number of proteins should be used for each XFEL beamtime. It is beneficial to spend more time to fully investigating one protein with several time-points to produce a more complete story, than to collect limited data from several proteins, but each with fewer time points. It is important that before the data collection starts that all parties agree on a priority list and schedule.

Microcrystal slurry should be well characterized in advance making sure that crystals are homogenous, in appropriate concentration density, and in enough quantity. The slurry is loaded in a Hamilton syringe to be connected to the ADE setup. The syringe is positioned on a syringe rocker to distribute the slurry evenly throughout the syringe. Usually, the first sample to be collected should be either a standard (like lysozyme crystals) or a resting state of one of the targets. It is important that the first sample gives good diffraction patterns. The first data collected is used to make sure everything is working fine (sample delivery, data acquisition, data processing, detector geometry (Brewster et al., 2018), and evaluate processing pipelines).

The first question in every time resolved experiment is *"What is the first time point?"*. The science team and beamline staff should discuss the appropriate technique to probe the relevant time points and collect data that is pertinent to the scientific goals. Preliminary experiments at synchrotrons should provide guidance. In many cases, all the scientific goals can be achieved with synchrotron sources. It is common to establish the "maximum" change in a system, typically at longer timepoints and higher ligand concentrations, and then to work toward shorter and shorter time delays that are likely to exhibit smaller differences. In some cases, XFEL experiments are likely to be critical because they introduce less radiation-induced alterations and yield good quality data from smaller samples. It is rather common that the "complete story" across a reaction coordinate comes from data collected at a variety of sources and includes complementary validations done in parallel.

The second question asked during beamtime is *"When should we stop collecting and move to the next time point?"*. To answer this question, near to real time feedback from on-line data processing is essential. It is important to keep in mind that between XRD and XES, data collection for sufficient signal over noise in either data type can be time-limiting. For instance, higher spatial resolution and/or lower symmetry space groups require more serial MX data compared to lower resolution results and/or higher symmetry systems. The XES signal for $K\alpha_{1,2}$ emission is much stronger and faster to build up compared to $K\beta_{1,3}$ and extraordinarily for $K\beta_{2,5}$ valence to core emission lines. Statistics such as multiplicity and $CC_{1/2}$ are monitored for completion of XRD data collection at a certain resolution (see Chapter 6). Signal over noise analyses of XES spectra from completed runs are used to assess data quality for the spectroscopic component. The isomorphous difference maps between resting state and first time point will give the direction for the next time point, especially when these manifest in the active site region of the macromolecule. For example, if no difference is visible, a longer time point and/or higher concentration is probably needed. Getting reliable difference maps may take a more time, and thus it may be a good idea to interleave projects if possible. While a member of the team is working to give one answer, another sample from a different project can be loaded on the system for another time-resolved, serial dataset.

In conclusion, expectation management and a good planning with respect to sample and/or beamtime availability and the priority list of timepoints will help yield a successful and harmonious beamtime.

3.3.4 Online/offline data processing
3.3.4.1 X-ray diffraction data collection
Serial crystallography merges single diffraction patterns from individual crystals presented at random orientations. The standard crystallographic workflow is data reduction — spot-finding, indexing and integration followed by scaling and merging reflections. Serial crystallography principally presents unique challenges for indexing as crystal orientation is not experimentally defined, for merging due to crystal variations and, for XFELs, the stochastic nature of the beam. There are currently a few options for serial data processing with DIALS (Brewster et al., 2018; Winter et al., 2018), CrystFEL (White et al., 2012), or XDS (Diederichs & Wang, 2017). For a more detailed and mathematical explanation of serial crystallography data processing, and implementation with DIALS, please see Chapter 6 in this book. Here we will briefly address practical dataprocessing considerations and the workflow at experiments generally, for the tape-drive, and resulting data storage requirements.

Due to the serial collection mode, each image needs to be individually assessed for diffraction, if present, indexed with the anticipated unit cell, and integrated. As discussed in Section 3.3.3 (above) normally a well behaving protein crystal sample is run at the start of the experiment to confirm the overall setup is working. For data-processing this is useful for checking appropriate spot-finding parameters and detector geometry refinement. Due to variations in intensities and experimental geometries re-optimizing appropriate parameters may be required between samples or at interventions to the sample delivery. Synchrotron sources typically provide a more stable input beam and detector geometry between requiring less re-refinement of experimental parameters. Regardless, as this data-reduction amounts to a search for diffraction through a haystack of images, it is this process that normally consumes the most computational resource and time. The question of how many crystals are required for a serial dataset is an active area of research, and we will not address this in detail here. However, as already covered, several merged statistics are used heuristically, and once more than several thousand lattices have been integrated it is worth merging the whole dataset to get an idea of the overall merge statistics and check electron density map quality.

A unique challenge for tape-drive setups where the tape face is parallel to the beam direction is that the tape itself shadows one side of the diffraction detector. This systematically lowers intensities on approximately half of the measured reflections thereby impacting scaling, and merging (Fuller et al., 2017). The Bragg reflections in the forward scattering signal are stronger and spread across whole surface of the large format 2D detector in comparison to the number of photons that yield the XES signals recorded in regions of interest on the XES detector(s). Consequently, the potentially deleterious impact of the parallel geometry manifests in the strongest Bragg signals, without measurable effect on the weaker XES signal. Therefore, a correction is applied to the pixels within the Bragg reflection shoeboxes during integration. A simple model describing the Kapton tape (thickness and width), the crystals' position on the tape, and the angle between the tape face and the X-ray beam is used to calculate the distance through the tape that an X-ray would have traveled for a given pixel. Using this, an attenuation factor is applied based on Beer's law and the mass attenuation coefficients for Kapton to determine the incident intensity. An error approximation for the correction is calculated by varying the input parameters for the tape-correction and determining the variation between the corrected intensity values. This is required for the appropriate weighting of the integrated intensities during merging. This correction is currently only implemented in DIALS in either the CCTBX.XFEL GUI or command line executions.

During a beamtime, the principal focus for 'online-processing' is to return information to the research scientist that is required for experimental decision making. Most importantly, these include the number of integrated lattices per image, the total number of high-quality lattices collected to a given point, upper resolution limit of the sample, and for time-resolved experiments whether there is an observed change for a given condition. The first three can be easily obtained while data-reduction and collection are taking place, however due to initial troubleshooting in indexing and integration, hit-rate (percentage of images with observable diffraction) is sometimes used in place of lattices per image. If integration-rate or hit-rate is too high (which may prevent indexing) or too low, then the experimenter can adjust the sample accordingly, while poor-resolution crystals can be replaced with a new batch or sample. Reporting on observed changes requires completing the data analysis workflow and inspecting 2Fo-Fc maps or Fo-Fc maps once sufficient data has been collected.

The volume of raw data generated is dependent on the total images collected and the storage cost of a single image. An upper limit can therefore be estimated by multiplying the size of a single image, the collection rate and total expected collection time. Moreover, recent examples have indicated that compression can be applied without losing information content (Leonarski et al., 2020). Diamond Light Source allows UK based research scientists to have their data safely stored and archived at Diamond. Finally, offline processing is done after the beamtime and allows several careful optimizations, such as the experimental geometry over the experiment, detector corrections and more time-consuming merging routines. This therefore encompasses routines that may improve data quality for further analysis and interpretation that is too time-consuming to provide useful additional feedback during the experiment.

3.3.4.2 XES data collection

The analyzer crystal array setup must be aligned to focus photons from the sample point source to an energy dispersive line on the 2D area detector located approximately in the plane and above or below the sample (Fig. 8). Obtaining XES spectra is achieved by integrating over the region of interest (ROI) and is corrected for background signal and panel gaps/dark pixels or hot pixels that are removed using masks or interpolation averaging. Due to the geometric requirements for a well-focused spectrum, a different ROI must be selected for each experiment. In theory, a setup with multiple banks of analyzer crystals could allow for simultaneous

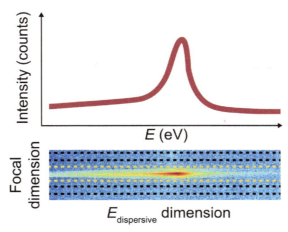

Fig. 8 An integrated X-ray emission spectroscopy (XES) signal obtained by integrating over the 2D position sensitive detector (PSD). The signal shown at the bottom, shows the photons counted on the detector with blue indicating low and red high number of counts. The region of interest (ROI) is between the two yellow lines, while background counts are obtained from the regions between the two black dashed lines above and below the ROI. The background signal is subtracted from the ROI signal to obtain the above spectrum.

detection of XES signals from different elements (Table 1). Even when recording the same XES signal on different occasions, it is likely that the signal will be on a different location on the detector and so a new ROI must be defined each time. It is important to keep in mind that to calculate difference spectra the setup needs to be as consistent as possible, and reliable calibrations are required to prevent systematic features in the difference spectra and ensure valid comparisons.

Although the drop-on-demand sample delivery system is equipped with a washing station, it is important to verify if metal contaminants build up on the tape. The potential for tape contamination should be checked with "empty" water droplets at an appropriate frequency and evaluated on a sample-by-sample basis.

Analyzing the spectra for changes between different states can be done in various ways. When sufficiently strong integrated signals are obtained and after normalization, changes in peak position can be used to indicate redox and/or spin state changes. For instance, an oxidized heme system (e.g. Fe(III)-OH$_x$, S = 5/2) may bind a strong ligand (e.g. NO, CO, CN$^-$, N$_3^-$, SCN$^-$, etc.) to yield an Fe(III)-ligand, S = 1/2 configuration that will elicit reasonable change in the K$\alpha_{1,2}$ XES signal, but perhaps reveal similar electron density features in maps with modest resolution, up to 2 Å. In addition, the Kα lines are sensitive to redox state and the XES will probably differ for Fe(II), Fe(III) and Fe(IV) species, but could also yield similar looking electron density maps. Thus, spectroscopic alterations are very informative and complementary data should also be explored in parallel because changes will also very likely manifest in electronic absorption, infrared, fluorescence, electron paramagnetic, resonance Raman, paramagnetically shifted NMR, and/or Mössbauer spectroscopies too. However, the magnitude of the XES changes can be harder to evaluate for weaker samples due to limited signal over noise. A moving average function may help to smooth a noisy difference spectrum. In even more dilute samples, comparing the integrated absolute difference (IAD) is a more robust alternative that provides a linear correlation between observed spectral changes and changes in spin state (Vanko et al., 2006):

$$IAD = \int |I_1(E) - I_2(E)| dE \tag{5}$$

Where $I_i(E)$ represents the intensity of the X-ray emission signals from two spectra over energy range E. Alternatively, the first moment of the XES peak can be used to determine small shifts in an obtained spectrum. The first moment uses statistics from the entire spectrum and is calculated as

$$\mu_1 = \frac{\sum_i E_i I_i}{\sum_i I_i} \tag{6}$$

with μ_1 as first moment, E_i as emitted photon energy (eV) and I_i as emission intensity (counts) per datapoint i, respectively. Mathematically, the first moment provides the total mass normalized "center of mass" of the spectrum. Comparison of the first moment of a spectrum of different perturbed states provides a shift in first moment that is directly correlated with spin state changes.

Post processing strategies are likely to mitigate this issue, especially by correlating XES data with diffraction images and filtering for images containing indexed diffraction patterns. Deploying such strategies helps ensure that the XES signal correlates with the atomic model refined from the XRD data. How much data is required for either method, and ultimately their correlation, is often of acute concern during experiments. The inclusion of XES signal from disordered metal atoms should be minimized for tr-SFX/tr-SSX + tr-XES studies because it can lead to incorrect conclusions. As the post-processing procedures improve, scientists will know sooner and during data acquisition if/when sufficient data has been collected and whether to move to the next dataset. It is of note that the tape-drive strategy for tr-XES can also be used for samples in solution for additional comparisons and validations.

3.4 Future developments

3.4.1 Complementary methods

Currently, the drop-on-demand sample delivery system has been demonstrated to be compatible with simultaneous XES and XRD data collection. The inherent versatility of this system allows for expansion and accommodation of other complementary spectroscopies such as UV-Vis and Raman spectroscopy, in addition to other X-ray spectroscopies. The current setup, equipped with von Hamos geometry multi-analyzer crystals, could theoretically be expanded to include X-ray Raman scattering, and resonant spectroscopies such as high-energy-resolution fluorescence-detected (HERFD) XAS and resonant inelastic X-ray scattering (RIXS), including magnetic circular dichroism (Sahle et al., 2023). A recent study demonstrated the feasibility of using the drop-on-demand system for XANES and EXAFS studies on the Mn cluster of PSII (Bogacz et al., 2023). However, this study did not include simultaneous XRD data acquisition, which remains a significant advantage of this system.

3.4.2 Diamond tape-drive for tr-SSX / tr-SFX ± tr-XES

At Diamond Light Source, we are commissioning a drop-on-demand tape-drive sample delivery system, inspired by extensive collaboration on the LBNL tape-drive. This sample delivery platform will allow for simultaneous collection of XRD and XES, while exploiting sufficient automation and feedback loops so that it can be operated by regular users, thereby requiring fewer interventions and staff members. Crystal slurry droplet formation will be done through PEI, which allows for ~100–300 pL droplet generation compared to ADE, reducing sample consumption by one order of magnitude. Piezo electric ejection of microcrystal slurry containing droplets was demonstrated using the PolyPico injector generating droplet size ranges from 30 - 100 s pL and routinely operating at 100–150 pL. If we stay conservative with a > 90 % droplet hit and 20 % lattice indexing hit ratios, then we will need less than 10 µL of sample for 10,000 indexed lattices compared to about 100 µL for ADE with the LBNL tape-drive. The first phase of development of our tape-drive will be to perform time-resolved mixing experiment with drop-on-drop set-up using

PEI to generate both slurry and ligand drops (Fig. 9A). The clean and dry unit allows continuous operation in contrast to the reel-to-reel tape-drive. The PEI ejecting the microcrystal slurry on the tape is positioned at 300 mm from the interaction region, while the ligand can be ejected by a PEI at two different positions: position 1 at 200 mm from the interaction region and position 2 is variable from 25 to 100 mm from the interaction region. The time points available depends on the tape speed; we estimate that we can run the tape-drive at the VMXi instrument at Diamond Light Source at up to about 300–600 mm/s and still get enough X-ray photons to produce an indexable diffraction pattern. At XFEL facilities the limiting

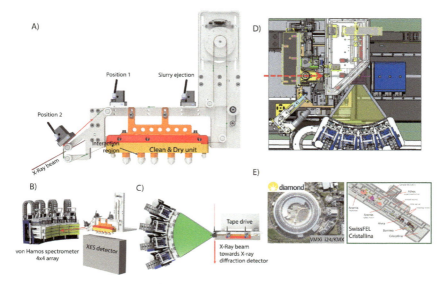

Fig. 9 Overview of the Diamond tape-drive. (A) The side view shows the 3 positions of the PEI. The microcrystal slurry ejection occurs at 300 mm away from the interaction region. Mixing with ligand is done at either position 1 (200 mm) or position 2 (25–100 mm). After the slurry is exposed to X-ray, the tape is cleaned and dried allowing continuous usage. The direction of the tape is noted with green arrows and is driven by a motor (top right) fitted with an encoder for finer control yielding higher precision. (B) Overall view of the tape-drive combined with the 4x4 array von Hamos spectrometer and XES detector. (C) Top view of the sample delivery and von Hamos spectrometer. (D) A top view schematic (top) for how the environmental/anaerobic chamber (translucent white polygon) containing the tape-drive is situated at the VMXi beamline at Diamond to allow the X-ray beam (red dashed line), tr-SSX diffraction detector (right, blue box) and the 400 mm radius von Hamos analyzer array with He cone (translucent yellow). The integrated instrumentation is designed to be portable and will be used for science programs at Diamond Light Source and SwissFEL, and potentially at other facilities too.

factor is the tape speed and the droplet stability at the interaction region. According to our test we can deliver a stable droplet at the interaction region at a speed of 4 m/s while maintaining a greater than 90 % droplet hit ratio. At synchrotrons, the time points for time-resolved experiments range from probably less than 100 ms to up to 20 s, while at XFELs we can reach single digit millisecond time-points and longer (Table 4).

The tape-drive will be enclosed within a high relative humidity chamber to avoid droplet dehydration, especially for the longer time-points. The interior of the chamber can be flushed with nitrogen gas to reduce the O_2 concentration to lower than 100 ppm. The PEIs in position 1 and 2 can be replaced by optic fiber coupled to a laser for pump probe experiments. A second phase of development will provide a gas reaction chamber (as shown in Fig. 2B) and improved anaerobic conditions with O_2 concentration below 10 ppm.

The tape-drive has been designed to be installed at the VMXi, but it may also be deployed at I24 / KMX at Diamond and Diamond II. We are also working closely with SwissFEL to install and operate the tape-drive at the Cristallina instrument to enable science programs for the user communities at Diamond and PSI. The portability of the system will also enable deployment at a variety of synchrotron and XFEL facilities.

Table 4 Possible reaction times according to the tape speed of the Diamond tape drive.

Δt reaction time point (s)

Tape velocity (mm/s)	Position 1–200 mm	Position 2–100 mm	Position 2–25 mm
10	20	10	2.5
25	8	4	1
50	4	2	0.5
100	2	1	0.25
150	1.3	0.67	0.167
200	1	0.5	0.125
400	0.5	0.25	0.062
800	0.25	0.125	0.031
1600	0.125	0.062	0.0156
3200	0.62	0.031	0.0078

We are commissioning our von Hamos spectrometer to perform XES experiment simultaneously with XRD experiment (Fig. 9B and C). The von Hamos spectrometer includes four banks of 4×1 crystal analyzer arrays. With the crystal analyzers currently available at Diamond, we will be able to routinely collect Fe Kα, Fe K$\beta_{1,3}$, Cu Kα, Cu K$\beta_{1,3}$ and Ni Kα emission spectra in various combinations simultaneously. We aim to start a user program at Diamond in 2025 (Fig. 9D and E).

References

Abdallah, B. G., Zatsepin, N. A., Roy-Chowdhury, S., Coe, J., Conrad, C. E., Dorner, K., ... Ros, A. (2015). Microfluidic sorting of protein nanocrystals by size for X-ray free-electron laser diffraction. *Structural Dynamics, 2*(4), 041719. https://doi.org/10.1063/1.4928688.

Alonso-Mori, R., Kern, J., Gildea, R. J., Sokaras, D., Weng, T. C., Lassalle-Kaiser, B., ... Bergmann, U. (2012). Energy-dispersive X-ray emission spectroscopy using an X-ray free-electron laser in a shot-by-shot mode. *Proceedings of the National Academy of Sciences of the United States of America, 109*(47), 19103–19107. https://doi.org/10.1073/pnas.1211384109.

Alonso-Mori, R., Kern, J., Sokaras, D., Weng, T. C., Nordlund, D., Tran, R., ... Bergmann, U. (2012). A multi-crystal wavelength dispersive x-ray spectrometer. *Review of Scientific Instruments, 83*(7), 073114. https://doi.org/10.1063/1.4737630.

Bachega, J. F. R., Maluf, F. V., Andi, B., Pereira, H. D., Carazzollea, M. F., Orville, A. M., ... Reboredo, E. H. (2015). The structure of the giant haemoglobin from. *Acta Crystallographica Section D-Structural Biology, 71*, 1257–1271. https://doi.org/10.1107/S1399004715005453.

Barends, T. R. M., Gorel, A., Bhattacharyya, S., Schiro, G., Bacellar, C., Cirelli, C., ... Schlichting, I. (2024). Influence of pump laser fluence on ultrafast myoglobin structural dynamics. *Nature*. https://doi.org/10.1038/s41586-024-07032-9.

Barends, T. R. M., Stauch, B., Cherezov, V., & Schlichting, I. (2022). Serial femtosecond crystallography. *Nature Reviews Methods Primers, 2*. https://doi.org/10.1038/s43586-022-00141-7.

Beale, J. H., Bolton, R., Marshall, S. A., Beale, E. V., Carr, S. B., Ebrahim, A., ... Owen, R. L. (2019). Successful sample preparation for serial crystallography experiments. *Journal of Applied Crystallography, 52*, 1385–1396. https://doi.org/10.1107/S1600576719013517.

Beitlich, T., Kuhnel, K., Schulze-Briese, C., Shoeman, R. L., & Schlichting, I. (2007). Cryoradiolytic reduction of crystalline heme proteins: analysis by UV–Vis spectroscopy and X-ray crystallography. *Journal of Synchrotron Radiation, 14*(Pt 1), 11–23. https://doi.org/10.1107/S0909049506049806.

Beyerlein, K. R., Dierksmeyer, D., Mariani, V., Kuhn, M., Sarrou, I., Ottaviano, A., ... Oberthuer, D. (2017). Mix-and-diffuse serial synchrotron crystallography. *IUCrJ, 4*(Pt 6), 769–777. https://doi.org/10.1107/S2052252517013124.

Bhowmick, A., Hussein, R., Bogacz, I., Simon, P. S., Ibrahim, M., Chatterjee, R., ... Yachandra, V. K. (2023). Structural evidence for intermediates during O(2) formation in photosystem II. *Nature, 617*(7961), 629–636. https://doi.org/10.1038/s41586-023-06038-z.

Bhowmick, A., Simon, P. S., Bogacz, I., Hussein, R., Zhang, M., Makita, H., ... Yano, J. (2023). Going around the Kok cycle of the water oxidation reaction with femtosecond X-ray crystallography. *IUCrJ, 10*(Pt 6), 642–655. https://doi.org/10.1107/S2052252523008928.

Bogacz, I., Makita, H., Simon, P. S., Zhang, M., Doyle, M. D., Chatterjee, R., ... Yano, J. (2023). Room temperature X-ray absorption spectroscopy of metalloenzymes with drop-on-demand sample delivery at XFELs. *Pure and Applied Chemistry, 95*(8), 891–897. https://doi.org/10.1515/pac-2023-0213.

Boutet, S., Lomb, L., Williams, G. J., Barends, T. R. M., Aquila, A., Doak, R. B., ... Schlichting, I. (2012). High-resolution protein structure determination by serial femtosecond crystallography. *Science, 337*(6092), 362–364. https://doi.org/10.1126/science.1217737.

Brewster, A. S., Waterman, D. G., Parkhurst, J. M., Gildea, R. J., Young, I. D., O'Riordan, L. J., ... Sauter, N. K. (2018). Improving signal strength in serial crystallography with geometry refinement. *Acta Crystallographica Section D-Structural Biology, 74*, 877–894. https://doi.org/10.1107/S2059798318009191.

Burgie, E. S., Clinger, J. A., Miller, M. D., Brewster, A. S., Aller, P., Butryn, A., ... Vierstra, R. D. (2020). Photoreversible interconversion of a phytochrome photosensory module in the crystalline state. *Proceedings of the National Academy of Sciences of the United States of America, 117*(1), 300–307. https://doi.org/10.1073/pnas.1912041116.

Butryn, A., Simon, P. S., Hinchliffe, P., Massad, R. N., Leen, G., ... Orville, A. M. (2021). An on-demand, drop-on-drop method for studying enzyme catalysis by serial crystallography. *Nature Communications, 12*(1), 4461. https://doi.org/10.1038/s41467-021-24757-7.

Carey, P. R. (2006). Raman crystallography and other biochemical applications of Raman microscopy. *Annual Review of Physical Chemistry, 57*, 527–554. https://doi.org/10.1146/annurev.physchem.57.032905.104521.

Chreifi, G., Baxter, E. L., Doukov, T., Cohen, A. E., McPhillips, S. E., Song, J., ... Poulos, T. L. (2016). Crystal structure of the pristine peroxidase ferryl center and its relevance to proton-coupled electron transfer. *Proceedings of the National Academy of Sciences of the United States of America, 113*(5), 1226–1231. https://doi.org/10.1073/pnas.1521664113.

Cohen, A. E., Doukov, T., & Soltis, M. S. (2016). UV-visible absorption spectroscopy enhanced X-ray crystallography at synchrotron and X-ray free electron laser sources. *Protein & Peptide Letters, 23*(3), 283–290. https://doi.org/10.2174/0929866523666160107115015.

Cole, K., Roessler, C. G., Mule, E. A., Benson-Xu, E. J., Mullen, J. D., Le, B. A., ... Soares, A. S. (2014). A linear relationship between crystal size and fragment binding time observed crystallographically: Implications for fragment library screening using acoustic droplet ejection. *PLoS One, 9*(7), e101036. https://doi.org/10.1371/journal.pone.0101036.

Colletier, J. P., Schiro, G., & Weik, M. (2018). Time-resolved serial femtosecond crystallography. *Towards molecular movies of biomolecules in action. X-ray free electron lasers, a revolution in structural biology, Springer Nature Switzerland*.

Cuttitta, C. M., Ericson, D. L., Scalia, A., Roessler, C. G., Teplitsky, E., Joshi, K., ... Soares, A. S. (2015). Acoustic transfer of protein crystals from agarose pedestals to micromeshes for high-throughput screening. *Acta Crystallographica. Section D, Biological Crystallography, 71*(Pt 1), 94–103. https://doi.org/10.1107/S1399004714013728.

Daughtry, K. D., Xiao, Y. L., Stoner-Ma, D., Cho, E. S., Orville, A. M., Liu, P. H., & Allen, K. N. (2012). Quaternary ammonium oxidative demethylation: X-ray crystallographic, resonance raman, and UV-Visible spectroscopic analysis of a rieske-type demethylase. *Journal of the American Chemical Society, 134*(5), 2823–2834. https://doi.org/10.1021/ja2111898.

De Groot, F. (2001). High-resolution X-ray emission and X-ray absorption spectroscopy. *Chemical Reviews, 101*(6), 1779–1808. https://doi.org/10.1021/cr9900681.

De La Mora-Rey, T., & Wilmot, C. M. (2007). Synergy within structural biology of single crystal optical spectroscopy and X-ray crystallography. *Current Opinion in Structural Biology, 17*(5), 580–586. https://doi.org/10.1016/j.sbi.2007.09.005.

DePonte, D. P., Weierstall, U., Schmidt, K., Warner, J., Starodub, D., Spence, J. C. H., & Doak, R. B. (2008). Gas dynamic virtual nozzle for generation of microscopic droplet streams. *Journal of Physics D-Applied Physics, 41*(19). https://doi.org/Artn19550510.1088/0022-3727/41/19/195505.

Diederichs, K., & Wang, M. (2017). Serial synchrotron X-ray crystallography (SSX). *Methods in Molecular Biology, 1607*, 239–272. https://doi.org/10.1007/978-1-4939-7000-1_10.

Dods, R., Bath, P., Arnlund, D., Beyerlein, K. R., Nelson, G., Liang, M., ... Neutze, R. (2017). From macrocrystals to microcrystals: A strategy for membrane protein serial crystallography. *Structure, 25*(9), 1461–1468.e1462. https://doi.org/10.1016/j.str.2017.07.002.

Dworkowski, F. S. N., Hough, M. A., Pompidor, G., & Fuchs, M. R. (2015). Challenges and solutions for the analysis of micro-spectrophotometric data. *Acta Crystallographica Section D-Structural Biology, 71*, 27–35. https://doi.org/10.1107/S1399004714015107.

Ebrahim, A., Moreno-Chicano, T., Appleby, M. V., Chaplin, A. K., Beale, J. H., Sherrell, D. A., ... Hough, M. A. (2019). Dose-resolved serial synchrotron and XFEL structures of radiation-sensitive metalloproteins. *IUCrJ, 6*(Pt 4), 543–551. https://doi.org/10.1107/S2052252519003956.

Echelmeier, A., Sonker, M., & Ros, A. (2019). Microfluidic sample delivery for serial crystallography using XFELs. *Analytical and Bioanalytical Chemistry, 411*(25), 6535–6547. https://doi.org/10.1007/s00216-019-01977-x.

Elrod, S. A. H. B., Khuri-Yakub, B. T., Rawson, E. G., & Richley, E. (1989). Nozzleless droplet formation with focused acoustic beams. *Journal of Applied Physics, 65*(9), 3441–3447. https://doi.org/10.1063/1.342663.

Engilberge, S., Caramello, N., Bukhdruker, S., Byrdin, M., Giraud, T., Jacquet, P., ... Royant, A. (2024). The TR-icOS setup at the ESRF: Time-resolved microsecond UV-Vis absorption spectroscopy on protein crystals. *Acta Crystallographica Section D: Structural Biology, 80*(Pt 1), 16–25. https://doi.org/10.1107/S2059798323010483.

Feehan, R., Franklin, M. W., & Slusky, J. S. G. (2021). Machine learning differentiates enzymatic and non-enzymatic metals in proteins. *Nature Communications, 12*(1), 3712. https://doi.org/10.1038/s41467-021-24070-3.

Fitzpatrick, P. F., Bozinovski, D. M., Héroux, A., Shaw, P. G., Valley, M. P., & Orville, A. M. (2007). Mechanistic and structural analyses of the roles of Arg409 and Asp402 in the reaction of the flavoprotein nitroalkane oxidase. *Biochemistry, 46*(48), 13800–13808. https://doi.org/10.1021/bi701557k.

Fransson, T., Chatterjee, R., Fuller, F. D., Gul, S., Weninger, C., Sokaras, D., ... Yano, J. (2018). X-ray emission spectroscopy as an in situ diagnostic tool for X-ray crystallography of metalloproteins using an X-ray free-electron laser. *Biochemistry, 57*(31), 4629–4637. https://doi.org/10.1021/acs.biochem.8b00325.

Fuller, F. D., Gul, S., Chatterjee, R., Burgie, E. S., Young, I. D., Lebrette, H., ... Yano, J. (2017). Drop-on-demand sample delivery for studying biocatalysts in action at X-ray free-electron lasers. *Nature Methods, 14*(4), 443–449. https://doi.org/10.1038/nmeth.4195.

Glatzel, P., & Bergmann, U. (2005). High resolution 1 s core hole X-ray spectroscopy in 3d transition metal complexes—Electronic and structural information. *Coordination Chemistry Reviews, 249*(1-2), 65–95.

Gog, T., Casa, D. M., Said, A. H., Upton, M. H., Kim, J., Kuzmenko, I., ... Khachatryan, R. (2013). Spherical analyzers and monochromators for resonant inelastic hard X-ray scattering: A compilation of crystals and reflections. *Journal of Synchrotron Radiation, 20*(Pt 1), 74–79. https://doi.org/10.1107/S0909049512043154.

Grunbein, M. L., Stricker, M., Nass Kovacs, G., Kloos, M., Doak, R. B., Shoeman, R. L., ... Schlichting, I. (2020). Illumination guidelines for ultrafast pump-probe experiments by serial femtosecond crystallography. *Nature Methods, 17*(7), 681–684. https://doi.org/10.1038/s41592-020-0847-3.

Hajdu, J., Neutze, R., Sjogren, T., Edman, K., Szoke, A., Wilmouth, R. C., & Wilmot, C. M. (2000). Analyzing protein functions in four dimensions. *Nature Structural Biology, 7*(11), 1006–1012. https://doi.org/10.1038/80911.

Henkel, A., Galchenkova, M., Maracke, J., Yefanov, O., Klopprogge, B., Hakanpaa, J., ... Oberthuer, D. (2023). JINXED: Just in time crystallization for easy structure determination of biological macromolecules. *IUCrJ, 10*(Pt 3), 253–260. https://doi.org/10.1107/S2052252523001653.

Henkel, A., & Oberthur, D. (2024). A snapshot love story: What serial crystallography has done and will do for us. *Acta Crystallographica Section D: Structural Biology, 80*(Pt 8), 563–579. https://doi.org/10.1107/S2059798324005588.

Héroux, A., Bozinovski, D. M., Valley, M. P., Fitzpatrick, P. F., & Orville, A. M. (2009). Crystal structures of intermediates in the nitroalkane oxidase reaction. *Biochemistry, 48*(15), 3407–3416. https://doi.org/10.1021/bi8023042.

Heymann, M., Opthalage, A., Wierman, J. L., Akella, S., Szebenyi, D. M., Gruner, S. M., & Fraden, S. (2014). Room-temperature serial crystallography using a kinetically optimized microfluidic device for protein crystallization and on-chip X-ray diffraction. *IUCrJ, 1*(Pt 5), 349–360. https://doi.org/10.1107/S2052252514016960.

Hough, M. A., & Owen, R. L. (2021). Serial synchrotron and XFEL crystallography for studies of metalloprotein catalysis. *Current Opinion in Structural Biology, 71*, 232–238. https://doi.org/10.1016/j.sbi.2021.07.007.

Huo, L., Davis, I., Liu, F., Andi, B., Esaki, S., Iwaki, H., ... Liu, A. (2015). Crystallographic and spectroscopic snapshots reveal a dehydrogenase in action. *Nature Communications, 6*, 5935. https://doi.org/10.1038/ncomms6935.

Hussein, R., Ibrahim, M., Bhowmick, A., Simon, P. S., Chatterjee, R., Lassalle, L., ... Yano, J. (2021). Structural dynamics in the water and proton channels of photosystem II during the S(2) to S(3) transition. *Nature Communications, 12*(1), 6531. https://doi.org/10.1038/s41467-021-26781-z.

Hutchison, C. D. M., Fadini, A., & Van Thor, J. J. (2022). Linear and non-linear population retrieval with femtosecond optical pumping of molecular crystals for the generalised uniaxial and biaxial systems. *Applied Sciences-Basel, 12*(9). https://doi.org/ARTN430910.3390/app12094309.

Ibrahim, M., Fransson, T., Chatterjee, R., Cheah, M. H., Hussein, R., Lassalle, L., ... Yano, J. (2020). Untangling the sequence of events during the S(2) – > S(3) transition in photosystem II and implications for the water oxidation mechanism. *Proceedings of the National Academy of Sciences of the United States of America, 117*(23), 12624–12635. https://doi.org/10.1073/pnas.2000529117.

Jirsa, J., Marcisovsky, M., Janoska, Z., & Jakovenko, J. (2024). Winne-leader-follower a novel hit allocation algorithm for pixel detectors. *Ieee Transactions on Nuclear Science, 71*(5), 1250–1256. https://doi.org/10.1109/Tns.2024.3378002.

Juers, D. H., & Ruffin, J. (2014). MAP_CHANNELS: A computation tool to aid in the visualization and characterization of solvent channels in macromolecular crystals. *Journal of Applied Crystallography, 47*(Pt 6), 2105–2108. https://doi.org/10.1107/S160057671402281X.

Kendrew, J. C. (1950). The crystal structure of horse met-myoglobin—I. General features: The arrangement of the polypeptide chains. *Proceedings of the Royal Society A, 201*(1064), 62–89.

Kern, J., Chatterjee, R., Young, I. D., Fuller, F. D., Lassalle, L., Ibrahim, M., ... Yachandra, V. K. (2018). Structures of the intermediates of Kok's photosynthetic water oxidation clock. *Nature, 563*(7731), 421–425. https://doi.org/10.1038/s41586-018-0681-2.

Lafuerza, S., Carlantuono, A., Retegan, M., & Glatzel, P. (2020). Chemical sensitivity of kbeta and kalpha X-ray emission from a systematic investigation of iron compounds. *Inorganic Chemistry, 59*(17), 12518–12535. https://doi.org/10.1021/acs.inorgchem.0c01620.

Lebrette, H., Srinivas, V., John, J., Aurelius, O., Kumar, R., Lundin, D., ... Högbom, M. (2023). Structure of a ribonucleotide reductase R2 protein radical. *Science, 382*(6666), 109–113. https://doi.org/10.1126/science.adh8160.

Lee, K., Kim, J., Baek, S., Park, J., Park, S., Lee, J. L., ... Nam, K. H. (2022). Combination of an inject-and-transfer system for serial femtosecond crystallography. *Journal of Applied Crystallography, 55*(Pt 4), 813–822. https://doi.org/10.1107/S1600576722005556.

Leonarski, F., Mozzanica, A., Brückner, M., Lopez-Cuenca, C., Redford, S., Sala, L., ... Wang, M. (2020). JUNGFRAU detector for brighter x-ray sources: Solutions for IT and data science challenges in macromolecular crystallography. *Struct. Dyn.* 7(014305), https://doi.org/10.1063/1.5143480.

Li, F., Burgie, E. S., Yu, T., Heroux, A., Schatz, G. C., Vierstra, R. D., & Orville, A. M. (2015). X-ray radiation induces deprotonation of the bilin chromophore in crystalline D. radiodurans phytochrome. *Journal of the American Chemical Society, 137*(8), 2792–2795. https://doi.org/10.1021/ja510923m.

Mahon, B. P., Kurian, J. J., Lomelino, C. L., Smith, I. R., Socorro, L., Bennett, A., ... McKenna, R. (2016). Microbatch mixing: "Shaken not stirred", a method for macromolecular microcrystal production for serial crystallography. *Crystal Growth & Design, 16*(11), 6214–6221. https://doi.org/10.1021/acs.cgd.6b00643.

Major, D. T., Heroux, A., Orville, A. M., Valley, M. P., Fitzpatrick, P. F., & Gao, J. L. (2009). Differential quantum tunneling contributions in nitroalkane oxidase catalyzed and the uncatalyzed proton transfer reaction. *Proceedings of the National Academy of Sciences of the United States of America, 106*(49), 20734–20739. https://doi.org/10.1073/pnas.0911416106.

Makita, H., Simon, P. S., Kern, J., Yano, J., & Yachandra, V. K. (2023). Combining on-line spectroscopy with synchrotron and X-ray free electron laser crystallography. *Current Opinion in Structural Biology, 80.* https://doi.org/ARTN10260410.1016/j.sbi.2023.102604.

Mehrabi, P., Schulz, E. C., Dsouza, R., Muller-Werkmeister, H. M., Tellkamp, F., Miller, R. J. D., & Pai, E. F. (2019). Time-resolved crystallography reveals allosteric communication aligned with molecular breathing. *Science, 365*(6458), 1167–1170. https://doi.org/10.1126/science.aaw9904.

Miller, K. R., Paretsky, J. D., Follmer, A. H., Heinisch, T., Mittra, K., Gul, S., ... Borovik, A. S. (2020). Artificial iron proteins: Modeling the active sites in non-heme dioxygenases. *Inorganic Chemistry, 59*(9), 6000–6009. https://doi.org/10.1021/acs.inorgchem.9b03791.

Monteiro, D. C. F., Amoah, E., Rogers, C., & Pearson, A. R. (2021). Using photocaging for fast time-resolved structural biology studies. *Acta Crystallographica Section D-Structural Biology, 77,* 1218–1232. https://doi.org/10.1107/S2059798321008809.

Nguyen, R. C., Davis, I., Dasgupta, M., Wang, Y., Simon, P. S., Butryn, A., ... Liu, A. (2023). In situ structural observation of a substrate- and peroxide-bound high-spin ferric-hydroperoxo intermediate in the P450 enzyme CYP121. *Journal of the American Chemical Society, 145*(46), 25120–25133. https://doi.org/10.1021/jacs.3c04991.

Ohmer, C. J., Dasgupta, M., Patwardhan, A., Bogacz, I., Kaminsky, C., Doyle, M. D., ... Ragsdale, S. W. (2022). XFEL serial crystallography reveals the room temperature structure of methyl-coenzyme M reductase. *Journal of Inorganic Biochemistry, 230.* https://doi.org/ARTN11176810.1016/j.jinorgbio.2022.111768.

Orville, A. M., Buono, R., Cowan, M., Héroux, A., Shea-McCarthy, G., Schneider, D. K., ... Sweet, R. M. (2011). Correlated single-crystal electronic absorption spectroscopy and X-ray crystallography at NSLS beamline X26-C. *Journal of Synchrotron Radiation, 18,* 358–366. https://doi.org/10.1107/S0909049511006315.

Pellegrini, G., Maiorino, M., Blanchot, G., Chmeissani, M., Garcia, J., Lozano, M., ... Ullan, M. (2007). Direct charge sharing observation in single-photon-counting pixel detector. *Nuclear Instruments & Methods in Physics Research Section a-Accelerators Spectrometers Detectors and Associated Equipment, 573*(1-2), 137–140. https://doi.org/10.1016/j.nima.2006.11.010.

Pletzer-Zelgert, J., Ehrt, C., Fender, I., Griewel, A., Flachsenberg, F., Klebe, G., & Rarey, M. (2023). LifeSoaks: A tool for analyzing solvent channels in protein crystals and obstacles for soaking experiments. *Acta Crystallographica Section D: Structural Biology, 79*(Pt 9), 837–856. https://doi.org/10.1107/S205979832300582X.

Pompidor, G., Dworkowski, F. S., Thominet, V., Schulze-Briese, C., & Fuchs, M. R. (2013). A new on-axis micro-spectrophotometer for combining Raman, fluorescence and UV/Vis absorption spectroscopy with macromolecular crystallography at the Swiss Light Source. *Journal of Synchrotron Radiation, 20*(Pt 5), 765–776. https://doi.org/10.1107/S0909049513016063.

Rabe, P., Kamps, J., Sutherlin, K. D., Linyard, J. D. S., Aller, P., Pham, C. C., ... Schofield, C. J. (2021). X-ray free-electron laser studies reveal correlated motion during isopenicillin N synthase catalysis. *Science Advances, 7*(34), https://doi.org/10.1126/sciadv.abh0250.

Roessler, C. G., Agarwal, R., Allaire, M., Alonso-Mori, R., Andi, B., Bachega, J. F. R., ... Zouni, A. (2016). Acoustic injectors for drop-on-demand serial femtosecond crystallography. *Structure, 24*(4), 631–640. https://doi.org/10.1016/j.str.2016.02.007.

Roessler, C. G., Kuczewski, A., Stearns, R., Ellson, R., Olechno, J., Orville, A. M., ... Heroux, A. (2013). Acoustic methods for high-throughput protein crystal mounting at next-generation macromolecular crystallographic beamlines. *Journal of Synchrotron Radiation, 20*(Pt 5), 805–808. https://doi.org/10.1107/S0909049513020372.

Ronda, L., Bruno, S., Bettati, S., Storici, P., & Mozzarelli, A. (2015). From protein structure to function via single crystal optical spectroscopy. *Frontiers in Molecular Biosciences, 2*, 12. https://doi.org/10.3389/fmolb.2015.00012.

Rose, S. L., Ferroni, F. M., Horrell, S., Brondino, C. D., Eady, R. R., Jaho, S., ... Hasnain, S. S. (2024). Spectroscopically Validated pH-dependent MSOX movies provide detailed mechanism of copper nitrite reductases. *Journal of Molecular Biology, 436*(18), 168706. https://doi.org/10.1016/j.jmb.2024.168706.

Sahle, C. J., Gerbon, F., Henriquet, C., Verbeni, R., Detlefs, B., Longo, A., ... Petitgirard, S. (2023). A compact von Hamos spectrometer for parallel X-ray Raman scattering and X-ray emission spectroscopy at ID20 of the European Synchrotron Radiation Facility. *Journal of Synchrotron Radiation, 30*, 251–257. https://doi.org/10.1107/S1600577522011171.

Sakaguchi, M., Kimura, T., Nishida, T., Tosha, T., Sugimoto, H., Yamaguchi, Y., ... Kubo, M. (2016). A nearly on-axis spectroscopic system for simultaneously measuring UV–visible absorption and X-ray diffraction in the SPring-8 structural genomics beamline. *Journal of Synchrotron Radiation, 23*(1), 334–338. https://doi.org/10.1107/S1600577515018275.

Schmidt, M. (2013). Mix and inject: Reaction initiation by diffusion for time-resolved macromolecular crystallography. *Advances in Condensed Matter Physics, 2013*. https://doi.org/Artn16727610.1155/2013/167276.

Shoeman, R. L., Hartmann, E., & Schlichting, I. (2023). Growing and making nano- and microcrystals. *Nature Protocols, 18*(3), 854–882. https://doi.org/10.1038/s41596-022-00777-5.

Simon, P. S., Makita, H., Bogacz, I., Fuller, F., Bhowmick, A., Hussein, R., ... Yano, J. (2023). Capturing the sequence of events during the water oxidation reaction in photosynthesis using XFELs. *FEBS Letters, 597*(1), 30–37. https://doi.org/10.1002/1873-3468.14527.

Soares, A. S., Engel, M. A., Stearns, R., Datwani, S., Olechno, J., Ellson, R., ... Orville, A. M. (2011). Acoustically mounted microcrystals yield high-resolution X-ray structures. *Biochemistry, 50*(21), 4399–4401. https://doi.org/10.1021/bi200549x.

Soares, A. S., Mullen, J. D., Parekh, R. M., McCarthy, G. S., Roessler, C. G., Jackimowicz, R., ... Sweet, R. M. (2014). Solvent minimization induces preferential orientation and crystal clustering in serial micro-crystallography on micro-meshes, in situ plates and on a movable crystal conveyor belt. *Journal of Synchrotron Radiation, 21*(Pt 6), 1231–1239. https://doi.org/10.1107/S1600577514017731.

Srinivas, V., Banerjee, R., Lebrette, H., Jones, J. C., Aurelius, O., Kim, I. S., ... Hogbom, M. (2020). High-resolution XFEL structure of the soluble methane monooxygenase hydroxylase complex with its regulatory component at ambient temperature in two oxidation states. *Journal of the American Chemical Society, 142*(33), 14249–14266. https://doi.org/10.1021/jacs.0c05613.

Stan, C. A., Milathianaki, D., Laksmono, H., Sierra, R. G., McQueen, T. A., Messerschmidt, M., ... Boutet, S. (2016). Liquid explosions induced by X-ray laser pulses. *Nature Physics, 12*(10), 966–971. https://doi.org/10.1038/Nphys3779.

Stoner-Ma, D., Skinner, J. M., Schneider, D. K., Cowan, M., Sweet, R. M., & Orville, A. M. (2011). Single-crystal Raman spectroscopy and X-ray crystallography at beamline X26-C of the NSLS. *Journal of Synchrotron Radiation, 18*, 37–40. https://doi.org/10.1107/S0909049510033601.

Stubbs, J., Hornsey, T., Hanrahan, N., Esteban, L. B., Bolton, R., Maly, M., ... West, J. (2024). Droplet microfluidics for time-resolved serial crystallography. *IUCrJ, 11*, 237–248. https://doi.org/10.1107/S2052252524001799.

Vanko, G., Neisius, T., Molnar, G., Renz, F., Karpati, S., Shukla, A., & de Groot, F. M. (2006). Probing the 3d spin momentum with X-ray emission spectroscopy: The case of molecular-spin transitions. *The Journal of Physical Chemistry B, 110*(24), 11647–11653. https://doi.org/10.1021/jp0615961.

Von Stetten, D., Giraud, T., Carpentier, P., Sever, F., Terrien, M., Dobias, F., ... Royant, A. (2015). In crystallo optical spectroscopy (icOS) as a complementary tool on the macromolecular crystallography beamlines of the ESRF. *Acta Crystallographica. Section D, Biological Crystallography, 71*(Pt 1), 15–26. https://doi.org/10.1107/S139900471401517X.

Weierstall, U., James, D., Wang, C., White, T. A., Wang, D. J., Liu, W., ... Cherezov, V. (2014). Lipidic cubic phase injector facilitates membrane protein serial femtosecond crystallography. *Nature Communications, 5*. https://doi.org/ARTN330910.1038/ncomms4309.

White, T. A., Kirian, R. A., Martin, A. V., Aquila, A., Nass, K., Barty, A., & Chapman, H. N. (2012). A software suite for snapshot serial crystallography. *Journal of Applied Crystallography, 45*, 335–341. https://doi.org/10.1107/S0021889812002312.

Wiedorn, M. O., Oberthür, D., Bean, R., Schubert, R., Werner, N., Abbey, B., ... Barty, A. (2018). Megahertz serial crystallography. *Nature Communications, 9*. https://doi.org/ARTN402510.1038/s41467-018-06156-7.

Winter, G., Waterman, D. G., Parkhurst, J. M., Brewster, A. S., Gildea, R. J., Gerstel, M., ... Evans, G. (2018). DIALS: Implementation and evaluation of a new integration package. *Acta Crystallographica Section D: Structural Biology, 74*(Pt 2), 85–97. https://doi.org/10.1107/S2059798317017235.

Yi, J., Orville, A. M., Skinner, J. M., Skinner, M. J., & Richter-Addo, G. B. (2010). Synchrotron X-ray-induced photoreduction of ferric myoglobin nitrite crystals gives the ferrous derivative with retention of the O-bonded nitrite ligand. *Biochemistry, 49*(29), 5969–5971. https://doi.org/10.1021/bi100801g.

Yin, X., Scalia, A., Leroy, L., Cuttitta, C. M., Polizzo, G. M., Ericson, D. L., ... Soares, A. S. (2014). Hitting the target: fragment screening with acoustic in situ co-crystallization of proteins plus fragment libraries on pin-mounted data-collection micromeshes. *Acta Crystallographica. Section D, Biological Crystallography, 70*(Pt 5), 1177–1189. https://doi.org/10.1107/S1399004713034603.

Young, I. D., Ibrahim, M., Chatterjee, R., Gul, S., Fuller, F. D., Koroidov, S., ... Yano, J. (2016). Structure of photosystem II and substrate binding at room temperature. *Nature, 540*(7633), 453. https://doi.org/10.1038/nature20161.

Zhao, F. Z., Zhang, B., Yan, E. K., Sun, B., Wang, Z. J., He, J. H., & Yin, D. C. (2019). A guide to sample delivery systems for serial crystallography. *Febs Journal, 286*(22), 4402–4417. https://doi.org/10.1111/febs.15099.

Zielinski, K. A., Prester, A., Andaleeb, H., Bui, S., Yefanov, O., Catapano, L., ... Oberthuer, D. (2022). Rapid and efficient room-temperature serial synchrotron crystallography using the CFEL TapeDrive. *IUCrJ, 9*(Pt 6), 778–791. https://doi.org/10.1107/S2052252522010193.

Zimmermann, P., Peredkov, S., Macarena Abdala, P., DeBeer, S., Tromp, M., Müller, C., & Van Bokhoven, J. A. (2020). Modern X-ray spectroscopy: XAS and XES in the laboratory. *Coordination Chemistry Reviews, 423*.

CHAPTER FOUR

Sample delivery for structural biology at the European XFEL

Katerina Dörner*, Peter Smyth, and Joachim Schulz

European XFEL, Schenefeld, Germany
*Corresponding author. e-mail address: katerina.doerner@xfel.eu

Contents

1. Introduction	106
2. Scientific instruments at the European XFEL	107
2.1 The Single Particles and Biomolecules Serial Femtosecond Crystallography (SPB/SFX) instrument	108
2.2 Structural biology at other instruments at the European XFEL	109
3. Sample delivery for SFX via liquid jets	109
3.1 Introduction	109
3.2 Sample properties for sample delivery via GDVN	114
4. High viscosity extrusion for SFX experiments	116
4.1 Introduction	116
4.2 Crystal properties and viscous media choices	118
5. Testing of sample delivery	120
6. Fixed-target sample delivery	122
7. Drop-on-demand sample delivery	124
8. Summary and outlook	125
References	126

Abstract

Serial femtosecond crystallography (SFX) at X-ray free electron lasers (XFELs) is a valuable technique for time-resolved structural studies on enzymes. This method allows for the collection of high-resolution datasets of protein structures at various time points during a reaction initiated by light or mixing. Experiments are performed under non-cryogenic conditions and allow the collection of radiation damage free structures. At the European XFEL (EuXFEL), SFX experiments are mainly performed with liquid jets produced by gas dynamic virtual nozzles (GDVNs) and less frequent with a high viscous extruder (HVE). In this chapter we describe these delivery methods, with the focus on GDVNs. Instrumentation, sample requirements, and preparation steps for SFX beamtimes are discussed. Other sample delivery methods available at the EuXFEL are briefly introduced at the end of this chapter.

Methods in Enzymology, Volume 709
ISSN 0076-6879, https://doi.org/10.1016/bs.mie.2024.10.007
Copyright © 2024 Elsevier Inc. All rights are reserved, including those for text and data mining, AI training, and similar technologies.

1. Introduction

The method of choice for structural biology at free electron lasers (FELs) is serial femto-second crystallography (SFX). Relying on the concept of diffraction before destruction, thousands of protein microcrystals are delivered to the X-ray beam and a high-resolution protein structure is reconstructed from a series of diffraction patterns (Chapman et al., 2011; Neutze et al., 2000). This allows structure determination at room temperature without radiation damage, and enables time-resolved experiments with high temporal resolution.

The very first SFX experiment was performed at the Linac Coherent Light Source (LCLS) (Chapman et al., 2011). The sample was delivered in a liquid jet via a gas dynamic virtual nozzle (GDVN), where protein crystals are exposed to the X-rays in a liquid stream of a few micrometers (DePonte et al., 2008). Time-resolved SFX experiments soon followed and opened up new possibilities for studying the structural mechanisms of light-sensitive proteins (Aquila et al., 2012; Tenboer et al., 2014). In these experiments, crystals in the liquid stream were exposed to an optical pump laser synchronized with the X-rays. Further development in GDVNs led to their use in the first SFX experiment at the European XFEL (EuXFEL), reaching jet speeds compatible with the megahertz X-ray pulse repetition rate (Fig. 1) (Grünbein et al., 2018; Wiedorn et al., 2018). Micro-mixers connected to the GDVNs were developed to enable time-resolved experiments with non–light sensitive proteins (Calvey et al., 2016; Vakili et al., 2022). Since 2011, static and time-resolved SFX experiments have been performed at XFEL facilities around the world. Over time, more sample delivery methods have been developed, with focus on reduction of the sample amount, the delivery of membrane proteins, and rapid mixing.

At the EuXFEL, liquid jets remain the preferred delivery method as it is the only method compatible with MHz X-ray repetition rates (Fig. 1). Therefore, sample delivery via GDVNs will be the main focus of this chapter. However, membrane protein crystals require a lipidic cubic phase (LCP) environment resulting in highly viscous delivery media which are not compatible with the GDVN. High viscosity extrusion (HVE) is offered at the EuXFEL as another standard delivery method to enable time-resolved SFX experiments on membrane proteins. This chapter will describe both GDVN and HVE methods covering sample properties, buffer compositions, sample delivery tests and general limitations. The chapter will also briefly introduce the development of fixed-target and drop-on-demand methods at the EuXFEL.

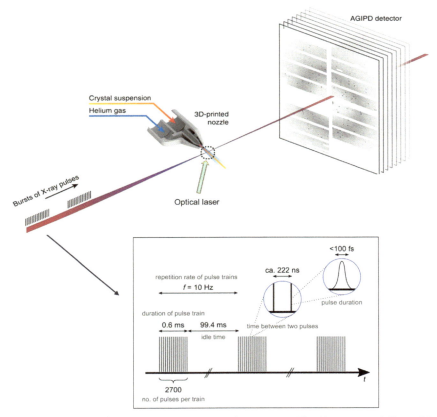

Fig. 1 Schematic of an SFX experiment at the EuXFEL. The X-rays are delivered in pulse trains and interact with the sample in a liquid jet produced by a 3D-printed GDVN. For pump-probe experiments, the jet is exposed to an optical laser in the interaction region and diffraction data are recorded with the AGIPD. *Picture re-printed and modified from Wiedorn, M. O. et al. (2018). Megahertz serial crystallography. Nature Communications, 9(1), 4025 and Vakili, M. et al. (2022). 3D printed devices and infrastructure for liquid sample delivery at the European XFEL. Journal of Synchrotron Radiation, 29(2), 331.*

2. Scientific instruments at the European XFEL

The European XFEL has seven scientific instruments on three beamlines. The superconducting accelerator of the EuXFEL delivers bunches of electrons with up to 17.5 GeV kinetic energy to the three beamlines, providing photon energies of 3–25 keV for the two hard X-ray undulators, and 0.26–3 keV for the soft X-ray undulator. The EuXFEL can deliver up to

27 000 photon pulses per second in 600 µs long pulse trains every 100 ms. Within the pulse trains, repetition rates up to 4.5 MHz are possible. The pulse pattern can be arbitrarily filled and multiplexed between the three beamlines. This enables parallel 10 Hz operation of three scientific instruments at any time (Decking et al., 2020; Tschentscher et al., 2017).

2.1 The Single Particles and Biomolecules & Serial Femtosecond Crystallography (SPB/SFX) instrument

The Single Particles and Biomolecules & Serial Femtosecond Crystallography (SPB/SFX) instrument is dedicated to diffraction of microcrystals, as well as single particle imaging experiments (Mancuso et al., 2019). Located on the SASE1 beamline SPB/SFX has two main interaction regions for experiments. The X-ray energy is tunable from 6–15 keV.

In the upstream interaction region (IRU), SPB/SFX provides optimal focus properties by windowless, in-vacuum beam transport using precise focusing mirrors in Kirkpatrick-Baez geometry. These mirrors provide micro- (~3 µm) and nanofocusing (~100 nm) capabilities (Bean et al., 2016). The megapixel Adaptive Gain Integrating Pixel Detector (AGIPD) with a sample-detector distance in the range 125–6000 mm (Allahgholi et al., 2016, 2019) can store up to 352 images per train repeating at 10 Hz (Tschentscher et al., 2017). The AGIPD operates in vacuum, which requires sample delivery in vacuum using a catcher with a differential pumping system (Fig. 3) (Mancuso et al., 2019). The majority of biological experiments at SPB/SFX are SFX experiments with liquid jets as the delivery method. Additionally, solution scattering via liquid jet sample delivery (Blanchet et al., 2023; Konold et al., 2024) and single particle experiments via aerosol injection can be performed (Bielecki et al., 2019).

The more modular downstream interaction region (IRD) currently uses a JUNGFRAU 4 M which operates in air instead of vacuum (Mozzanica et al., 2018). This detector has 16 memory cells (Mozzanica et al., 2016) which can be filled during the 600 µs pulse train, with the data read out occurring between trains. Various sample environments can be used in the downstream interaction region, including liquid jets, HVE, and fixed targets in a helium atmosphere. The majority of biological experiments are SFX experiments with sample delivery via HVE, which will be discussed in more detail below.

The SPB/SFX instrument is equipped with several laser systems for time-resolved pump-probe experiments (Koliyadu et al., 2022) which cover the wavelength range 400–2600 nm at MHz repetition rates. A tunable system between 210–2200 nm at up to 20 Hz is also available with pulse duration and pulse energy depending on the laser system. For a detailed description please see Koliyadu et al (2022). Laser parameters need to be determined for each specific experiment, considering factors such as reaction times, jet speed and the distance from position of pump laser and jet overlap to the X-ray focus point.

2.2 Structural biology at other instruments at the European XFEL

The other instruments that can be used for structural biology at the European XFEL are the hard X-ray instruments Material Imaging and Dynamics (MID) and Femtosecond X-ray Experiments (FXE), and the soft X-ray instrument Small Quantum Systems (SQS). MID provides an AGIPD in varying geometries to allow small angle X-ray scattering (SAXS), wide angle X-ray scattering (WAXS), and X-ray photon correlation spectroscopy (XPCS). The MID sample environment can be run in air or in vacuum. FXE is optimized for time resolution and can combine forward scattering with X-ray fluorescence and resonant X-ray scattering in a single shot. Both MID and FXE can be configured for fixed target or liquid jet sample delivery. For FXE, the main sample delivery method is a liquid jet in helium atmosphere. SQS can be equipped with an aerodynamic lens system to inject nanoparticles from an aerosol into vacuum. In this configuration, single particle imaging of biological samples is possible.

3. Sample delivery for SFX via liquid jets
3.1 Introduction

The main delivery method for SFX at the EuXFEL is via liquid jets using gas dynamic virtual nozzles (GDVNs)(DePonte et al., 2008). In these nozzles, the stream of protein crystal slurry is compressed by a concentric gas sheath (Fig. 2), resulting in a fast jet of a few micrometers in diameter and allowing nozzle orifices large enough to avoid clogging. Currently, this is the only crystal delivery method for high-resolution SFX which is compatible with MHz X-ray repetition rates enabling data collection at 1.1 MHz at the EuXFEL allowing the collection of complete data sets within 20–60 min. The method is compatible with pump-probe studies as well as mixing experiments. This allows the collection of several time points in one user beamtime and therefore is an excellent method for time-resolved structural studies.

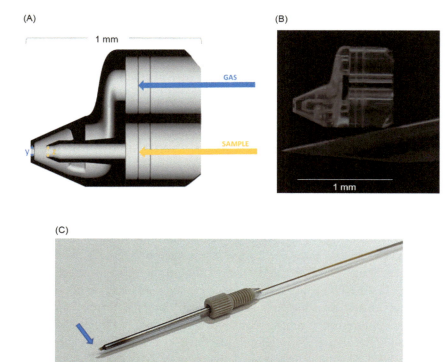

Fig. 2 Gas Dynamic Virtual Nozzle (GDVN). Schematic of a GDVN (A) showing the incoming gas and liquid channels as well as the gas orifice (y) and liquid orifice (z). Micrograph of a 3D-printed GDVN with a syringe needle for comparison (B). Nozzle assembly with the nozzle marked with an arrow (C). Liquid and gas connections run through a metal tube and are glued to the nozzle. The fitting which connects the nozzle with the SPB/SFX nozzle rod is shown.

GDVNs are available in several different designs, originally built out of glass but now predominantly 3D-printed using the two-photon polymerization (2PP)-based method. An overview of 3D-printed nozzles and devices at the EuXFEL can be found in Vakili et al. (2022). A typical nozzle for SFX experiments has a sample line with an inner diameter of 100 μm, with the sample passing through an orifice of 75 μm within the tip of the nozzle. At that point the sample is compressed by helium gas running in an outer channel and exiting the nozzle through a 60 μm orifice (Fig. 2). The result is a fast jet (20 to 50 m·s^{-1}) with a typical diameter of 3 to 5 μm. The jet speed and diameter are adjustable and depend on the gas and liquid flow rates as well as diameter of the gas orifice. Double-flow focusing nozzles (DFFNs) have an additional liquid line creating a liquid sheath, usually ethanol or water, around the sample slurry (Oberthuer et al., 2017). For some samples, jetting stability can be improved by using a DFFN, which generally requires a lower sample flow rate.

Samples are provided in syringe-like sample reservoirs driven by an HPLC pump (Schulz et al., 2019). A plunger in the reservoir divides the water side from the sample side. The water from the HPLC pump pushes the plunger forward and drives the sample into the sample line. The sample line is connected to the nozzle via a switching valve that enables switching between water and sample flowing to the nozzle. This way, sample flow can be quickly stopped in case data collection needs to be interrupted, and then started again easily. More valves can be included to switch between several sample reservoirs. To avoid settling of the crystals, the reservoir is rotated during sample delivery. Temperature in the reservoir holder can be controlled in the range between 4 °C to room temperature. The nozzles are mounted on a nozzle rod with 2.2 m lines connected to the sample reservoir and pumps in order to reach the X-ray interaction region in the relatively large SPB/SFX vacuum chamber (Fig. 3), and at the same time enable a fast nozzle exchange when needed (Schulz et al., 2019).

During the experiment, the jet is imaged via a digital side-microscope (Koliyadu et al., 2022; Vakili et al., 2022) with pulsed laser illumination enabling for monitoring of the X-ray-induced jet explosion and resulting gap in the jet (Fig. 3). In this way, sufficient jet speed and stability can be confirmed. A minimum flow rate is required to ensure replenishment of sample between pulses, so each exposure is on a fresh undamaged crystal. This applies to the X-ray beam as well as to pump illumination for time-resolved experiments, which often use a larger ($\sim$100 μm) focal spot (Grünbein et al., 2020). The required linear speed of a liquid jet (in $m{\cdot}s^{-1}$) depends on the intended X-ray repetition rate. The jet must move fast enough for the sample damaged by one XFEL pulse to be replaced with fresh sample, and for the jet to reform, in the time interval between pulses. For the highest intra-train repetition rate of 1.128 MHz used at SPB/SFX (887 ns between pulses), the jet speed should be at least $40\ m{\cdot}s^{-1}$, although at 0.5 MHz, a minimum of $20\ m{\cdot}s^{-1}$ is required. At repetition rates higher than approximately 1.1 MHz, sample degradation may occur due to shockwaves traveling along the jet into the unexposed sample which can impact crystal integrity and protein structure as observed at 4.5 MHz (Grünbein et al., 2021; Grünbein et al., 2021; Grünbein et al., 2021).

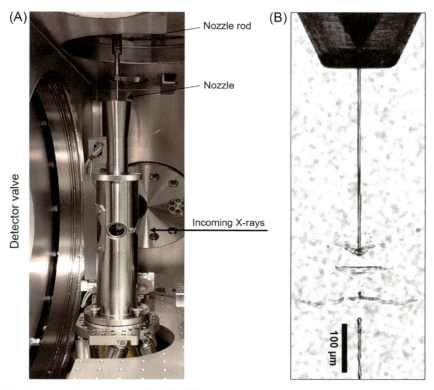

Fig. 3 Interaction region at the SPB/SFX instrument. A view of the catcher in the SPB/SFX vacuum chamber with X-rays entering from the right (A). Once the experiment is running, the valve to the detector on the left is opened. The picture shows the insertion of the nozzle mounted on the nozzle rod. After insertion, the nozzle tip is a few hundred micrometers above the X-ray beam and is visible via the side-view microscope (not shown) mounted close to the catcher window. (B) An image taken with the side-view microscope of a jet in the SPB/SFX chamber, showing the gaps resulting from the explosion after X-ray exposure.

Faster and thinner jets are obtained with a combination of increased GDVN sheath gas pressure and a smaller orifice. Maximum flow rates of liquid and gas into the IRU chamber are limited by the vacuum requirement of the AGIPD (Allahgholi et al., 2019). The detector, which is mounted in a vacuum chamber adjoining the sample chamber, cannot operate above 2×10^{-4} mbar due to the high voltage electronics. The catcher in which the nozzle sits and the sample is injected, is connected to its own turbo pump and acts as a differential pumping system, keeping the pressure in the main chamber below the required 2×10^{-4} mbar (Mancuso et al., 2019).

Choice of nozzles affects the jet parameters, with smaller gas orifice diameters providing a faster jet at the same volumetric flow rates. Smaller nozzles, however, have an increased risk of clogging due to protein crystals that are too large or have aggregated, or from other precipitated material. Hence, prior characterization of sample behavior in a jet is important for efficient use of limited beamtime.

As mentioned above the SPB/SFX instrument is equipped with several laser system to enable pump-probe experiment for time-resolved studies. Different geometries of the GDVN and adjustments of gas and liquid flow rate allow specific jet speeds compatible with desired timepoints. The jet imaging via the side-microscope enables the detection of jet speeds during the experiment. The used sample reservoirs are light-tight and liquid connections can be covered in the hutch to ensure a dark environment.

GDVNs can be combined with micro-mixers for rapid mixing of protein crystals with substrates and therefore enable time-resolved studies of non-light dependent enzymatic reactions. At the EuXFEL mixers are 3D-printed using the two-photon polymerization (2PP) method and available in different designs (Vakili et al., 2022). The designs are based on glass mixers developed at Cornell University (Calvey et al., 2016). Reaction times down to 5 ms can be reached.

With sample delivery via GDVNs, the minimum flow rate required to maintain a fast jet means many crystals pass through the interaction region during gaps between pulses, without being exposed to the X-ray beam (DePonte et al., 2008; Weierstall, 2014). Solutions are needed to minimize the waste of sample and provide sample only during the pulse train. In general, drop-on-demand, fixed-target methods, and HVE sample delivery address this problem, but these methods are not compatible with MHz repetition rates of the X-ray pulses.

A promising approach to reduce sample waste while utilizing MHz repetition rate of the X-ray pulses is the segmented-flow method (Echelmeier et al., 2020). Samples are delivered via GDVNs, but the jet consist of oil segments without sample, and buffer segments with protein crystals. Droplets of sample are produced in a capillary within a continuous oil stream. When exiting the GDVN, the droplets are squeezed and elongated, resulting in segments in the jet. The droplets can be electrically triggered and synchronized with the X-ray pulse trains, ensuring that sample delivery only occurs during X-ray pulses. Between the pulse trains, only oil flows through the jet. This would lead to a 10-fold reduction in sample consumption. Furthermore, this method is compatible with time-resolved experiments via pump-probe as well as mixing.

3.2 Sample properties for sample delivery via GDVN

Since SFX relies on collection of a single diffraction pattern from each crystal, thousands of crystals are needed to build a complete dataset with sufficient multiplicity for high quality and high resolution protein structure determination (Kirian et al., 2010; Moreno-Chicano et al., 2019). This is especially important in time-resolved studies, where small differences between conformations must be resolved, often in a mix of partial occupancies (Caramello & Royant, 2024; Gorel et al., 2021). For optimal sample consumption, small crystals are preferred, so less protein is used in each crystal. Specifically, for time-resolved experiments, small and uniformly sized crystals are essential to ensure consistent reaction starting times and light intensities. Small and homogeneous crystals are also better suited for liquid sample delivery as they are less likely to clog the small diameter nozzles used to produce jets. This is the most common cause for unplanned interruptions in data collection. Crystal concentration is also an important factor to consider when preparing samples. If the concentration of crystals in the slurry is too low, then the hit rate will be low requiring more time to collect a full dataset, resulting in wasted beamtime. If too high a concentration of crystals is used, then more protein is required, as well as increasing the probability of nozzle clogging, and increased chance of multiple hits, which are harder to process, and can affect laser illumination for time-resolved experiments. In addition, the composition of the delivery buffer needs careful examination since viscosity is an important factor for jet stability.

In general, SFX data analysis and experimental set-up leads to some crucial sample parameters to enable smooth sample delivery and effective data collection. The most important are the following, and will be discussed in more detail below.

1. Protein crystal size and size distribution
2. Composition of the delivery buffer
3. Sample amount and final crystal density

3.2.1 Protein crystal size and size distribution

One factor for the upper limit of the crystal size is the diameter of the nozzle orifices. The GDVNs provide thin jets and at the same time have orifices large enough to handle microcrystals without clogging. However, to achieve fast enough jets the nozzle orifice has to be below $100\,\mu m$ in diameter, limiting the crystal size to approximately $20\,\mu m$. Larger crystals have a higher risk of clogging the nozzle, and therefore optimal crystal size for delivery is 10 µm or less. Size distribution of the crystals should be narrow, as differences in crystal size lead to higher inhomogeneity of the sample and therefore instability in flow behavior of the resulting jets.

From a sample delivery perspective, smaller crystals are the better. However, previous SFX experiments indicate that there might be a protein-specific optimal crystal size for highest resolution of diffraction (Shoeman et al., 2023). In time-resolved experiments, additional limitations on the crystal size apply. For mixing experiments, the diffusion times of substrates within the crystals need to be considered and must be shorter than the turnover time of the enzyme. To minimize diffusion times, crystals should be only a few micrometers in size (Pandey et al., 2021; Schmidt, 2013). A more detailed discussion can be found in Schmidt (2013). In pump-probe experiments, crystal size should not exceed the light penetration depth, if the observation only of single-photon-induced processes is desired (Grünbein et al., 2020). If crystals are too large, milling or filtration steps can be tested to achieve smaller crystals (Shoeman et al., 2023).

3.2.2 Composition of the delivery buffer

The buffer components of a sample greatly affect the ability to form a stable jet. Since the upstream interaction region is in a vacuum, liquids have the tendency to evaporate upon introduction to the chamber. This can result in precipitation of compounds in the sample, such as highly concentrated salts in the crystallization buffer. Precipitated compounds and other solids in the sample can clog the nozzle, disrupting jet flow. Ice formation on the nozzle due to evaporative cooling is also an issue. Solid or liquid build-up around the nozzle has the potential to enter the X-ray beam, and cause large amounts of scatter which can damage the detector. If the viscosity of the sample is too high, a jet is not formed. This leads to limitations in the concentration of crystallization agents such as polyethylene glycol and ammonium sulfate. Since protein crystal type, size, and density and buffer composition contribute to the viscosity of the sample, it is difficult to give general statements on concentrations of compounds. For each sample, jetting tests have to be performed before the experiment to identify conditions for stable jetting.

3.2.3 Sample amount and final crystal density

With liquid jets, the hit rate depends on concentration of crystals in the sample. Higher crystal concentrations increase the likelihood of a crystal being in the beam during an X-ray pulse. However, too high a density of crystals can prevent stable jetting or lead to clogging of the nozzles. In general, the concentration of settled crystals in the delivery buffer should be around 10–15 % (v/v). The hit rate should be determined at the beginning of each experiment. By varying the sample concentration, it can be optimized during further data collection. As a rough estimate for one data set at the EuXFEL, 1 mL of crystal slurry with 15 % (v/v) of settled crystals is needed.

4. High viscosity extrusion for SFX experiments
4.1 Introduction

Another sample delivery method available at the SPB/SFX instrument is high viscosity extrusion (HVE), which involves extruding a viscous crystal-containing medium into the X-ray beam. Membrane proteins are often crystallized in lipidic cubic phase (LCP), known as in meso crystallization (Aherne et al., 2012; Liu et al., 2014). As LCP is naturally highly viscous, HVE is an ideal method for delivery of membrane protein crystals. This method is suitable for pump-probe experiments with an optical laser, enabling time-resolved studies. With HVE, sample consumption is much lower than with liquid jets. The high viscosity of the sample allows a stable extrusion to be formed at lower flow rates, thereby increasing the proportion of crystals hit by the beam, and thus requiring a lower volume of sample. However, the low flow rates limit data collection to 10 Hz at the EuXFEL, making data collection time consuming. One advantage of using HVE is the lower flow rate providing a reduced linear sample speed (measured in $m \cdot s^{-1}$), so longer timepoints can be accessed with laser pump-probe before breakup of the sample flow. Due to the lower maximum repetition rate, HVE is not the default delivery method at the EuXFEL, but is useful for structural studies of membrane proteins.

The viscous extruder used at SPB/SFX is based on the Arizona State University design (Fig. 4) (Weierstall et al., 2014), which is also in use at other facilities. The sample reservoir is made of steel, with a volume of 20–220 µL, and its cylindrical bore is sealed at one end using a PTFE ball. The crystal-containing medium is inserted from the front using a gas-tight syringe connected via an adapter. The reservoir is then connected to a fused silica capillary with an inner diameter of 50–125 µm, which has been ground with an external taper to a point at one end. The capillary is held inside a ceramic outer nozzle through which helium flows as a sheath around the sample to stabilize the extrusion (Fig. 4). The helium prevents the extrusion from fluctuating in position relative to the X-ray

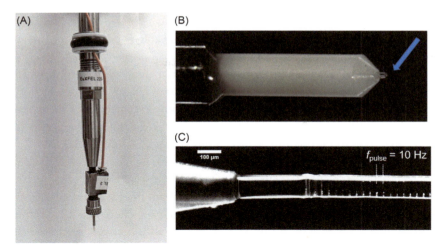

Fig. 4 High viscosity extrusion (HVE) device. The assembly of the extruder (A) and a micrograph of the nozzle (B). The tip with inner diameter of 100 μm is marked with an arrow. (C) A micrograph of the tip with the sample extruding at the SPB/SFX instrument. Interaction with the X-rays pulses at 10 Hz repetition rate cause a regular pattern to be visible in the sample.

and laser beams, but does not provide flow focusing as with a GDVN. Extrusion is powered by an HPLC pump. The pump forces water into the upper hydraulic section of the extruder, which contains a syringe plunger with a connected rod (Weierstall et al., 2014). The rod extends into the reservoir and pushes on the rear of the PTFE ball to force the sample out of the front. The differing internal diameters of the water syringe and the sample reservoir result in the sample having a reduced flow rate and increased pressure compared to the water, proportional to the difference in cross-sectional area. Typical flow rates for the extruded sample are on the order of 1 $\mu L \cdot min^{-1}$, which result in linear velocities of 2 $mm \cdot s^{-1}$. This flow rate allows, the extrusion to move 200 μm in the 10 ms between pulse trains, which is enough to remove damaged sample and bring new crystals into the beam, ensuring data free of the effects of beam-induced damage.

An alternative extruder nozzle design is available. It is 3D-printed in the same manner as GDVN nozzle (Vakili et al., 2022). Instead of the sample being extruded directly from a short-tapered capillary connected to the reservoir, a capillary is used to connect the 3D-printed nozzle to the reservoir. The 3D-printed piece includes both the externally tapered sample nozzle and the surrounding gas orifice. An advantage of the one-piece design is the reproducible geometry, which is not achievable with hand-ground and assembled capillary nozzles.

To enable time-resolved SFX experiments based on mixing, a nozzle was designed that incorporates an additional capillary for delivering a reaction substrate, which is mixed with the sample upstream of the nozzle exit. A helical static mixer is incorporated into the length between the inlets and the tapered section before the orifice. Because of the low flow rate of the viscous medium and the limited rate of diffusion, mixing times are in the range 2–20 s (Vakili et al., 2023).

The viscous extruder is used in the downstream interaction region of the SPB/SFX instrument, contained within a helium-filled cube to reduce background scatter. The assembled extruder with sample reservoir and nozzle is inserted into a steel cone at the top of the cube. An O-ring on its outside forms a seal inside the cone. The nozzle and cube are aligned with the X-ray beam using an on-axis camera immediately upstream, which incorporates a drilled lens and mirror to pass the X-ray beam. The downstream face of the cube contains a Kapton (polyimide) window to allow the diffracted X-rays to exit, with a short distance to the detector through air. The direct beam passes out through a hole in the Kapton window, then through a metal straw to block air scatter before the beam reaches the central hole in the detector. The side windows of the cube are transparent to allow orthogonal viewing of the jet and pump-probe illumination.

To catch the extrusion upon breakup, a catcher is installed below the interaction region. This helps to keep the extrusion flowing vertically at a stable rate. Depending on the properties and flow rate of the sample, the catcher can consist of a wide hole, a static cone, or be connected to a vacuum pump to draw the extrusion downwards. When using a vacuum pump, care must be taken not to pump too fast, as this could remove the helium atmosphere in the sample cube and draw air into the cube through gaps, causing increased background scatter.

4.2 Crystal properties and viscous media choices

The best viscous medium to use depends on the protein sample being investigated. Membrane proteins are often crystallized in LCP, which is naturally highly viscous (Aherne et al., 2012; Liu et al., 2014). Membrane proteins are first reconstituted in LCP by passing the protein and lipids between two coupled gas-tight syringes to mix them. The protein can then be crystallized by mixing the LCP with a syringe of precipitant solution.

Another method of protein crystallization in LCP has been developed, which enables high quantities of crystals to be grown in sealed vials (Birch et al., 2023). Protein-laden LCP can be dispensed into vials of precipitant solution, which are incubated until crystals grow. The LCP with crystals can then be lifted from the liquid in the vial and loaded into the extruder for sample delivery. The viscosity of LCP can be tuned for optimal sample delivery by changing the lipid content and concentrations, subject to protein stability in the lipid mixture used. It is also possible to crystallize proteins under normal batch conditions, and then embed them in LCP prior to sample delivery (Nam, 2019).

Another option for extrusion media when using non-membrane proteins is embedding crystals in various types of viscous substances. A common medium is grease, which is hydrophobic and encases the crystals with minimal amounts of remaining mother liquor surrounding them. Various types of grease (Sugahara et al., 2015, 2020) and animal and plant fats (Nam, 2020a, 2020c, 2022) have been used successfully. Viscous media for embedding crystals can also be made from polyacrylamide (Park et al., 2019) and water-soluble polysaccharides such as agarose and hydroxyethyl cellulose (Conrad et al., 2015; Nam, 2020b; Sugahara et al., 2015), which are dissolved in mother liquor prior to crystal embedding. The concentration of the medium can be adjusted to control its viscosity. Embedding of crystals in media is usually performed by first concentrating the crystals with gentle centrifugation and removal of excess mother liquor from the top. The crystals can then be loaded into a gas-tight syringe and coupled to another syringe containing a larger volume of viscous medium, often a volume ratio of 1:4, then mixed by passing the contents back and forth until homogeneous. Embedding can also be performed by mixing the crystals and medium together on a glass plate prior to loading into the sample reservoir (Nam, 2019; Sugahara et al., 2016).

For HVE, crystals similar to those used for liquid jets are required. Crystal preparation needs to consider the homogeneity and size of the crystals. To prevent clogging, crystals must be smaller than the nozzle orifice. Due to the lack of flow focusing, the extrusion has the same diameter as the nozzle (50–125 μm), which results in higher background scattering from the carrier medium than a focused (~5 μm) liquid jet. This means that small sub-micrometer crystals are unsuitable for HVE due to the limited signal to noise ratio.

As with liquid jets, crystal concentration within the delivery medium is important to control hit rate. Unlike in low-viscosity suspension, it is hard to increase concentration of crystals after embedding in viscous medium. Therefore, correct initial preparation of crystals is important. The concentration of crystal is controlled during sample preparation by altering the volume ratio of concentrated crystal slurry added to the viscous medium.

5. Testing of sample delivery

The ability to successfully deliver a sample in a liquid jet or viscous extrusion is highly dependent on the properties of the sample. To make efficient use of precious XFEL beamtime, the optimal conditions for sample delivery should be known beforehand. To this end, the EuXFEL provides facilities and assistance for performing offline tests of sample jetting prior to beamtime. Jetting test stations are set up in the user laboratories, located above the experimental hall. The setups include similar equipment to that on the instrument, including several HPLC pumps, flow meters, switching valves, a sample rotator, and regulated pressurized helium. Sample reservoirs and connections to the nozzle are identical to those at the beamline. The jet can be examined using a high-speed camera and microscope lens (Fig. 5). GDVNs are tested in small vacuum chambers, and HVE is tested in air.

One of the test stations is a replica of the IRU vacuum sample chamber of the SPB/SFX instrument. This system allows testing of entire sample delivery setups in conditions as close as possible to those at the instrument. The replica chamber also provides space for developing and testing upgrades to the IRU chamber, without disrupting use of the instrument itself. The chamber and associated equipment can be used to train staff and users on loading nozzles and delivering samples into the interaction region.

The first test for liquid jets involves checking the functioning of the GDVNs themselves. After 3D printing, fused silica capillaries are glued to the input ports of the nozzle, and the nozzle is glued into a steel tube for support. These connections are tested for leaks or blockages by jetting water. The reliable formation of a jet at the flow rates and gas pressures intended for use with the samples is examined. Manufacturing defects in the nozzles can result in them behaving erratically above certain values.

For sample testing, a minimum of 300 µL of crystal slurry with crystal density of 10–20 % (v/v) is needed. As well as testing real samples and consuming precious protein crystals, testing of jetting can be performed on the buffer in which the crystals are suspended. This can provide some

Sample delivery for structural biology at the European XFEL

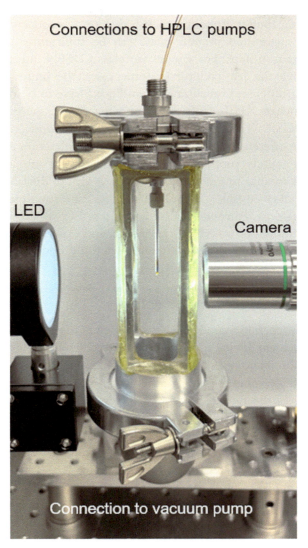

Fig. 5 Jetting test chamber. A GDVN is shown in the vacuum test chamber in the EuXFEL user laboratories. A microscope lens connected to camera (not shown), and an LED light source, enable monitoring of the jet. Gas and liquid lines for the sample are coming in from the top and are connect to a HPLC pump and a pressure regulator (not shown).

information on suitability for jetting at a given speed, and the required nozzle and flow parameters. A stable jet should be straight, run continuously with a certain length (more than 150 μm) and without disruptions. In most cases, failure is due to an inhomogeneous sample or too high viscosity of the buffer. Inhomogeneity, due to a broad crystal size distribution or aggregated protein often can be sufficiently improved by additional filtration steps. For viscous samples, a DFFN instead of GDVN might improve the jet stability. In some cases, the buffer needs to be exchanged or different crystallization conditions have to be obtained.

The European XFEL offers beamtime specifically for protein crystal screening (PCS). Within the scope of a PCS proposal, a few hours beamtime are offered to test sample delivery and diffraction quality of the samples. This includes support in the user laboratory for sample preparation and testing of conditions for sample injection. A visit to the labs prior to the beamtime is also possible. Early testing gives time to optimize sample conditions prior to returning for the beamtime. PCS is aimed at research groups with limited or no XFEL experience, and experienced groups who wish to test new samples in preparation for a full proposal.

6. Fixed-target sample delivery

At high repetition rate X-ray sources such as the European XFEL, fast sample replacement is vital to make good use of the available X-ray pulses. This calls for fast jets as described in previous chapters. Unfortunately, fast jets need large sample volumes that are not always available. Fixed-target sample delivery enables the measurement of samples which are only available in small amounts. In addition, limitations of viscosity and salt concentration of the crystal buffer do not apply. All instruments at the EuXFEL provide fast and accurate scanning stages for fixed target sample delivery. With this technology, biological targets such as crystals are fixed on a two-dimensional surface and scanned through the X-ray focus. At the EuXFEL, this scanning happens at the 10 Hz pulse-train repetition rate. The train length of 600 μs is too short to consider replacing the sample within a pulse train.

In addition to scanning in two dimensions, some instruments also allow rotating the sample stage around the vertical axis. For crystallography, however, this option is normally not needed as the crystals are randomly oriented on the surface, and are destroyed with a single pulse of the XFEL. In a fixed target SFX approach, each measured crystal samples a random angle and a full data set is taken in the same manner as in liquid jet SFX. Only exceptionally, crystals might lie preferentially specific in specific angles on the surface because of their shape. In this case, scanning in the stage at different angles might be necessary.

At the EuXFEL, two types of fixed targets for macromolecular crystallography are available; the Roadrunner II goniometer (Roedig et al., 2016; Tolstikova et al., 2019) or the HARE (hit-and-return) chips used at the PETRA III P14 T-REXX endstation (Mehrabi et al., 2020; Von Stetten et al., 2019).

The Roadrunner II goniometer uses an aluminum or polyimide frame to hold chips that contain randomly distributed crystals. There are several layouts of chip, with different patterns of perforations, including larger chips with several compartments capable of holding independent samples to reduce sample exchange time. The chips are made of either polyimide or monocrystalline silicon. For silicon chips, the data collection positions must be aligned to the apertures so that the X-ray pulses pass through and avoid causing strong Bragg peaks from silicon diffraction, which can damage a detector with an XFEL pulse.

The goniometer includes a humidified sample chamber with Kapton exit window, into which the chips can be loaded. A heated or cooled water bath and gas bubbler are used to control sample chamber temperature and humidity (Tolstikova et al., 2019). As well as sample translation between positions on the chip, the chip can also be rotated during data collection, to compensate for potential preferential orientation of crystals.

The European XFEL also has the fast solid sample scanner (García-Tabarés Valdivieso et al., 2024), which is capable of supporting the HARE chip design (Mehrabi et al., 2020). These chips contain tapered apertures into which crystals are drawn as vacuum pressure removes mother liquor from the rear of the chip. Fiducials are used to align the apertures with the beam, so that it passes through the aperture, and pulses are more likely to hit a crystal. These chips are sealed either side with polymer foils to prevent evaporation.

Another fixed target method is to place a thin layer of crystals in mother liquor onto a foil, sealed with another foil layer on top. This method sandwiches the sample between two protective layers and thereby prevents drying. The additional layer, however, creates background and the sample is irregularly distributed. This requires either a pre-investigation of target positions, or the sample is scanned in a regular pattern, limiting the hit rate and thereby the efficiency both of sample and beam usage (Doak et al., 2018).

At the European XFEL, an approach to automatically search and qualify samples has been developed. Using deep-learning algorithms, it is possible to pre-characterize samples and even distinguish different sample types. This approach can be used for pre-characterization of a sample chip in the lab. The positions of good targets are then stored relative to fiducial marks on the chip, and can be automatically retrieved during the X-ray experiment after positional calibration using the fiducial marks (Kardoost et al., 2024). First tests show that algorithms can be fast enough for real-time target finding in 10 Hz. More development in hard and software is still required.

7. Drop-on-demand sample delivery

The pulsed operation of XFELs opens the possibility to increase sample efficiency of liquid samples. Samples can be delivered on demand, synchronized with the XFEL pulses. Drop-on-demand technology is widely available because ink-jet printers use a similar technique to apply a drop of ink to the paper at the moment the print head scans over the to-be-colored pixel (Su et al., 2021).

The aim of drop-on-demand sample delivery is to increase sample efficiently by only injecting the sample when an X-ray pulse is approaching. If the emitted droplets are sufficiently small to contain only one sample crystal each, the sample efficiency can be close to 100 %. This means that every injected crystal is hit by an X-ray pulse and creates a measurable diffraction pattern. If additionally, the droplet delivery rate can be fast enough to deliver one droplet per X-ray pulse, and the particle density in the crystal slurry is high enough to provide a crystal in most droplets, the X-ray pulses can be used very efficiently as well.

Drop-on-demand sample delivery systems can operate at repetition rates on the order of tens of kilohertz. It is therefore not possible to match the 4.5 MHz maximum intra-train repetition rate of the EuXFEL, however the arbitrary filling pattern of the EuXFEL allows matching of the pulse rate to that attainable by drop-on-demand methods. A 30 kHz repetition rate, for example, allows the delivery of 18 pulses in the 600 µs pulse train, leading to an effective droplet hit rate of 180 shots per second.

Another option is to produce elongated droplets that can be hit sequentially by several X-ray pulses in a pulse train. The elongated droplets are short segments of a liquid jet containing a number of crystals, just as in a fast jet. Therefore, hitting the jet several times with X-ray pulses leads to hit rates similar to liquid jet operation. Because these elongated drops are only provided when pulses are expected, the sample consumption can be reduced by at least two orders of magnitude, close to the duty cycle of the bunched XFEL.

This droplet pattern has been demonstrated recently by Perrett et al. (2024) at the FXE instrument of the European XFEL. Each pulse train was filled with 16 pulses at a repetition rate of 47 kHz. During these 340 µs long bunch trains, four elongated droplets were produced so that every droplet was hit by four consecutive photon pulses (Perrett et al., 2024).

Drop-on-demand technology cannot be used in vacuum environments. Evaporation of solvent at the nozzle would cool the nozzle between shots and thereby freeze it, or the increased solute concentration caused by solvent evaporation could cause precipitation and nozzle clogging. Therefore, for liquid sample delivery in a vacuum, a continuous flow of liquid is preferable. A combination of drop-on-demand and liquid jet delivery is the segmented flow method mentioned in the GDVN paragraph above. Droplets are formed in the liquid line resulting in a segmented jet compatible with vacuum, while reducing the sample amount at the same time.

8. Summary and outlook

GDVNs and HVE are crucial sample delivery methods for time-resolved SFX experiments at the European XFEL and complement each other in respect of sample viscosity and x-ray repetition rates. They are compatible with pump-probe and mixing set-ups, enabling time-resolved studies to reveal structural mechanisms of proteins. While preparing samples for an SFX experiment at XFELs, attention needs to be paid to crystal size, sample homogeneity and buffer composition. Pre-testing of sample delivery is a necessity for successful beamtime and efficient data collection. GDVNs enable data collection at MHz repetition rates and therefore the collection of many time-points of an enzymatic reaction within one beamtime at the EuXFEL. HVE is a valuable alternative for time-resolved studies on membrane proteins.

A remaining challenge is high sample consumption, and more effective sample delivery methods are needed. Drop-on-demand methods only require small volumes of sample and will be developed further to be compatible with higher X-ray repetition rates. The segmented-flow method, a combination of GDVN and drop-on-demand, is a promising method to make use of the full EuXFEL pulse train while reducing the sample amount needed substantially. It can be combined with pump-probe set-ups as well as mixing enabling time-resolved structural studies.

References

Aherne, M., Lyons, J. A., & Caffrey, M. (2012). A fast, simple and robust protocol for growing crystals in the lipidic cubic phase. *Journal of Applied Crystallography, 45*(6), 1330–1333. https://doi.org/10.1107/S0021889812037880.

Allahgholi, A., Becker, J., Bianco, L., Bradford, R., Delfs, A., Dinapoli, R., ... Zhang, J. (2016). The adaptive gain integrating pixel detector. *Journal of Instrumentation, 11*(02), C02066. https://doi.org/10.1088/1748-0221/11/02/C02066.

Allahgholi, A., Becker, J., Delfs, A., Dinapoli, R., Goettlicher, P., Greiffenberg, D., ... Graafsma, H. (2019). The adaptive gain integrating pixel detector at the European XFEL. *Journal of Synchrotron Radiation, 26*(1), 74–82. https://doi.org/10.1107/S1600577518016077.

Aquila, A., Hunter, M. S., Doak, R. B., Kirian, R. A., Fromme, P., White, T. A., ... Chapman, H. N. (2012). Time-resolved protein nanocrystallography using an X-ray free-electron laser. *Optics Express, 20*(3), 2706–2716. https://doi.org/10.1364/OE.20.002706.

Bean, R. J., Aquila, A., Samoylova, L., & Mancuso, A. P. (2016). Design of the mirror optical systems for coherent diffractive imaging at the SPB/SFX instrument of the European XFEL. *Journal of Optics, 18*(7), 074011. https://doi.org/10.1088/2040-8978/18/7/074011.

Bielecki, J., Hantke, M. F., Daurer, B. J., Reddy, H. K. N., Hasse, D., Larsson, D. S. D., ... Maia, F. R. N. C. (2019). Electrospray sample injection for single-particle imaging with x-ray lasers. *Science Advances, 5*(5), eaav8801. https://doi.org/10.1126/sciadv.aav8801.

Birch, J., Kwan, T. O. C., Judge, P. J., Axford, D., Aller, P., Butryn, A., ... Moraes, I. (2023). A versatile approach to high-density microcrystals in lipidic cubic phase for room-temperature serial crystallography. *Journal of Applied Crystallography, 56*(5), 1361–1370. https://doi.org/10.1107/S1600576723006428.

Blanchet, C. E., Round, A., Mertens, H. D. T., Ayyer, K., Graewert, M., Awel, S., ... Svergun, D. (2023). Form factor determination of biological molecules with X-ray free electron laser small-angle scattering (XFEL-SAS). *Communications Biology, 6*(1), 1057. https://doi.org/10.1038/s42003-023-05416-7.

Calvey, G. D., Katz, A. M., Schaffer, C. B., & Pollack, L. (2016). Mixing injector enables time-resolved crystallography with high hit rate at X-ray free electron lasers. *Structural Dynamics (Melville, N. Y.), 3*(5), 054301. https://doi.org/10.1063/1.4961971.

Caramello, N., & Royant, A. (2024). From femtoseconds to minutes: Time-resolved macromolecular crystallography at XFELs and synchrotrons. *Acta Crystallographica Section D, 80*(2), 60–79. https://doi.org/10.1107/S2059798323011002.

Chapman, H. N., Fromme, P., Barty, A., White, T. A., Kirian, R. A., Aquila, A., ... Spence, J. C. H. (2011). Femtosecond X-ray protein nanocrystallography. *Nature, 470*(7332), 73–77. https://doi.org/10.1038/nature09750.

Conrad, C. E., Basu, S., James, D., Wang, D., Schaffer, A., Roy-Chowdhury, S., ... Gati, C. (2015). A novel inert crystal delivery medium for serial femtosecond crystallography. *IUCrJ, 2*(4), 421–430.

Decking, W., Abeghyan, S., Abramian, P., Abramsky, A., Aguirre, A., Albrecht, C., ... Bak, P. (2020). A MHz-repetition-rate hard X-ray free-electron laser driven by a super-conducting linear accelerator. *Nature Photonics, 14*(6), 391–397. https://doi.org/10.1038/s41566-020-0607-z.

DePonte, D., Weierstall, U., Schmidt, K., Warner, J., Starodub, D., Spence, J., & Doak, R. (2008). Gas dynamic virtual nozzle for generation of microscopic droplet streams. *Journal of Physics D: Applied Physics, 41*(19), 195505. https://doi.org/10.1088/0022-3727/41/19/195505.

Doak, R. B., Nass Kovacs, G., Gorel, A., Foucar, L., Barends, T. R. M., Grünbein, M. L., ... Schlichting, I. (2018). Crystallography on a chip - without the chip: Sheet-on-sheet sandwich. *Acta Crystallographica Section D: Structural Biology, 74*(Pt 10), 1000–1007. https://doi.org/10.1107/S2059798318011634.

Echelmeier, A., Cruz Villarreal, J., Messerschmidt, M., Kim, D., Coe, J. D., Thifault, D., ... Ros, A. (2020). Segmented flow generator for serial crystallography at the European X-ray free electron laser. *Nature Communications, 11*(1), 4511. https://doi.org/10.1038/s41467-020-18156-7.

García-Tabarés Valdivieso, A., Deiter, C., Gelisio, L., Göde, S., Hauf, S., Kardoost, A. K., ... Sohn, F. (2024). The solid sample scanning workflow at the European XFEL. *Proceedings, 19th International Conference on Accelerator and Large Experimental Physics Control Systems (ICALEPCS'23), 19*, 78–83. https://doi.org/10.18429/JACoW-ICALEPCS2023-MO2AO03.

Gorel, A., Schlichting, I., & Barends, T. R. M. (2021). Discerning best practices in XFEL-based biological crystallography—Standards for nonstandard experiments. *IUCrJ, 8*(4), 532–543. https://doi.org/10.1107/S205225252100467X.

Grünbein, M. L., Bielecki, J., Gorel, A., Stricker, M., Bean, R., Cammarata, M., ... Schlichting, I. (2018). Megahertz data collection from protein microcrystals at an X-ray free-electron laser. *Nature Communications, 9*(1), 3487. https://doi.org/10.1038/s41467-018-05953-4.

Grünbein, M. L., Foucar, L., Gorel, A., Hilpert, M., Kloos, M., Nass, K., ... Stan, C. A. (2021). Observation of shock-induced protein crystal damage during megahertz serial femtosecond crystallography. *Phys. Rev. Res. 3*(1), 013046. https://doi.org/10.1103/PhysRevResearch.3.013046.

Grünbein, M. L., Gorel, A., Foucar, L., Carbajo, S., Colocho, W., Gilevich, S., ... Schlichting, I. (2021). Effect of X-ray free-electron laser-induced shockwaves on haemoglobin microcrystals delivered in a liquid jet. *Nature Communications, 12*(1), 1672. https://doi.org/10.1038/s41467-021-21819-8.

Grünbein, M. L., Stricker, M., Nass Kovacs, G., Kloos, M., Doak, R. B., Shoeman, R. L., ... Schlichting, I. (2020). Illumination guidelines for ultrafast pump–probe experiments by serial femtosecond crystallography. *Nature Methods, 17*(7), 681–684. https://doi.org/10.1038/s41592-020-0847-3.

Kardoost, A., Schönherr, R., Deiter, C., Redecke, L., Lorenzen, K., Schulz, J., & de Diego, I. (2024). Convolutional neural network approach for the automated identification of in cellulo crystals. *Journal of Applied Crystallography, 57*(Pt 2), 266–275. https://doi.org/10.1107/S1600576724000682.

Kirian, R. A., Wang, X., Weierstall, U., Schmidt, K. E., Spence, J. C. H., Hunter, M., ... Holton, J. (2010). Femtosecond protein nanocrystallography—Data analysis methods. *Optics Express, 18*(6), 5713–5723. https://doi.org/10.1364/OE.18.005713.

Koliyadu, J. C. P., Letrun, R., Kirkwood, H. J., Liu, J., Jiang, M., Emons, M., ... Sato, T. (2022). Pump-probe capabilities at the SPB/SFX instrument of the European XFEL. *Journal of Synchrotron Radiation, 29*(Pt 5), 1273–1283. https://doi.org/10.1107/S1600577522006701.

Konold, P. E., Monrroy, L., Bellisario, A., Filipe, D., Adams, P., Alvarez, R., ... Westenhoff, S. (2024). Microsecond time-resolved X-ray scattering by utilizing MHz repetition rate at second-generation XFELs. *Nature Methods.* https://doi.org/10.1038/s41592-024-02344-0.

Liu, W., Ishchenko, A., & Cherezov, V. (2014). Preparation of microcrystals in lipidic cubic phase for serial femtosecond crystallography. *Nature Protocols, 9*(9), 2123–2134. https://doi.org/10.1038/nprot.2014.141.

Mancuso, A. P., Aquila, A., Batchelor, L., Bean, R. J., Bielecki, J., Borchers, G., ... Tschentscher, T. (2019). The single particles, clusters and biomolecules and serial femtosecond crystallography instrument of the European XFEL: Initial installation. *Journal of Synchrotron Radiation, 26*(Pt 3), 660–676. https://doi.org/10.1107/S1600577519003308.

Mehrabi, P., Müller-Werkmeister, H. M., Leimkohl, J.-P., Schikora, H., Ninkovic, J., Krivokuca, S., ... Owen, R. L. (2020). The HARE chip for efficient time-resolved serial synchrotron crystallography. *Journal of Synchrotron Radiation, 27*(2), 360–370.

Moreno-Chicano, T., Ebrahim, A., Axford, D., Appleby, M. V., Beale, J. H., Chaplin, A. K., ... Hough, M. A. (2019). High-throughput structures of protein-ligand complexes at room temperature using serial femtosecond crystallography. *IUCrJ, 6*(Pt 6), 1074–1085. https://doi.org/10.1107/S2052252519011655.

Mozzanica, A., Andrä, M., Barten, R., Bergamaschi, A., Chiriotti, S., Brückner, M., ... Leonarski, F. (2018). The JUNGFRAU detector for applications at synchrotron light sources and XFELs. *Synchrotron Radiation News, 31*(6), 16–20.

Mozzanica, A., Bergamaschi, A., Brueckner, M., Cartier, S., Dinapoli, R., Greiffenberg, D., ... Tinti, G. (2016). Characterization results of the JUNGFRAU full scale readout ASIC. *Journal of Instrumentation, 11*(02), C02047. https://doi.org/10.1088/1748-0221/11/02/C02047.

Nam, K. H. (2019). Sample delivery media for serial crystallography. *International Journal of Molecular Sciences, 20*(5), https://doi.org/10.3390/ijms20051094.

Nam, K. H. (2020a). Lard injection matrix for serial crystallography. *International Journal of Molecular Sciences, 21*(17), 5977.

Nam, K. H. (2020b). Polysaccharide-based injection matrix for serial crystallography. *International Journal of Molecular Sciences, 21*(9), https://doi.org/10.3390/ijms21093332.

Nam, K. H. (2020c). Shortening injection matrix for serial crystallography. *Scientific Reports, 10*(1), 107. https://doi.org/10.1038/s41598-019-56135-1.

Nam, K. H. (2022). Beef tallow injection matrix for serial crystallography. *Scientific Reports, 12*(1), 694. https://doi.org/10.1038/s41598-021-04714-6.

Neutze, R., Wouts, R., van der Spoel, D., Weckert, E., & Hajdu, J. (2000). Potential for biomolecular imaging with femtosecond X-ray pulses. *Nature, 406*(6797), 752–757. https://doi.org/10.1038/35021099.

Oberthuer, D., Knoška, J., Wiedorn, M. O., Beyerlein, K. R., Bushnell, D. A., Kovaleva, E. G., ... Bajt, S. (2017). Double-flow focused liquid injector for efficient serial femtosecond crystallography. *Scientific Reports, 7*, 44628. https://doi.org/10.1038/srep44628.

Park, J., Park, S., Kim, J., Park, G., Cho, Y., & Nam, K. H. (2019). Polyacrylamide injection matrix for serial femtosecond crystallography. *Scientific Reports, 9*(1), 2525. https://doi.org/10.1038/s41598-019-39020-9.

Perrett, S., Fadini, A., Hutchison, C. D. M., Bhattacharya, S., Morrison, C., Turkot, O., ... Van Thor, J. J. (2024). Kilohertz droplet-on-demand serial femtosecond crystallography at the European XFEL station FXE. *Structural Dynamics, 11*(2), 024310. https://doi.org/10.1063/4.0000248.

Roedig, P., Duman, R., Sanchez-Weatherby, J., Vartiainen, I., Burkhardt, A., Warmer, M., ... Meents, A. (2016). Room-temperature macromolecular crystallography using a micro-patterned silicon chip with minimal background scattering. *Journal of Applied Crystallography, 49*(Pt 3), 968–975. https://doi.org/10.1107/S1600576716006348.

Schmidt, M. (2013). Mix and inject: Reaction initiation by diffusion for time-resolved macromolecular crystallography. *Advances in Condensed Matter Physics, 2013.*

Schulz, J., Bielecki, J., Doak, R. B., Dörner, K., Graceffa, R., Shoeman, R. L., ... Mancuso, A. P. (2019). A versatile liquid-jet setup for the European XFEL. *Journal of Synchrotron Radiation, 26*(Pt 2), 339–345. https://doi.org/10.1107/S1600577519000894.

Shoeman, R. L., Hartmann, E., & Schlichting, I. (2023). Growing and making nano- and microcrystals. *Nature Protocols, 18*(3), 854–882. https://doi.org/10.1038/s41596-022-00777-5.

Su, Z., Cantlon, J., Douthit, L., Wiedorn, M., Boutet, S., Kern, J., ... DePonte, D. (2021). Serial crystallography using automated drop dispensing. *Journal of Synchrotron Radiation, 28*(Pt 5), 1386–1392. https://doi.org/10.1107/S1600577521006160.

Sugahara, M., Mizohata, E., Nango, E., Suzuki, M., Tanaka, T., Masuda, T., ... Iwata, S. (2015). Grease matrix as a versatile carrier of proteins for serial crystallography. *Nature Methods, 12*(1), 61–63. https://doi.org/10.1038/nmeth.3172.

Sugahara, M., Motomura, K., Suzuki, M., Masuda, T., Joti, Y., Numata, K., ... Ishikawa, T. (2020). Viscosity-adjustable grease matrices for serial nanocrystallography. *Scientific Reports, 10*(1), 1371. https://doi.org/10.1038/s41598-020-57675-7.

Sugahara, M., Song, C., Suzuki, M., Masuda, T., Inoue, S., Nakane, T., ... Iwata, S. (2016). Oil-free hyaluronic acid matrix for serial femtosecond crystallography. *Scientific Reports, 6*(1), 24484. https://doi.org/10.1038/srep24484.

Tenboer, J., Basu, S., Zatsepin, N., Pande, K., Milathianaki, D., Frank, M., ... Koglin, J. E. (2014). Time-resolved serial crystallography captures high-resolution intermediates of photoactive yellow protein. *Science (New York, N. Y.), 346*(6214), 1242–1246. https://doi.org/10.1126/science.1259357.

Tolstikova, A., Levantino, M., Yefanov, O., Hennicke, V., Fischer, P., Meyer, J., ... Meents, A. (2019). 1 kHz fixed-target serial crystallography using a multilayer monochromator and an integrating pixel detector. *IUCrJ, 6*(5), 927–937. https://doi.org/10.1107/S205225251900914X.

Tschentscher, T., Bressler, C., Grünert, J., Madsen, A., Mancuso, A. P., Meyer, M., ... Zastrau, U. (2017). Photon beam transport and scientific instruments at the European XFEL. *Applied Sciences, 7*(6), https://doi.org/10.3390/app7060592.

Vakili, M., Bielecki, J., Knoska, J., Otte, F., Han, H., Kloos, M., ... Schulz, J. (2022). 3D printed devices and infrastructure for liquid sample delivery at the European XFEL. *Journal of Synchrotron Radiation, 29*(2), 331–346. https://doi.org/10.1107/S1600577521013370.

Vakili, M., Han, H., Schmidt, C., Wrona, A., Kloos, M., de Diego, I., ... Schulz, J. (2023). Mix-and-extrude: High-viscosity sample injection towards time-resolved protein crystallography. *Journal of Applied Crystallography, 56*(Pt 4), 1038–1045. https://doi.org/10.1107/S1600576723004405.

Von Stetten, D., Agthe, M., Bourenkov, G., Polikarpov, M., Horrell, S., Yorke, B., ... Gehrmann, T. (2019). TREXX: A new endstation for serial time-resolved crystallography at PETRA III. *Acta Crystallographica Section A: Foundations and Advances, 75*, e26.

Weierstall, U. (2014). Liquid sample delivery techniques for serial femtosecond crystallography. *Philosophical Transactions of the Royal Society B: Biological Sciences, 369*(1647), 20130337. https://doi.org/10.1098/rstb.2013.0337.

Weierstall, U., James, D., Wang, C., White, T. A., Wang, D., Liu, W., ... Cherezov, V. (2014). Lipidic cubic phase injector facilitates membrane protein serial femtosecond crystallography. *Nature Communications, 5*(1), 3309. https://doi.org/10.1038/ncomms4309.

Wiedorn, M. O., Oberthür, D., Bean, R., Schubert, R., Werner, N., Abbey, B., ... Barty, A. (2018). Megahertz serial crystallography. *Nature Communications, 9*(1), 4025. https://doi.org/10.1038/s41467-018-06156-7.

CHAPTER FIVE

Experimental approaches for time-resolved serial femtosecond crystallography at PAL-XFEL

Jaehyun Park[a,*] and Ki Hyun Nam[b,*]
[a]Pohang Accelerator Laboratory, Pohang University of Science and Technology, Pohang, Republic of Korea
[b]College of General Education, Kookmin University, Seoul, Republic of Korea
*Corresponding authors. e-mail address: jaehyun.park@postech.ac.kr; structure@kookmin.ac.kr

Contents

1. Introduction	132
2. NCI experimental hutch	134
2.1 Beamline instruments	135
2.2 Optical lasers	136
3. Sample environments for TR-SFX at the NCI experimental hutch	137
3.1 Sample chamber	137
3.2 Sample injectors	141
3.3 Fixed target scanning	144
4. Sample preparation	146
4.1 Sample preparation laboratory	146
4.2 Sample preparation	146
5. Experimental setup for TR-SFX at NCI experimental hutch	149
5.1 Procedure	149
6. Data processing	153
7. Safety considerations and others	155
7.1 Preliminary discussion for experiment	155
7.2 Laser	156
7.3 Hutch access	156
7.4 Data access	156
Acknowledgments	156
References	157

Abstract

Understanding the structures and dynamics of biomolecules and chemical compounds is crucial for deciphering their molecular functions and mechanisms. Serial femtosecond crystallography (SFX) using X-ray free-electron lasers (XFELs) is a useful technique for determining structures at room temperature, while minimizing radiation damage. Time-resolved serial femtosecond crystallography (TR-SFX), which uses an optical laser or a mixing device, allows molecular dynamic visualization during a

Methods in Enzymology, Volume 709
ISSN 0076-6879, https://doi.org/10.1016/bs.mie.2024.10.005
Copyright © 2024 Elsevier Inc. All rights are reserved, including those for text and data mining, AI training, and similar technologies.

reaction at specific time points. Because the XFEL beamline has unique properties for beams and instruments, understanding the beamline system is essential to conduct TR-SFX experiments and develop related technologies. In this study, we introduce an experimental system for performing TR-SFX using a Nano Crystallography and Coherent Imaging (NCI) experimental hutch at the Pohang Accelerator Laboratory XFEL (PAL-XFEL). Specifically, we present the XFEL properties of the PAL-XFEL and the main instruments in the NCI experimental hutch. In addition, the characteristics and uses of the sample delivery methods for TR-SFX and the general sample preparation process are discussed. Furthermore, the general time schedule and experimental procedures for TR-SFX during the beam time are outlined, along with data analysis programs. This chapter contributes to understanding the performance of TR-SFX experiments conducted at the PAL-XFEL NCI experimental hutch.

Abbreviations

PAL	Pohang Accelerator Laboratory.
XFEL	X-ray Free Electron Laser.
SFX	Serial Femtosecond Crystallography.
NCI	Nano Crystallography and Coherent Imaging.
MICOSS	Multivariable Injection Chamber for Molecular Structure Study.
CMD	Carrier Matrix Delivery.
PSD	Particle Solution Delivery.
MLV	Micro-Liter Volume.
OPA	Optical Parametric Amplifier.
OPO	Optical Parametric Oscillator.
CPM	Chamber Position Manipulator.
HPLC	High-Performance Liquid Chromatography.
OH	Optical Hutch.
DCM	Double Crystal Monochromator.

1. Introduction

To accurately comprehend the properties, functions, and molecular mechanisms of molecules, acquiring structural information and insights into their molecular dynamics within a spatiotemporal framework is necessary (Di Rienzo et al., 2016; Hekstra, 2023; Nam, 2021; Sponer et al., 2018). In targeted molecule reactions, various observable motions occur, including electronic motion, photodissociation, photoionization, bond vibrations, molecular collisions, vibration relaxation, solvation, proton transfer, temperature-jump Raman spectroscopy, fluorescence, molecular rotations, large-molecule relaxation, molecular diffusion, and phosphorescence (Farr et al., 2018; Fisette et al., 2012; Orville, 2020; Sahoo et al., 2011). These motions span different timescales, ranging from a few seconds to femtoseconds (Farr et al., 2018; Fisette et al., 2012; Orville, 2020; Sahoo et al., 2011). Various spectroscopy-, electron-, and X-ray-based analytical methods have been developed to investigate the molecular motions occurring at these diverse time scales (Farr et al., 2018; Fisette et al., 2012; Orville, 2020; Sahoo et al., 2011).

Among them, serial femtosecond crystallography (SFX) using X-ray free electron laser (XFEL) is an emerging technique to determine the room-temperature structures of target molecule without radiation damage based on the "diffraction before destruction" regime (Chapman et al., 2014; Neutze et al., 2000). In particular, the ultrashort pulse width characteristics of XFELs are valuable for time-resolving the molecular dynamics of fast reaction mechanisms (Chapman, 2017, 2019). Time-resolved SFX (TR-SFX) using optical lasers offers valuable insights into the molecular mechanisms and characteristics of various photoactive proteins and small molecules (Poddar et al., 2021; Westenhoff et al., 2022). Furthermore, the time resolution achieved by studying the protein–ligand binding process using a mixing device enhances our understanding of the molecular mechanism and offers insights into potential industrial applications (Nango & Iwata, 2023).

The XFEL facilities currently capable of performing time-resolved serial femtosecond crystallography (TR-SFX) include the Linac Coherent Light Source (LCLS) (Emma et al., 2010), SPring-8 Angstrom Compact free electron LAser (SACLA) (Ishikawa et al., 2012), Pohang Accelerator Laboratory XFEL (PAL-XFEL) (Kang et al., 2017; Ko et al., 2017; Nam et al., 2021), European XFEL (EuXFEL) (Decking et al., 2020), and Swiss X-ray FreeElectron Laser (SwissFEL) (Prat et al., 2020). TR-SFX observes the reaction process in a crystal sample by external stimuli, such as light or reactant mixing into the crystal sample using an optical laser or mixing device. The basic concept is that there are almost the same XFEL beamlines capable of performing all TR-SFX (Hekstra, 2023). However, the unique characteristics of the XFEL facilities (repetition rate, photon flux, available X-ray energy, and beam size) and the specifications of the experimental instruments (optical laser system, sample delivery system, and sample chamber) used by the beamline may vary the XFEL beamline application (Park & Nam, 2023). Accordingly, to successfully collect TR-SFX data, understanding the characteristics of the XFEL facility and beamline that will be used, the available equipment, and the experimental approach specific to the beamline is critical.

TR–SFX experiments in PAL-XFEL were performed in the Nano Crystallography and Coherent Imaging (NCI) experimental hutch. Here, we present information on the XFEL, optics, beamline properties, and sample delivery methods for performing TR-SFX in the NCI experimental hutch.

2. NCI experimental hutch

The TR-SFX experiments were performed at the NCI experimental hutch on a hard X-ray beamline at PAL-XFEL (Park et al., 2016). The NCI experimental hutch is designed with a forward scattering geometry. Research on SFX and coherent diffraction imaging (CDI) has mainly been conducted in NCI experiments (Park et al., 2018; Sung et al., 2021). Additionally, experiments that share this scattering geometry, such as small- and wide-angle X-ray scattering (SAXS, WAXS), can also be performed (Kim et al., 2017; Späh et al., 2019). The energy of the XFEL transported to the NCI experimental hutch ranges between 6–15 keV. Depending on the research purpose, self-amplified spontaneous emission (SASE) or self-seeding X-rays can be used with bandwidths of 0.2 % and 0.002 % at 9.7 keV, respectively. The pulse duration, maximum repetition rate, and photon flux (at 9.7 keV, SASE) of XFELs are <25 fs, 60 Hz, and >5 × 10^{11} photons per pulse, respectively. The specifications of XFEL beam at the NCI experimental hutch are summarized in Table 1.

Table 1 Specification of the X-ray free-electron laser (XFEL) beam at the NCI experimental hutch.

Parameter	Specification
X-ray Energy (keV)	6–15
Energy bandwidth (%)	SASE: ~0.2 (at 9.7 keV) Self-seeding: ~0.002 (at 9.7 keV)
Pulse duration (fs)	<25
Repetition rate	Up to 60 Hz
Photon flux (photons/pulse)	>5 × 10^{11} (at 9.7 keV, SASE mode)
Beam size (μm^2 FWHM)	2 × 2 (horizontal X vertical, at 12.4 keV)

2.1 Beamline instruments

The dimensions of the NCI experimental hutch were 8.2 × 4.0 × 22 m³ (width × height × length) (Fig. 1). Various instruments, such as a reference laser, XFEL beam diagnostics, X-ray focusing mirror, sample chamber, and detector are installed in the beamline. The XFEL diagnostics and X-ray focusing mirror are placed in fixed positions on the beamline and used in all the experiments. The sample chamber and detector are replaced according to the purpose of the research. For example, when conducting an SFX experiment, a sample chamber is used to maintain the SFX sample delivery environment and a large-area detector. In contrast, the CDI experiments require a dedicated vacuum chamber and a high-resolution detector with a small pixel size.

Two beam profile monitors (pop-ins) consisting of a fluorescent screen and camera are installed at the beamline, and are mainly used to check the XFEL beam entering the NCI experimental hutch and the position of the XFEL beam. Two quadrant beam position monitors (QBPMs) are used to monitor the intensity and position of the X-rays transmitted to the beamline. This instrument is used for data collection, and provides real-time information on the XFEL flux used in the experiment.

The XFEL pulses are focused horizontally and vertically using a Kirkpatrick-Baez mirror to provide a high photon flux density at the sample position (Kim et al., 2018). The K-B mirror consists of two quartz pieces (600 × 50 × 50 mm³) with two 10 mm-wide optical strip lines, one bare side

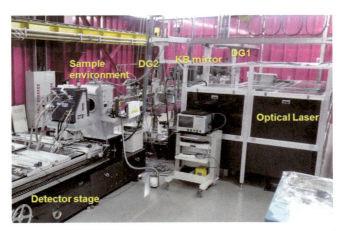

Fig. 1 Nano Crystallography and Coherent Imaging (NCI) experimental hutch at Pohang Accelerator Laboratory X-ray free-electron laser (PAL-XFEL).

and the other side coated with rhodium for reflecting X-rays at higher energies. The K-B mirror is designed with a beam size of approximately $2 \times 2\,\mu m^2$ at 9.7 keV at the sample position. A commercially available MX225-HS detector (Rayonix, LLC) is used to collect the diffraction data for the SFX experiment. The detector consists of an array of 3×3 tiled frame-transmitting CCD modules with an effective area of $225 \times 225\,mm^2$ tiled without gaps in the imaging area. The detection quantum efficiency is 0.8, ranging between 8–12 keV. The pixel size in the detector is 39 µm with an overall resolution of 5760×5760 pixels. The readout speed of this detector bins pixels to collect all the data on the XFEL repetition rate used in the experiment. For the 4×4 and 6×6 binding modes, the pixel sizes are 156 and 234 µm, with readout speeds of 33 frames/second and 62 frames/second, respectively. The pixel binning of the detector is set according to the XFEL repetition rate and the purpose of the study. The MX225-HS detector is operated at $-80\,°C$ to reduce the thermal background signals on the sensors. The XFEL passing through the sample position is transmitted toward the detector, and a beam stopper is installed in the active area of the detector to avoid physical damage to the detector. The diameter of the beam stopper is approximately 6 mm, and the stopper is connected near the center of the upper edge. The detector moves back and forth from the sample position according to the diffraction pattern of the crystal. When performing other experiments such as CDI, or when the MX225-HS detector is not in use, the detector can be moved horizontally to lie outside the XFEL path. The detector position is manipulated using 3-axis linear stages on a granite table. The sample chamber is located in front of the detector and placed on a chamber position manipulator (CPM) with the 2-axes high-load stages for vertical and horizontal motions. The CPM sits on a granite block to minimize the vibrational motion on the sample. On the CPM, a breadboard (M6 tap at a 25 mm distance) is attached to fix the sample chamber. The temperature inside the hutch is controlled by an air-conditioning unit that is installed near the end wall. The nominal temperature is maintained at approximately 20–23 °C by reducing the effect of heat sources, such as detector servers and vacuum pumps.

2.2 Optical lasers

Both femtosecond (fs) and nanosecond (ns) lasers can be used to perturb solid and molecular samples in TR-SFX. The fundamental wavelength of Ti:sapphire-based femtosecond lasers is approximately 800 nm. The maximum pulse-repetition rate and energy are 120 Hz and 10 mJ, respectively. The pulse duration is approximately 40 fs. The available wavelengths are 400 nm and 266 nm for the harmonics. In addition, continuous wavelengths ranging

Table 2 Specification of laser system for time-resolved serial femtosecond crystallography (TR-SFX).

fs-Laser	Specification
Wavelength	800 nm (Fundamental)
	400, 266 nm (Harmonics)
	* 240–20,000 nm (OPA)
Pulse energy	<3 mJ @ 800 nm (depending on wavelength)
Pulse width	>40 fs (depending on wavelength)
Repetition rate	<120 Hz
Timing diagnostics (OXC)	Available
ns-Laser	
Wavelength	210–2600 nm (OPO)
Pulse energy	1–45 mJ
Pulse width	2–5 ns
Repetition rate	<60 Hz
Option	Fiber coupling for 350–2000 nm

between 240–2600 nm are provided using an optical parametric amplifier (OPA). Because of their short pulse duration, femtosecond lasers can be applied to ultrafast dynamics studies. In contrast, nanosecond laser systems provide continuous wavelengths from 210 to 2600 nm using an optical parametric oscillator (OPO) system. The pulse duration is approximately 2–5 ns. The optical lasers are selected and used according to the purpose of the experiment. The specifications are summarized in Table 2.

3. Sample environments for TR-SFX at the NCI experimental hutch

3.1 Sample chamber

At the NCI experimental hutch, the sample delivery system for SFX experiments can accommodate both injector and fixed-target scanning methods. In the injector-based sample delivery method, crystal samples are surrounded by a solution or viscous material. If a sample is delivered at a low flow rate, its condition may change depending on the temperature or humidity. For example, when a sample containing a crystallization solution with a high salt concentration is delivered at a low flow rate, salt crystals can form at the end of the nozzle, altering the path of the injection stream, or exposing the grown crystals to the XFEL, resulting in unwanted salt diffraction patterns. Additionally, the lipidic cubic phase (LCP) that uses monoolein, which is poplular as a crystal delivery medium (Weierstall et al., 2014), may change its phase depending on the temperature or humidity, affecting the injection stream or optical laser transmittance (Van Dalsen et al., 2020; Wells et al., 2022). Factors such as temperature, humidity, and air conditions can influence the crystal structure. Therefore, maintaining a constant sample environment is crucial for reliable and efficient data collection during SFX experiments.

To date, for SFX experiments in the NCI experimental hutch, the MICOSS (Multivariable Injection Chamber for Molecular Structure Study) and fixed-target (FT) chamber have been developed to use injector and fixed-target scanning methods. MICOSS and the fixed-target chamber are isolated from the atmosphere and collected in a helium environment to reduce background scattering from the air during data collection. The injector or FT sample holder is mounted on the manipulator inside the chamber, which aligns the sample with the X-ray path. The specifications of the sample chamber are summarized in Table 3.

3.1.1 MICOSS

MICOSS was the first platform dedicated to SFX experiments (Park et al., 2018). The chamber is designed to operate various sample delivery instruments under a liquid flow scheme (Fig. 2). MICOSS can operate three representative sample injectors: particle solution delivery (PSD) (Park et al., 2018), carrier matrix delivery (CMD) (Park et al., 2018), and microliter volume (MLV) (Park & Nam, 2023). The positions of all injector systems were aligned to the X-ray beam path by position manipulation comprising three linear stages combined with stepping motors. The travel lengths of the linear stages for position manipulation in the X, Y, and Z directions are 300, 50, and 51 mm, respectively. Within MICOSS, injector position and sample delivery status are monitored using two ultra-long-zoom microscopes (UWZ-300F, Union Optical Co., Ltd.). The working distance of these microscopes is 300 mm, which not only

Table 3 Specification of sample chamber.

Specification	MICOSS	Fixed-target chamber
Dimension	Rectangle shape 400(W) × 300(L) × 400(H)	L-shape 400(W) × 220(L1) × 433(H1) 400(W) × 160(L2) × 262(H2)
Delivery method	Injection	Translation
Sample delivery	PSD injector CMD injector MLV injector	1D fixed target 2D fixed target
Sample environment	Ambient pressure Ambient temperature (Helium gas purging)	Ambient pressure Ambient temperature (Helium gas purging)
Position manipulation	X, Y, Z motion with linear stages	X, Y, Z motion with linear stages

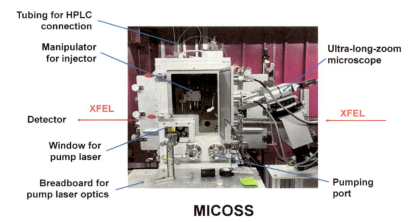

MICOSS

Fig. 2 Multivariable Injection Chamber for Molecular Structure Study (MICOSS) for time-resolved serial femtosecond crystallography (TR-SFX) experiments at the Nano Crystallography and Coherent Imaging (NCI) experimental hutch at Pohang Accelerator Laboratory X-ray free-electron laser (PAL-XFEL).

avoids positional interference with other instruments, but also facilitates various experimental setups at the sample position in MICOSS. Two microscopes monitor the front and side directions of the X-rays and are used to accurately align the extruded injection stream from the injector to the XFEL pulse path and focal point. For TR-SFX using an optical laser, two quartz windows are prepared so that the optical laser can reach the sample location at two angles of incidence for the X-rays (15° and 90°). In the case of the sample injector, the perpendicular incidence of the optical laser promotes easy overlapping of the X-rays with the optical laser through the microscope at the side. MICOSS is operated in a helium gas environment at ambient pressure to reduce background scattering from air. The helium purging speed is accelerated using an attached vacuum scroll pump. After the chamber door is closed, the purging process is completed within approximately 5 min. The purging level is monitored using an oxygen concentration sensor inside MICOSS. In addition, a temperature/humidity sensor is installed near the sample position to monitor the temperature and humidity levels during the experiment.

3.1.2 FT chamber

The FT chamber is employed for SFX data collection by scanning a fixed target sample delivery system (Fig. 3). Two-dimensional (2D) scanning using a nylon mesh-based sample holder (Lee et al., 2019) and one-dimensional (1D) scanning using polyimide tubing (Lee et al., 2020) are currently the primary methods used in FT chambers. The sample holder or microtubing containing the crystal samples are initially installed in the acryl-based holder to install the manipulator in the FT chamber. The 2D and 1D sample chips containing the crystals are raster scanned using piezo stages (SmarAct SLC-24180; stroke: 180 mm). The speed of the stages is

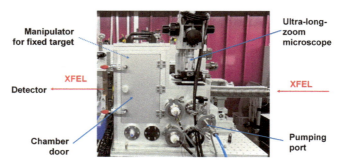

FT chamber

Fig. 3 Fixed-target serial femtosecond crystallography (FT-SFX) chamber for time-resolved serial femtosecond crystallography (TR-SFX) experiments at the Nano Crystallography and Coherent Imaging (NCI) experimental hutch at Pohang Accelerator Laboratory X-ray free-electron laser (PAL-XFEL).

sufficient to cover repetition rates up to 60 Hz with a 50 μm step size. For sample monitoring, a long-working distance microscope was installed on the top surface of the lower stairs, with a transparent glass window. A flat mirror with a centered hole for X-ray transmission was installed at an angle of 45° to reflect the sample image onto the microscope. A collimator (stainless steel, approximately 50 mm) is installed at the central hole to block noisy signals from upstream and parasitic scattering from the air. The FT chamber is operates in a helium gas environment to minimize X-ray background scattering from air. The FT chamber has a vacuum port connected to a roughing pump and a He gas injection port for quick purging, similar to MICOSS.

3.2 Sample injectors

3.2.1 PSD injector

The PSD injector is designed to deliver a crystal suspension or solution sample (Park et al., 2018). The PSD injector nozzle consists of a tapered capillary for sample suspension delivery and a glass capillary for gas focusing (Fig. 4A). The crystal suspension is injected through the fused silica tubing with an inner diameter of 75 or 100 μm. The liquid beam, injected through the fused silica, tubing is focused using a homemade gas dynamic virtual nozzle (GDVN) lens with a diameter of a few microns. The beamline uses two types of sample reservoirs: a borosilicate glass column that can store 10 mL of the sample and a stainless-steel column that can store 2 mL of the sample. An injector nozzle is connected to the sample reservoir, and the other side is connected to a high-

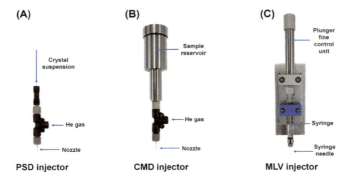

Fig. 4 Injector devices for time-resolved serial femtosecond crystallography (TR-SFX) experiment at Nano Crystallography and Coherent Imaging (NCI) experimental hutch at Pohang Accelerator Laboratory X-ray free-electron laser (PAL-XFEL). (A) Particle solution delivery (PSD), (B) carrier matrix delivery (CMD), and (C) micro-liter volume (MLV) injectors. The original figures were obtained from a previous study (Park & Nam, 2023).

performance liquid chromatography (HPLC) pump. Sample delivery is based on the principle that the sample in the HPLC pump reservoir is delivered to the nozzle when the sample transfers the water pressure to the sample reservoir. When a solution sample is delivered from the PSD injector at a low flow rate, a drop forms at the tip of the nozzle. Consequently, high flow rates ($>20\,\mu L/min$) are typically required to generate stable liquid beams from the PSD injector. Considering that the repetition rate of the XFEL at the PAL-XFEL can reach 60 Hz, there is an exceeding number of samples that are not exposed to X-rays than those that are exposed. Given the beam repetition rate, PSD injectors can have a drawback in terms of sample consumption. Generally, the sample volume required to determine the protein structure may vary depending on the quality and density of the crystal; however, it is approximately 50–100 mL.

3.2.2 CMD injector

The CMD injector is designed to deliver crystals embedded in viscous media, such as monoolein (LCP) and other viscous chemicals (Park et al., 2018; Park et al., 2019), which have the advantage of creating a stable injection stream even at low flow rates, thereby reducing sample consumption (Nam, 2019). The CMD injector consists of an injector body containing the sample reservoir and an injector nozzle, where the sample is extruded (Fig. 4B). The volume of the sample reservoir is $40\,\mu L$, which is sufficient to deliver samples continuously for 6 h at a flow rate of $110\,nL/min$. The sample in the reservoir is extruded using a Teflon ball and a metal plunger. The metal plunger is placed between the Teflon ball and the water plunger (material: Teflon), which delivers pressure from an HPLC pump. The nozzle for extrusion is a tapered capillary made of fused silica, with an inner diameter of 75 or $100\,\mu m$. To fill the viscous sample in the reservoir, a prepared Hamilton syringe is connected to the CMD injector body. Initially, a Teflon ball is inserted into the sample reservoir and a Teflon ball plunger placed in the CMD injector body. The plunger is then pushed until the Teflon ball reaches the lowest part of the reservoir. After removing the coupler, the prepared Hamilton syringe is connected directly to the CMD injector body. Finally, the Hamilton syringe plunger is used to fill the sample inside the reservoir until the desired volume is reached.

3.2.3 MLV syringe injector

In the sample preparation procedure using LCP and viscous media to deliver microcrystals, Hamilton syringes are used to crystallize or distribute the crystals in the media. The main idea behind this injector is that a Hamilton syringe can be used as the injector (Eom et al., 2022). In general, the RN type syringe with a 100 micro-liter volume is feasible with an adaptive needle option. The MLV syringe injector consists of three main parts (Fig. 4C). A microliter volume Hamilton syringe (RN 1700 series, volume: 100 μL), a needle, and a plunger fine control unit (PFCU) (Eom et al., 2022). The PFCU delivers pressure to the syringe plunger with precise and slow motion that matches the intended flow rate of the sample. The PFCU is an integrated device composed of a metal cylinder with a through-hole at the center, a water plunger, and a fitting on one side of the cylinder to seal the water from the HPLC pump. The through-hole has a diameter of 5 mm and a length of 102 mm, allowing sufficient elongation of the syringe plunger to hold 100 μL inside the barrel. The water plunger consists of Teflon and seals the water from the HPLC pump. The PEEK tubing (OD: 1/16 in.) is connected to the PFCU with a 10–32 tap. The plunger speed is controlled by varying the flow rate of the HPLC pump. For sample delivery to the X-ray interaction point, a commercial SUS needle can be used (22s-gauge, ID approximately 150 μm, Hamilton Company).

A specialized feature of the MLV syringe injector is its ability to use two optical lasers at different wavelengths during time-resolved experiments. Because the syringe barrel is made of borosilicate glass, which is transparent

Table 4 Specification of the injectors for time-resolved serial femtosecond crystallography (TR-SFX).

Injector	PSD	CMD	MLV
Sample environment	Solvent	Viscous medium	Viscous medium
Injector type	Nozzle	Nozzle	Commercial syringe
Injection pressure	HPLC	HPLC	HPLC
Reservoir volume	10 mL	40 μL	100 μL
Flow rate (μL/min)	>50	>2.8	>6.0
Data collection time for single injection (FWHM$_{Laser}$ ~ 100 μm)	~200 min	~14 min	~16 min

to the pumping nanosecond laser, the molecular state can be initiated to a specific chemical state when a single optical laser illuminates the glass barrel. This initially conserved state can then be perturbed by a second optical laser to drive fast dynamics of the molecules. The specifications of different types of injectors are summarized in Table 4.

3.3 Fixed target scanning
3.3.1 2D fixed target
A 2D fixed target scanning mesh-based sample holder is primarily used (Fig. 5A) (Lee et al., 2019). The crystal sample is distributed on a nylon mesh placed on a polyimide film and then covered with another film to prevent evaporation. The nylon mesh prevents the crystal sample from sinking due to gravity when the sample holder is positioned vertically. A thin polyimide film is preferred to minimize background scattering caused by the film. Additionally, the polyimide film can be replaced with another film such as Mylar. The crystal samples are mainly located in the pores of the nylon mesh, with some on top of the nylon material. Because the nylon mesh allows X-rays to pass through it, precise alignment of the XFEL to a specific position, such as a nylon mesh pore, is unnecessary, thereby enabling fast data collection. During data collection, XFELs can penetrate both the nylon mesh pores and the nylon mesh material. The background scattering from nylon typically exhibits a fiber-scattering form and shows negligible background scattering during data processing. The specifications of 2D fixed taget sample holders are summarized in Table 5.

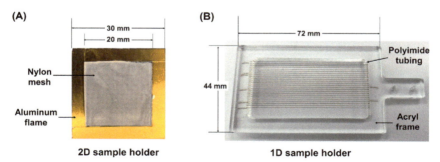

Fig. 5 Sample holder for fixed-target time-resolved serial femtosecond crystallography (TR-SFX) at the Nano Crystallography and Coherent Imaging (NCI) experimental hutch at Pohang Accelerator Laboratory X-ray free-electron laser (PAL-XFEL). (A) nylon-mesh based sample holder for 2D scan and (B) polyimide microtubing for 1D scan.

Table 5 Specification of fixed target sample holders.

Method	1D target	2D target
Sample holder materials	Polyimide-tubing	Nylon mesh Polyimide film
Sample holder Dimension (standard)	100 µm (inner diameter) 126 µm (outer diameter) 500 mm (length)	20–50 mm (vertical) 20–50 mm (horizontal)
Sample environment	Hydrated	Hydrated
Volume of crystal suspension	16 µL (single chip−4 tubes)	60–80 µL (single chip)
Translation of sample holder	Horizontal direction	Vertical and Horizontal direction
Scanning range	56 mm/line	18 mm × 18 mm
Translation speed	1.5 mm/sec (horizontal @ 30 Hz)	1.5 mm/sec (vertical & horizontal @ 30 Hz)
Data acquisition time	15 min (24 lines, 50 µm interval)	72 min (18 ×18 mm^2, 50 µm interval)

3.3.2 1D fixed target

A significant feature of the 1D fixed target is the reduction in sample consumption (Fig. 5B). The spatial confinement of the sample along the scanning trajectory enhances sample consumption efficiency compared to a 2D fixed target with a continuous crystal distribution. Microtubing is used to achieve this goal (Lee et al., 2020). Microtubing serves as a tiny container for microcrystals while maintaining a perfectly hydrated environment. The microtubing material was polyimide, which is commonly used as an X-ray window in X-ray facilities. The inner diameter, outer diameter, and length of the microtubing are 100, 126, and 500 mm, respectively. A frame with multiple holes to thread tubes was fabricated for raster scanning. In total, the plate can hold four tubes simultaneously. The contained sample volume is approximately 4 µL per tube and 16 µL per chip. The results of SFX experiments with a 1D fixed target demonstrated that reasonable data collection is feasible even with a single chip. One of the advantages of a 1D fixed target is that the solution within the tubing does not evaporate, allowing the crystals to be stored for a long time. The

1D fixed-target tubing, when exposed to XFEL, showed no solution evaporation and maintained crystal diffraction even after six months. The specifications of 1D fixed taget sample holders are summarized in Table 5.

4. Sample preparation

4.1 Sample preparation laboratory

Two sample preparation laboratories (SPLs) for the SFX experiments are located near the NCI experimental hutch of the PAL-XFEL. User groups can use these facilities to prepare samples before or during XFEL beamtime. The SPLs are equipped with a temperature-controlled incubator, several centrifuges of various tube sizes, a high-resolution microscope, and second-order nonlinear imaging of chiral crystals (SONICC) to monitor and detect the nanocrystals and microcrystals. The SPL on the first floor has a dark room for the preparation of light-sensitive samples, which is particularly useful for TR-SFX experiments involving photoactive enzymes.

The sample preparation room is equipped with the following:
- Temperature-controlled incubator
- Centrifugation for various tube sizes
- High-resolution microscope (KH-8700, Hirox Co. Ltd.)
- SONICC (Formulatrix)
- Dark room for light-sensitive samples

4.2 Sample preparation

The selection of sample delivery and preparation methods is tailored to the specific research objectives and the quantity and quality of the samples involved. The sample preparation process varies depending on the type of sample and the preferred approach of the researcher. Here, the general sample preparation methods used during method development at the NCI beamline are introduced.

4.2.1 PSD injector

Crystal samples with diameters larger than the inner diameter of the nozzle may cause clogging problems in the tubing or nozzle during sample delivery. If clogging due to large crystals occurs in the tubing or nozzle, high pressure may be applied to the sample, or the nozzle may be damaged. Accordingly, the size of the prepared crystals is first determined using a high-resolution microscope (SONICC). If the crystal size is larger than that of the injector nozzle, a filter must be used to remove the crystals. The crystal suspension is then transferred to a sample reservoir column using a syringe. Sample delivery is ready when both the sample reservoir and injector nozzle are connected directly through fittings for fluidics and PEEK tubes. Crystals in the sample reservoir settle owing to gravity, which can potentially result in uneven delivery of the crystals or clogging of the tubing. Accordingly, the reservoir column is continuously moved and shaken using a rocker (anti-settler) to prevent the crystal samples from settling during sample delivery. The experiment continues as an HPLC system and delivers the crystal suspension from the sample reservoir to the injector nozzle. Once the sample comes out (confirmed using a camera), the flow rate at which a stable liquid beam is generated is determined, continuously delivered, and diffraction data collected.

4.2.2 CMD injector

A sample delivery medium that does not physically or chemically damage the crystals during mixing or storage is initially selected. Crystal samples and viscous materials are mixed using a dual-syringe setup. The crystal suspension is transferred to one syringe, and the viscous material transferred to another syringe. The syringes are then connected to a syringe coupler. The syringe plungers are gently moved back and forth to mix the crystals and the viscous material. The crystalline sample embedded in the viscous material is then transferred to a single syringe. The syringe coupler is disconnected and the crystal embedded in the viscous medium is transferred from the syringe to the sample reservoir in the body of the CMD injector using commercial fittings for 10–32 thread. The height of the water plunger in the CMD injector increases as the reservoir fills with the sample until the plunger hit the end surface of the injector body. The sample syringe is separated from the injector body. The CMD injector nozzle is connected to the sample and the HPLC tubing is connected to the other side of the CMD injector body. By operating the HPLC system, crystals embedded in a viscous medium emerge from the CMD injector body into the injector nozzle. Initially, the injector nozzle is empty; therefore, if the sample is delivered at a low flow rate, it takes a long time to see the sample exiting the nozzle. Accordingly, the beam time can be used efficiently by initially delivering the sample at a high flow rate and then switching to a lower flow rate once it is confirmed that the sample is exiting the nozzle. The injection stream is aligned with the XFEL, and diffraction data is collected.

4.2.3 MLV injector

The sample preparation method is the same as that used for the CMD injectors. After mixing the crystal sample and viscous material, the syringe needle is connected to a syringe containing the crystals embedded in the viscous medium and installed in the MLV syringe holder. The crystal sample is extruded from the syringe using the pressure from an HPLC pump. The injection stream is aligned with the XFEL, and diffraction data collected.

4.2.4 2D fixed target

A sample holder is required for 2D fixed-target scanning, which can be manually manufactured in a sample preparation laboratory. The standard frame size is 30 mm × 30 mm with a frame width of 5 mm, providing an area of 20 mm × 20 mm for sample distribution. A polyimide film is fixed to the frame using double-sided adhesive polyimide tape. Inside the frame, a nylon mesh is placed on the polyimide film, and the crystal suspension is transferred onto the nylon mesh. The samples are spread evenly using a pipette tip. Another frame containing a polyimide film is immediately placed over the sample on a nylon mesh to prevent evaporation of the solution. The two frames are attached using double-sided adhesive poly-imide tape. The sample holder containing the crystal suspension is mounted on a manipulator using FT-SFX. After aligning the sample holder, a 2D raster scan is performed to collect diffraction data.

4.2.5 1D fixed target

An RN-type Hamilton syringe (1710 gas-tight series) containing the crystal suspension is connected to a standard metal needle. The micro-tubing is inserted into PEEK tubing (OD: 1/16-inch, length: ~ 30 mm) and fixed with a fast-drying glue, such as Loctite 401 adhesive. The microtubing is fixed securely with PEEK tubing for approximately 10 min. A union fitting (P-704; IDEX Health and Science) connects both parts of the syringe needle and PEEK tubing using a nut (F-331; IDEX Health and Science) accompanied by a ferrule. When the connection is completed, the solution is injected into the microtube by pushing the plunger of the Hamilton syringe. Sample injection is completed if the solution extrudes from the micro-tubing end on the other side. After cutting the microtubing attached to the PEEK tubing, both ends are closed with fast-drying glue to avoid dehydration. This process is repeated for a total of four tubes. The prepared tubes are then stitched into a metal frame with a comb-shaped groove and fixed to the surface of the metal frame with adhesive tape. The sample holder containing the crystal suspension is mounted on a manipulator using FT-SFX. After aligning the sample holder, a 1D scan is performed to collect the diffraction data.

5. Experimental setup for TR-SFX at NCI experimental hutch

5.1 Procedure

5.1.1 Day < 0: Installation and inspection of the SFX instrument

Prior to beamtime, beamline instruments are installed and inspected. Various studies, including SFX and CDI, have been conducted at the NCI beamline, each using sample chambers and detectors tailored for their research purposes. Consequently, the sample chamber is replaced with a MICOSS or FT chamber to position it on the CPM, and an MX225-HS detector is installed on the detector stage. The beamline instruments are roughly aligned using a reference laser system located on the front side of the NCI experimental hutch. The path of the reference laser follow almost the same trajectory as that of the XFEL beam. In addition, a preliminary inspection is performed on the sample environment used in the SFX research, which typically includes checking the sample viewing system, inspecting the injector nozzle, conducting sample injection tests, and confirming movement of the translator. These processes are generally conducted by the beamline staff, and all instrument installations and rough alignments are completed before the user beamtime. Especially, the following steps are basically described for the case of the beamtime allocated in 6 shifts for 6 days.

5.1.2 Day 1: NCI and SFX instrument alignment

5.1.2.1 Step 1. XFEL transport

The generated XFEL at the undulator passes through an optical hutch (OH) containing optical instruments, such as plate mirrors, a double crystal monochromator (DCM), and an inline spectrometer to manipulate the X-rays for optimal conditions. For SFX experiments, the XFEL beam typically operates in the SASE mode without using DCM. Two pairs of flat mirrors are used to reject higher harmonics, with rejections occurring below and above 10 keV. The vertical offset is set at 30 mm. The first step in aligning the beamline optics begin with a beam-position check. Two reference points are observed at the OH and NCI using pop-in instruments, which determine the pointing vector of the incident X-ray beam. The positions of the flat mirrors (M1 − M2 or M1 − M3) are finely adjusted to achieve the designated trajectory, which is also in close proximity to the optimized beam position on the K-B mirrors.

5.1.2.2 Step 2. XFEL focusing using a K-B mirror system

The K-B mirrors are aligned to achieve a higher photon flux density at the sample position. To determine the optimal setup, the position of the focusing point based on the Foucault test and the beam size are checked. This procedure involves knife-edge scanning with a crosswire made of a tungsten rod with a diameter of $150 \, \mu m$. The beam shape is monitored using an X-ray detector (C12849–101U, Hamamatsu Corp.) and the intensity is measured using a photodiode installed downstream of the detector stage. The beam size is determined by differentiating the measured intensity during the knife-edge scanning. The translation and pitch of the K-B mirrors are finely adjusted to achieve the desired beam size. Under normal circumstances, this process can be completed within two hours if there are no significant issues with the beamline.

5.1.2.3 Step 3. Scattering instrument alignment

The position alignment process is conducted for the experimental instruments, including collimators, a 90° reflection mirror with a hole at the center (used for sample monitoring), a sample chamber, and an X-ray detector. In particular, the detector position is crucial because it directly affects the correct placement of the beam stop.

5.1.2.4 Step 4: Detector setup

Before the data collection, a dark image is obtained without X-ray exposure on the detector side. These data helped remove initial background noise signals. Typically, three dark images are used for background reduction and serve as backup data to monitor the detector status.

5.1.2.5 Step 5: Sample chamber fine alignment

The MICOSS or FT chamber is installed above the CPM and aligned with the incident X-rays. A temporal beam stop is installed at the outgoing port facing the detector inside the chamber. This indicates the central location of the port, which ensures symmetrical data collection.

5.1.2.6 Step 6: Sample-to-detector distance (SDD) measurement

The calibration of the distance between the sample and the detector is conducted using powder diffraction with a standard powder sample (e.g., LaB_6, CeO_2). The powder chip is prepared in a sandwiched form between polyimide thin films. The powder chip is then placed on the sample plane, which can either be a liquid flow from the injectors or a fixed target chip. Software dedicated to 1D and 2D data analyses, such as FIT2D, is used to calculate the distances. The software determines the initial distance and coordinates of the beam center, providing this information to users for data processing.

5.1.3 Day 2: Preliminary process and SFX data collection

5.1.3.1 Step 1: XFEL energy scan

The energy spectra of the incident X-rays are measured using scanning DCM. The step size of the energy scan is approximately 1–2 eV. The nominal energy bandwidths of the SASE and self-seeded XFEL are approximately 20 and 0.2 eV, respectively. Owing to the limitations in energy resolution, an inline spectrometer is used to determine the energy spectrum of the self-seeded XFEL.

5.1.3.2 Step 2: Initial injection test

PAL-XFEL staff or users can perform an offline test of the sample injector in a sample injector room located on the first floor of the sample preparation room. A test bench is set up with a microscope (KH-8700, Hirox Co. Ltd.) to monitor the operational status of the injector.

5.1.3.3 Step 3: Optical laser setup and data collection

The NCI hutch can deliver beams of various wavelengths to femtosecond lasers using an optical parametric amplifier (OPA) and nanosecond lasers using an optical parametric oscillator (OPO) system. Beamline staff, specializing in optical lasers, set up the laser system according to the desired specifications of the optical pump (wavelength, energy density, and beam size) provided by the user for the TR-SFX experiment. Depending on the experimental plan, the diffraction data can be collected for the native structures without the pump lasers or for the perturbed structures at several delay points with the pump lasers after completing the laser ovelapping process described below.

5.1.4 Day 3–6: SFX data collection

5.1.4.1 Step 1: Overlapping X-rays and optical lasers

For TR-SFX experiments, overlapping the X-rays and optical lasers is crucial. As previously mentioned, sample-monitoring microscopes are used to identify the illumination position. The X-ray is positioned at the center of the crosshair, and the large red spot shows where the optical laser illuminates the sample (Fig. 6).

5.1.4.2 Step 2: Determining time zero and data collection at different time delay points

The graphical user interface (GUI) allows for easy adjustment of the time delay between the X-rays and optical lasers. There are different methods to set a time delay depending on the time range of interest. For delays ranging from a few femtoseconds to picoseconds, the transmission signal of the incident optical laser is measured by changing the relative arrival time between the X-ray and optical laser. The time delay is controlled by the motion of the linear stage, which results in a small difference in the optical path length. The change in the refractive index caused by the X-ray leads to intensity variation, which defines time zero, marked by an abrupt

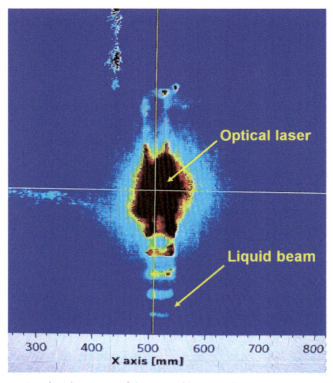

Fig. 6 Monitoring the alignment of the optical laser to the injection stream during the overlapping process.

intensity variation in the time domain. However, an electronic control is used for longer time delays. In this case, time zero can be defined using an avalanche photodiode (APD) that measures the intensity of both beams using an oscilloscope. Time zero and time delay can be adjusted electronically using a delay generator. Diffraction images are collected at several time delay points.

6. Data processing

During TR-SFX experiments, data collection is monitored in real-time using the OnDA program (Mariani et al., 2016) or a detector image viewer provided by the MX225-HS detector (Fig. 7). The OnDA GUI enables diffraction pattern accumulation monitoring, which helps assess data completeness. Moreover, the program calculates the intensity of the Bragg peak in the image and displays the hit rate providing insight into the diffraction intensity of the crystal sample used in the experiment and helps to determine whether the density of the crystal sample is suitable. This real-time information provided by the OnDA is invaluable for guiding the preparation of crystal samples for subsequent experiments.

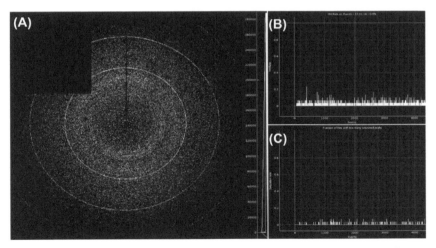

Fig. 7 Graphical user interface (GUI) of the OnDA program for real time diffraction pattern monitoring. (A) accumulated XFEL diffraction patterns (B) hit rate (C) fraction of shots containing more than three saturated peaks. The original figures for (A) were obtained from a previous study (Nam et al., 2022).

The collected data include images with either no crystal diffraction or a weak crystal diffraction intensity. A data filtering program is used to exclude images lacking sufficient diffraction intensity, which are deemed unnecessary for further data processing. Within the PAL-XFEL NCI control room, Cheetah (Barty et al., 2014), Psocake (Yoon, 2020), and NanoPeakCell (Coquelle et al., 2015) are available as software tools for filtering non-hit images. Users have the flexibility to select their preferred program for data filtering. Typically, in an NCI control room, Cheetah programs are used for data filtering during SFX data processing. The default parameters for data filtering are listed in Table 6.

Researchers have the flexibility to adjust the hit finder parameters based on their research objectives or desired quality of data. In addition, an image acquired by a user through a data-filtering program is referred to as a hit. Typically, hit images exhibit an intensity and number of Bragg peaks that satisfy the criteria established by the data-filtering program. The proportion of images obtained via

Table 6 Data filtering parameter in Cheetah program at Nano Crystallography and Coherent Imaging (NCI) control room.

pal.ini

hitfinderADC	150
hitfinderNpeaks	15
hitfinderNpeaksMax	5000
hitfinderMinPixCount	2
hitfinderMaxPixCount	50
hitfinderMinSNR	5
[front]	
detectorType	mx225hs-4x
detectorName	mx225hs-4x
pixelSize	0.000156
detectorZpvname	86
cameraLengthOffset	750
saveCXI	1

the filtering program relative to the total dataset was termed the hit rate. This hit rate serves as a benchmark for assessing the density and quality of the crystal samples exposed to XFEL. The parameters used in data filtering can be modified to align with the specific research goals of the investigator.

The CrystFEL program is available for further data processing, such as indexing, integrating, and scaling of the hit images (White, 2019; White et al., 2016). Various algorithms, such as MOSFLM (Battye et al., 2011), XDS (Kabsch, 2010), DIRAX (Duisenberg, 1992), and XGANDALF (Gevorkov et al., 2019) can be applied during data processing in CrystFEL. Optimization of the detector geometry using a geoptimiser is essential for achieving high indexing efficiency (Yefanov et al., 2015). Fully integrated intensities are generated from partially integrated intensities using Monte Carlo integration. Structure factors are used for structure determination. To solve the phasing problem and determine the structure, other crystallographic programs such as PHENIX (Liebschner et al., 2019), CCP4 (Agirre et al., 2023), COOT (Emsley & Cowtan, 2004), and PyMOL are installed. The available crystallographic programs for SFX are listed in Table 7.

7. Safety considerations and others
7.1 Preliminary discussion for experiment

When the beamtime allocation is completed and the beamtime of a user group approaches, discussions regarding the required experimental conditions,

Table 7 Crystallographic program for serial femtosecond crystallography (SFX) data processing.

Program	PAL-XFEL
Real time monitoring	OnDA
	MX225-HS (detector image viewer)
Image viewer	HDFsee, ADXV
Data filtering program	Cheetah, Pesocake, NanoPeakCell
Data processing	CrystFEL
Structure determination and refinement protein	Phenix, CCP4
Structure Building	COOT
Structure viewer	PyMOL

hardware, and software for data analysis are conducted with beamline scientists and engineers prior to beamtime. To conduct an SFX study, researchers should discuss data collection parameters, such as the preferred experimental environment and setup, with beamline personnel in advance. For instance, users must inform the beamline manager regarding the XFEL energy they wish to use, the XFEL repetition rate, and whether the XFEL mode they prefer is SASE or self-seeded. This information is communicated to the accelerator department prior to beamtime and is incorporated into the XFEL tuning process by the accelerator department on the day of beamtime.

7.2 Laser

The optical laser system is managed and optimized exclusively by laser experts in PAL-XFEL. Access inside the hutch is prohibited during the optical laser setup. In addition, for safety purposes, user groups intending to use the optical laser system for their beamtime must undergo laser safety training beforehand.

7.3 Hutch access

Accessing the interior of a hutch requires several steps. Generally, an interlock system is established to prevent accidental access when X-rays and optical lasers are active in the area of the hutch.

7.4 Data access

All the XFEL data collected during the experiment are subjected to data analysis and processing in an NCI control room or data processing room. However, owing to the data security policy of PAL-XFEL, off-facility online access to the PAL-XFEL data server is not permitted. Instead, all the data are transferred to the Korea Institute of Science and Technology Information (KISTI). Data transferred to KISTI can be accessed and processed by beamtime users at any time in accordance with KISTI's security policy.

Acknowledgments

We would like to express our gratitude to Yunje Cho, Jong-Lam Lee, Wan Kyun Chung, Donghyun Lee, Keondo Lee, Sangwon Baek, and Jihan Kim from POSTECH for their valuable contributions in developing the sample delivery techniques. We would like to thank Sehan Park and Sang Jae Lee at PAL-XFEL. This work was funded by the National Research Foundation of Korea (NRF) (NRF-2017M3A9F6029736 and NRF-2021R1I1A1A01050838).

References

Agirre, J., Atanasova, M., Bagdonas, H., Ballard, C. B., Baslé, A., Beilsten-Edmands, J., ... Yamashita, K. (2023). The CCP4 suite: Integrative software for macromolecular crystallography. *Acta Crystallographica Section D Structural Biology, 79*(6), 449–461. https://doi.org/10.1107/s2059798323003595.

Barty, A., Kirian, R. A., Maia, F. R., Hantke, M., Yoon, C. H., White, T. A., & Chapman, H. (2014). Cheetah: Software for high-throughput reduction and analysis of serial femtosecond X-ray diffraction data. *Journal of Applied Crystallography, 47*(Pt 3), 1118–1131. https://doi.org/10.1107/S1600576714007626.

Battye, T. G., Kontogiannis, L., Johnson, O., Powell, H. R., & Leslie, A. G. (2011). iMOSFLM: A new graphical interface for diffraction-image processing with MOSFLM. *Acta Crystallographica Section D, Biological Crystallography, 67*(Pt 4), 271–281. https://doi.org/10.1107/S0907444910048675.

Chapman, H. N. (2017). Structure determination using X-ray free-electron laser pulses. *Protein Crystallography, 295*–324. https://doi.org/10.1007/978-1-4939-7000-1_12.

Chapman, H. N. (2019). X-ray free-electron lasers for the structure and dynamics of macromolecules. *Annual Review of Biochemistry, 88*(1), 35–58. https://doi.org/10.1146/annurev-biochem-013118-110744.

Chapman, H. N., Caleman, C., & Timneanu, N. (2014). Diffraction before destruction. *Philosophical Transactions of the Royal Society B, 369*(1647), 20130313. https://doi.org/10.1098/rstb.2013.0313.

Coquelle, N., Brewster, A. S., Kapp, U., Shilova, A., Weinhausen, B., Burghammer, M., & Colletier, J. P. (2015). Raster-scanning serial protein crystallography using micro- and nano-focused synchrotron beams. *Acta Crystallographica. Section D, Biological Crystallography, 71*(Pt 5), 1184–1196. https://doi.org/10.1107/S1399004715004514.

Decking, W., Abeghyan, S., Abramian, P., Abramsky, A., Aguirre, A., Albrecht, C., ... Zybin, D. (2020). A MHz-repetition-rate hard X-ray free-electron laser driven by a superconducting linear accelerator. *Nature Photonics, 14*(6), 391–397. https://doi.org/10.1038/s41566-020-0607-z.

Di Rienzo, C., Gratton, E., Beltram, F., & Cardarelli, F. (2016). Spatiotemporal fluctuation analysis: A powerful tool for the future nanoscopy of molecular processes. *Biophysical Journal, 111*(4), 679–685. https://doi.org/10.1016/j.bpj.2016.07.015.

Duisenberg, A. J. M. (1992). Indexing in single-crystal diffractometry with an obstinate list of reflections. *Journal of Applied Crystallography, 25*(2), 92–96. https://doi.org/10.1107/S0021889891010634.

Emma, P., Akre, R., Arthur, J., Bionta, R., Bostedt, C., Bozek, J., ... Galayda, J. (2010). First lasing and operation of an angstrom-wavelength free-electron laser. *Nature Photonics, 4*(9), 641–647. https://doi.org/10.1038/Nphoton.2010.176.

Emsley, P., & Cowtan, K. (2004). Coot: model-building tools for molecular graphics. *Acta Crystallographica. Section D, Biological Crystallography, D60,* 2126–2132. https://doi.org/10.1107/S0907444904019158.

Eom, I., Chun, S. H., Lee, J. H., Nam, D., Ma, R., Park, J., ... Kim, C. (2022). Recent progress of the PAL-XFEL. *Applied Sciences, 12*(3), 1010. https://doi.org/10.3390/app12031010.

Farr, E. P., Quintana, J. C., Reynoso, V., Ruberry, J. D., Shin, W. R., & Swartz, K. R. (2018). Introduction to time-resolved spectroscopy: Nanosecond transient absorption and time-resolved fluorescence of eosin B. *Journal of Chemical Education, 95*(5), 864–871. https://doi.org/10.1021/acs.jchemed.7b00941.

Fisette, O., Lague, P., Gagne, S., & Morin, S. (2012). Synergistic applications of MD and NMR for the study of biological systems. *Journal of Biomedicine & Biotechnology, 2012,* 254208. https://doi.org/10.1155/2012/254208.

Gevorkov, Y., Yefanov, O., Barty, A., White, T. A., Mariani, V., Brehm, W., ... Chapman, H. N. (2019). XGANDALF - extended gradient descent algorithm for lattice finding. *Acta Crystallographica Section A: Foundations and Advances, 75*(Pt 5), 694–704. https://doi.org/10.1107/S2053273319010593.

Hekstra, D. R. (2023). Emerging time-resolved X-ray diffraction approaches for protein dynamics. *Annual Review of Biophysics, 52*(1), 255–274. https://doi.org/10.1146/annurev-biophys-111622-091155.

Ishikawa, T., Aoyagi, H., Asaka, T., Asano, Y., Azumi, N., Bizen, T., ... Kumagai, N. (2012). A compact X-ray free-electron laser emitting in the sub-ångström region. *Nature Photonics, 6*(8), 540–544. https://doi.org/10.1038/nphoton.2012.141.

Kabsch, W. (2010). Xds. *Acta Crystallographica. Section D, Biological Crystallography, 66*(Pt 2), 125–132. https://doi.org/10.1107/S0907444909047337.

Kang, H. S., Min, C. K., Heo, H., Kim, C., Yang, H., Kim, G., ... Ko, I. S. (2017). Hard X-ray free-electron laser with femtosecond-scale timing jitter. *Nature Photonics, 11*(11), 708–713. https://doi.org/10.1038/s41566-017-0029-8.

Kim, J., Kim, H. Y., Park, J., Kim, S., Kim, S., Rah, S., ... Nam, K. H. (2018). Focusing X-ray free-electron laser pulses using Kirkpatrick-Baez mirrors at the NCI hutch of the PAL-XFEL. *Journal of Synchrotron Radiation, 25*, 289–292. https://doi.org/10.1107/S1600577517016186.

Kim, K. H., Späh, A., Pathak, H., Perakis, F., Mariedahl, D., Amann-Winkel, K., ... Nilsson, A. (2017). Maxima in the thermodynamic response and correlation functions of deeply supercooled water. *Science (New York, N. Y.), 358*(6370), 1589–1593. https://doi.org/10.1126/science.aap8269.

Ko, I. S., Kang, H. S., Heo, H., Kim, C., Kim, G., Min, C. K., ... Lee, K. B. (2017). Construction and commissioning of PAL-XFEL facility. *Applied Sciences, 7*(5), 479. https://doi.org/10.3390/app7050479.

Lee, D., Baek, S., Park, J., Lee, K., Kim, J., Lee, S. J., ... Nam, K. H. (2019). Nylon mesh-based sample holder for fixed-target serial femtosecond crystallography. *Scientific Reports, 9*(1), 6971. https://doi.org/10.1038/s41598-019-43485-z.

Lee, D., Park, S., Lee, K., Kim, J., Park, G., Nam, K. H., ... Park, J. (2020). Application of a high-throughput microcrystal delivery system to serial femtosecond crystallography. *Journal of Applied Crystallography, 53*(Pt 2), 477–485. https://doi.org/10.1107/S1600576720002423.

Liebschner, D., Afonine, P. V., Baker, M. L., Bunkoczi, G., Chen, V. B., Croll, T. I., ... Adams, P. D. (2019). Macromolecular structure determination using X-rays, neutrons and electrons: Recent developments in Phenix. *Acta Crystallographica Section D Structural Biology. 75*(Pt 10), 861–877. https://doi.org/10.1107/S2059798319011471.

Mariani, V., Morgan, A., Yoon, C. H., Lane, T. J., White, T. A., O'Grady, C., ... Chapman, H. N. (2016). OnDA: Online data analysis and feedback for serial X-ray imaging. *Journal of Applied Crystallography, 49*(3), 1073–1080. https://doi.org/10.1107/s1600576716007469.

Nam, I., Min, C.-K., Oh, B., Kim, G., Na, D., Suh, Y. J., ... Kang, H.-S. (2021). High-brightness self-seeded X-ray free-electron laser covering the 3.5 keV to 14.6 keV range. *Nature Photonics, 15*(6), 435–441. https://doi.org/10.1038/s41566-021-00777-z.

Nam, K. H. (2019). Sample delivery media for serial crystallography. *International Journal of Molecular Sciences, 20*(5), 1094. https://doi.org/10.3390/ijms20051094.

Nam, K. H. (2021). Molecular dynamics—From small molecules to macromolecules. *International Journal of Molecular Sciences, 22*(7), 3761. https://doi.org/10.3390/ijms22073761.

Nam, K. H., Park, S., & Park, J. (2022). Preliminary XFEL data from spontaneously grown endo-1,4-β-xylanase crystals from *Hypocrea virens. Acta Crystallographica Section F: Structural Biology Communications, 78*(Pt 6), 226–231. https://doi.org/10.1107/S2053230X22005118.

Nango, E., & Iwata, S. (2023). Recent progress in membrane protein dynamics revealed by X-ray free electron lasers: Molecular movies of microbial rhodopsins. *Current Opinion in Structural Biology, 81*. https://doi.org/10.1016/j.sbi.2023.102629.

Neutze, R., Wouts, R., van der Spoel, D., Weckert, E., & Hajdu, J. (2000). Potential for biomolecular imaging with femtosecond X-ray pulses. *Nature, 406*(6797), 752–757. https://doi.org/10.1038/35021099.

Orville, A. M. (2020). Recent results in time resolved serial femtosecond crystallography at XFELs. *Current Opinion in Structural Biology, 65*, 193–208. https://doi.org/10.1016/j.sbi.2020.08.011.

Park, J., Kim, S., Kim, S., & Nam, K. H. (2018). Multifarious injection chamber for molecular structure study (MICOSS) system: Development and application for serial femtosecond crystallography at Pohang Accelerator Laboratory X-ray Free-Electron Laser. *Journal of Synchrotron Radiation, 25*(Pt 2), 323–328. https://doi.org/10.1107/S160057751800022X.

Park, J., Kim, S., Nam, K. H., Kim, B., & Ko, I. S. (2016). Current status of the CXI beamline at the PAL-XFEL. *Journal of the Korean Physical Society, 69*(6), 1089–1093. https://doi.org/10.3938/jkps.69.1089.

Park, J., & Nam, K. H. (2023). Sample delivery systems for serial femtosecond crystallography at the PAL-XFEL. *Photonics, 10*(5), 557. https://doi.org/10.3390/photonics10050557.

Park, J., Park, S., Kim, J., Park, G., Cho, Y., & Nam, K. H. (2019). Polyacrylamide injection matrix for serial femtosecond crystallography. *Scientific Reports, 9*(1), 2525. https://doi.org/10.1038/s41598-019-39020-9.

Poddar, H., Heyes, D. J., Schirò, G., Weik, M., Leys, D., & Scrutton, N. S. (2021). A guide to time-resolved structural analysis of light-activated proteins. *The FEBS Journal, 289*(3), 576–595. https://doi.org/10.1111/febs.15880.

Prat, E., Abela, R., Aiba, M., Alarcon, A., Alex, J., Arbelo, Y., ... Zimoch, E. (2020). A compact and cost-effective hard X-ray free-electron laser driven by a high-brightness and low-energy electron beam. *Nature Photonics, 14*(12), 748–754. https://doi.org/10.1038/s41566-020-00712-8.

Sahoo, S. K., Umapathy, S., & Parker, A. W. (2011). Time-resolved resonance Raman spectroscopy: Exploring reactive intermediates. *Applied Spectroscopy, 65*(10), 1087–1115. https://doi.org/10.1366/11-06406.

Späh, A., Pathak, H., Kim, K. H., Perakis, F., Mariedahl, D., Amann-Winkel, K., ... Nilsson, A. (2019). Apparent power-law behavior of water's isothermal compressibility and correlation length upon supercooling. *Physical Chemistry Chemical Physics: PCCP, 21*(1), 26–31. https://doi.org/10.1039/c8cp05862h.

Sponer, J., Bussi, G., Krepl, M., Banas, P., Bottaro, S., Cunha, R. A., ... Otyepka, M. (2018). RNA structural dynamics as captured by molecular simulations: A comprehensive overview. *Chemical Reviews, 118*(8), 4177–4338. https://doi.org/10.1021/acs.chemrev.7b00427.

Sung, D., Nam, D., Kim, M.-J., Kim, S., Kim, K. S., Park, S.-Y., ... Kim, S. (2021). Single-shot coherent X-ray imaging instrument at PAL-XFEL. *Applied Sciences, 11*(11), 5082. https://doi.org/10.3390/app11115082.

Van Dalsen, L., Weichert, D., & Caffrey, M. (2020). In meso crystallogenesis. Compatibility of the lipid cubic phase with the synthetic digitonin analogue, glyco-diosgenin. *Journal of Applied Crystallography, 53*(2), 530–535. https://doi.org/10.1107/s1600576720002289.

Weierstall, U., James, D., Wang, C., White, T. A., Wang, D., Liu, W., ... Cherezov, V. (2014). Lipidic cubic phase injector facilitates membrane protein serial femtosecond crystallography. *Nature Communications, 5*, 3309. https://doi.org/10.1038/ncomms4309.

Wells, D. J., Berntsen, P., Balaur, E., Kewish, C. M., Adams, P., Aquila, A., ... Darmanin, C. (2022). Observations of phase changes in monoolein during high viscous injection. *Journal of Synchrotron Radiation, 29*(3), 602–614. https://doi.org/10.1107/s1600577522001862.

Westenhoff, S., Meszaros, P., & Schmidt, M. (2022). Protein motions visualized by femtosecond time-resolved crystallography: The case of photosensory vs photosynthetic proteins. *Current Opinion in Structural Biology, 77*, 102481. https://doi.org/10.1016/j.sbi.2022.102481.

White, T. A. (2019). Processing serial crystallography data with CrystFEL: A step-by-step guide. *Acta Crystallographica Section D Structural Biology, 75*(Pt 2), 219–233. https://doi.org/10.1107/S205979831801238X.

White, T. A., Mariani, V., Brehm, W., Yefanov, O., Barty, A., Beyerlein, K. R., ... Chapman, H. N. (2016). Recent developments in CrystFEL. *Journal of Applied Crystallography, 49*(Pt 2), 680–689. https://doi.org/10.1107/S1600576716004751.

Yefanov, O., Mariani, V., Gati, C., White, T. A., Chapman, H. N., & Barty, A. (2015). Accurate determination of segmented X-ray detector geometry. *Optics Express, 23*(22), 28459–28470. https://doi.org/10.1364/OE.23.028459.

Yoon, C. H. (2020). Psocake: GUI for making data analysis a piece of cake. *In Handbook on Big Data and Machine Learning in the Physical Sciences*, 169–178. https://doi.org/10.1142/9789811204579_0010.

CHAPTER SIX

Time-resolved IR spectroscopy for monitoring protein dynamics in microcrystals

Wataru Sato, Daichi Yamada, and Minoru Kubo*

Graduate School of Science, University of Hyogo, Kouto, Kamigori, Ako, Hyogo, Japan
*Corresponding author. e-mail address: minoru@sci.u-hyogo.ac.jp

Contents

1. Introduction	162
2. Overview of instrument	163
3. Experimental procedure	165
3.1 Instrument preparation	165
3.2 Alignment of the pump beam	165
3.3 Collection of a background spectrum	166
3.4 Sample preparation	167
3.5 Sample packing within the FTIR cell	167
3.6 Selection of the sample-containing wells	168
3.7 Adjustment of the pump irradiation timing and rapid-scan measurements	168
3.8 Setting measurement conditions	170
3.9 Data analysis	170
4. Results and discussion	171
4.1 Observation of CO photolysis in CcO	171
4.2 Observation of P450nor intermediate	173
5. Conclusion	174
Acknowledgments	175
References	175

Abstract

Analysis of protein dynamics is crucial for understanding the molecular mechanisms underlying protein function. To gain insights into the structural changes in proteins, time-resolved X-ray crystallography has been greatly advanced by the development of X-ray free-electron lasers. This tool has the potential to trace structural changes at atomic resolution; however, data interpretation and extrapolation to the solution state is often not straightforward as the *in crystallo* environment is not the same as it is in solution. On the other hand, time-resolved spectroscopy techniques, which have long been used for tracking protein dynamics, offer the advantage of being applicable irrespective of whether the target proteins are in crystalline or solution phase. Time-resolved IR spectroscopy is a particularly powerful technique, as it can be used on

Methods in Enzymology, Volume 709
ISSN 0076-6879, https://doi.org/10.1016/bs.mie.2024.10.006
Copyright © 2024 Elsevier Inc. All rights are reserved, including those for text and data mining, AI training, and similar technologies.

161

various proteins, including those that are colorless, and provides information on the chemical structures of functional sites of proteins and ligands which complements X-ray crystallography. This chapter presents the protocol for time-resolved IR microspectroscopic measurements of protein microcrystals. It includes an overview of the measurement system assembly, sample preparation, setting of experimental conditions, and time-resolved data analysis. It also describes, with examples, the usefulness of time-resolved IR measurements for comparing the dynamics between crystalline and solution conditions.

1. Introduction

Proteins undergo fluctuations and structural changes during the course of their function. Time-resolved (TR) X-ray crystallography using an X-ray free-electron laser (XFEL) is a novel technique that enables the direct observation of such protein dynamics at ambient temperature with atomic resolution. However, there remain some technical difficulties due to the complexities associated with handling protein microcrystals. One of the critical challenges in interpreting TR X-ray crystallography data is determining the extent to which protein dynamics in microcrystals reflect those in aqueous solutions. To address this issue, we previously developed a TR UV–visible microspectroscopic system that can measure protein microcrystals and investigated the similarities and differences in the reaction dynamics between microcrystal and solution conditions for three microbial rhodopsins (Nango et al., 2016; Oda et al., 2021; Hosaka et al., 2022). We found that all three proteins failed to exhibit functionally important late intermediates in the microcrystals, probably due to the crystal packing. Our *in crystallo* spectroscopic observations showed the importance of the combined use of TR UV–visible microspectroscopy with TR X-ray crystallography; importantly the spectroscopic data were practically used to select appropriate delay times for TR X-ray crystallography to correctly assign intermediates. However, UV–visible microspectroscopy suffers from limited applicability, being effective only for the analysis of colored proteins that possess chromophores. To overcome this shortcoming, the use of TR-IR microspectroscopy is desirable, as it is applicable to any proteins, including colorless ones.

IR spectroscopy is a powerful technique that offers deep insight into molecular structures through the observation of molecular vibrations (Barth, 2007; Kötting and Gerwert, 2005). In this context the method reports on the vibrations associated with polar bonds in amino acid side chains, cofactors, and ligands. Given the pivotal roles of these in the chemical reactions in proteins, the use of IR spectroscopy is crucial for exploring the reaction kinetics in the crystal environment as well as for accessing specific chemical information at the functional sites of proteins, which is complementary to the structural information provided by X-ray crystallography. To enhance the integration of TR-IR spectroscopy with TR X-ray crystallography, we recently developed a TR-IR microspectroscopy system that can handle protein microcrystals. This chapter explains the methodology of *in crystallo* TR-IR microspectroscopy measurements, as exemplified by the photolysis of CO in cytochrome *c* oxidase (C*c*O). It then presents an application of this technique to probe a key intermediate of cytochrome P450 NO reductase (P450nor).

2. Overview of instrument

The experimental apparatus is depicted in Fig. 1. An interferometer coupled to an FTIR microscope (Vertex 70 and Hyperion 2000, Bruker) was used for TR-IR microspectroscopy. Protein microcrystals were placed within the microwells (Ø500 μm) of a fixed-target-type FTIR cell, which was set on the stage of the microscope. The IR beam diameter was *ca.* 500 μm, but the sample area used for IR spectral measurements was defined by a knife-edge aperture (spatial filter) in front of a mercury cadmium telluride (MCT) detector. The contribution of water vapor to the spectrum was reduced by purging the apparatus with N_2 gas.

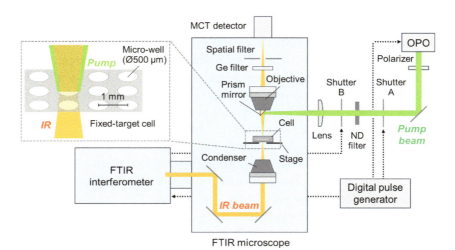

Fig. 1 Experimental setup for TR-IR microspectroscopy.

To initiate the dynamics in proteins, a 6-ns pump pulse from an optical parametric oscillator (OPO; NT230, Ekspla) was used. The wavelength of the OPO output was tunable from 300 to 2600 nm, and thus enzymatic reactions with caged compounds, which are often activated by UV light, can be triggered and observed by our system. The pump beam was introduced to the sample in an on-axis manner with the IR beam *via* a prism mirror attached to a Cassegrain objective mirror (15 ×, N.A. 0.4, W.D. 24 mm) and focused on the sample with Ø600 μm. The pump irradiation to the sample was controlled with a beam shutter (TH05, Thorlabs). A germanium (Ge) filter (WG90530-G, Thorlabs) was placed at the rear of the Cassegrain objective to protect the MCT detector from the pump light scattering.

TR-IR spectra were acquired in the transmission mode in a rapid scan manner. The time resolution was < 35 ms with a spectral resolution of $8\ cm^{-1}$, using a double-sided, forward-backward data acquisition mode and a scan speed of 160 kHz. The pump illumination to the sample and the start of the rapid scan were synchronized and controlled using digital pulse generators (DG645, Stanford Research Systems). The dark data (before pump illumination) and TR-IR spectral data (after pump illumination) were recorded for each well of the FTIR cell on the sample stage, which was automatically repositioned following each rapid-scan measurement. The TR-IR minus dark difference spectra were calculated and averaged over the measured wells to improve the signal-to-noise ratio.

The assembly of our FTIR cell is shown in Fig. 2. The cell was composed of a stainless steel spacer, two CaF_2 window plates, two Teflon spacers, and a cell holder that can be connected to a cooling water chiller. CaF_2 is a suitable material for IR measurements owing to its low solubility, good resistance to a wide range of acids and alkalis, and transparency in the mid-IR region. The stainless steel spacer contained 100 holes with a diameter of 500 μm, distributed at 1 mm intervals. The thickness of the stainless steel spacer, which corresponds to the IR path length, is alterable; in the current case we set it to be 30 μm because this is comparable to the typical microcrystal size. As shown in Fig. 2, the stainless steel spacer was sandwiched between the two CaF_2 plates, which were further sandwiched by the Teflon spacers serving as cushions. All of these components were mounted and fixed in the cell holder with screws. The detailed procedure for performing a TR-IR experiment on protein microcrystals is described in the following section, using CO photolysis in C*c*O as a representative example.

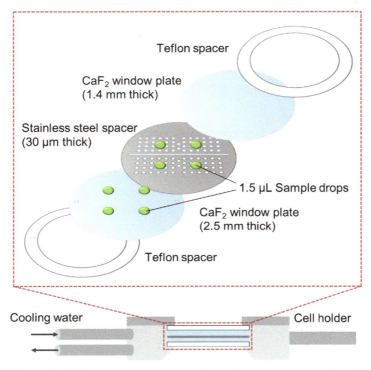

Fig 2 FTIR cell for TR-IR microspectroscopy. The stainless spacer contains 100 holes, designed for fixed-target measurements of microcrystal samples. A long channel is also included, enabling flow measurements of solution samples.

3. Experimental procedure

3.1 Instrument preparation

The air conditioning and a dehumidification system were turned on in the experimental room. The room was maintained at 20 °C. The N_2 gas purging system was activated to remove water vapor from the FTIR apparatus. Liquid nitrogen was added to the MCT detector dewar to cool down the detector.

3.2 Alignment of the pump beam

The OPO was turned on and the output wavelength was set to 532 nm to induce CO photolysis in CcO. To avoid the pump beam entering the MCT detector, a Ge filter was inserted at the rear of the Cassegrain objective. The pump beam was aligned to the sample stage on axis with the IR beam using a mirror placed outside the FTIR microscope (Fig. 1). The beam was focused on the sample cell with Ø600 μm using a lens. The pump light intensity was adjusted to be 50 μJ at the sample point using a polarizer and a neutral-density (ND) filter.

3.3 Collection of a background spectrum

To obtain a background spectrum, we set the empty FTIR cell (without sample) on the stage of the FTIR microscope. The cooling water chiller was connected to the cell holder to maintain a cell temperature of 4 °C. We then placed a plastic tube between the Cassegrain objective mirror and the cell, which allowed the N_2 gas leaking from the Cassegrain objective mirror to remain in place, thereby facilitating the removal of water vapor from the IR optical path.

The IR focal point was adjusted on a randomly chosen well of the cell as follows (Fig. 3): (1) Using the "visible light reflectance" mode (Hyperion 2000, Bruker), the height of the sample stage (along the z axis) was adjusted such that the well image was clear through the binocular eyepiece or video

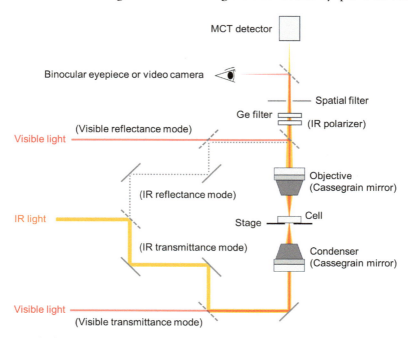

Fig. 3 Light beam path in the FTIR microscope.

camera. (2) Using the "visible light transmittance" mode, the height of the condenser Cassegrain mirror below the stage (along the z axis) was adjusted to obtain a clear well image. At this point, the focus of the IR beam should be on the well. (3) Using the "IR light transmittance" mode, the IR light intensity was measured to confirm that it had been maximized.

After the above instrument alignment, the background IR light intensity spectrum of the empty well was measured with 32 scans over the wavenumber range of 1000–2500 cm^{-1} with a spectral resolution of 8 cm^{-1}.

3.4 Sample preparation

We prepared the sample inside a glove box (Vinyl Anaerobic Chamber Type A, Coy Laboratory Products) as follows: (1) Microcrystals of bovine heart CcO, suspended in 40 mM sodium phosphate (pH 6.5) containing 1 % polyethylene glycol 4000 (PEG 4000), 2 % ethylene glycol, and 0.2 % n-decyl-β-D-maltoside (DM), were centrifuged for a few seconds and the supernatant was removed. (2) 1 mL of CO-saturated reduction buffer (40 mM sodium phosphate (pH 5.7) containing 8 % PEG 4000, 40 % ethylene glycol, and 0.2 % DM, supplemented with 2.5 mM dithionite, 0.5 μM catalase, 50 mM D-glucose, and 1 μM glucose oxidase) was added to the microcrystals and incubated for 30 min at 4 °C. (3) The CcO microcrystals in the CO-saturated reduction buffer were centrifuged for a few seconds and the supernatant was removed. (4) Steps (2) and (3) were repeated several times to ensure complete binding of CO to the microcrystalline CcO.

3.5 Sample packing within the FTIR cell

The CO-bound CcO microcrystals were packed within the FTIR cell (Fig. 2) inside a glove box as follows: (1) Four 1.5 μL drops of the sample containing CO-bound CcO microcrystals were deposited onto the lower CaF$_2$ window plate (2.5 mm thickness). (2) The stainless spacer of 30 μm thickness was placed on the CaF$_2$ plate, followed by the addition of another four drops of the sample onto the spacer. (3) The upper CaF$_2$ window plate (1.4 mm thickness) was carefully placed on the stainless steel spacer to enclose the sample, such that the spacer was sandwiched between the two CaF$_2$ plates. (4) The CaF$_2$ plates holding the stainless steel spacer were further sandwiched by Teflon spacers serving as cushions and finally mounted and fixed in the cell holder using screws.

The assembled cell holder was set on the stage of the FTIR microscope and connected to the cooling water chiller to maintain a cell temperature of 4 °C.

Using the "visible light reflectance" mode, we confirmed the placement of the C*c*O microcrystals within the wells of the cell using the video camera attached to the FTIR microscope (Fig. 3). It should be noted that an IR polarizer can be inserted above the Cassegrain objective mirror, although we did not use it because the protein microcrystals were randomly oriented within the wells.

3.6 Selection of the sample-containing wells

After confirming the optimal beam alignment of the FTIR microscope as described in Step 3, the stage was scanned in the xy plane and the center of each well was assigned a specific address (x and y values). Next, the IR light intensity spectrum of each well was measured by performing a few scans in the wavenumber range of $1000-2500\,\text{cm}^{-1}$ with a spectral resolution of $8\,\text{cm}^{-1}$. Here, the sample stage was automatically moved to place each well in the IR beam path using the previously designated well addresses. The strong IR absorption of water at $1645\,\text{cm}^{-1}$ permitted straightforward differentiation between the wells that contained the sample and those that did not, with the former exhibiting a marked decrease in the IR light intensity at this wavenumber (Fig. 4A). Thus, the sample-containing wells were identified by calculating the absorbance spectrum for each well using the background spectrum obtained in Step 3. In addition to the peak at $1645\,\text{cm}^{-1}$, the sample-containing wells exhibited an Amide II peak at $1550\,\text{cm}^{-1}$ derived from the protein polypeptide chain (Fig. 4B).

3.7 Adjustment of the pump irradiation timing and rapid-scan measurements

Upon pump light irradiation to CO-bound C*c*O, the CO undergoes photodissociation and then rebinds to C*c*O within 1 s (Einarsdóttir et al., 1993; Shimada et al., 2017). For reversible reactions such as this, repetitive

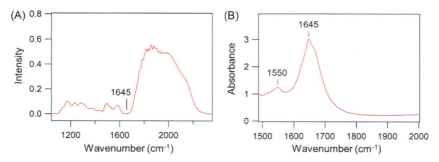

Fig. 4 A typical example of the obtained spectral data. (A) IR light intensity spectrum and (B) IR absorption spectrum of a well containing the C*c*O microcrystals.

measurements of a single well are possible by extending the interval between pump irradiations to 1 s. A time chart for the TR-FTIR measurements of CO-bound C*c*O is presented in Fig. 5A. The pump laser (OPO) in our system can only operate at 30 Hz (33 ms interval). Thus, we installed a mechanical shutter (shutter A; SH05, Thorlabs) to allow pump irradiation of the sample at 1 Hz (every 1 s).

The pump laser was operated using the TTL signal from a digital pulse generator (DG645, Stanford Research Systems) at a frequency of 30 Hz, whereas the TTL signal for opening shutter A was input to the shutter control box *via* another digital pulse generator, with the frequency reduced from 30 to 1 Hz (Fig. 5B). The same digital pulse generator also sent the TTL at 1 Hz to the FTIR interferometer trigger box to initiate the rapid

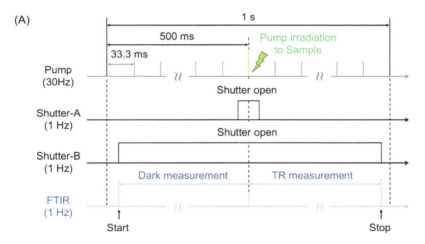

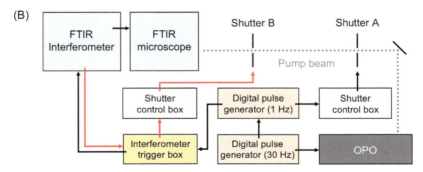

Fig. 5 Timing control for the TR-FTIR measurements. (A) Time chart of the measurements. (B) Schematic of the timing control using digital pulse generators.

scan measurement. It should be noted that the time required for the movable mirror to complete one scan in the FTIR interferometer was *ca.* 77 ms, allowing 12 scans to be completed within 1 s. In the current work, the first six scans were used to record the dark reference (before pump irradiation) and the next six scans were used to obtain TR-IR spectra (after pump irradiation), as shown in Fig. 5A. In this experimental setting, the time window for TR-IR measurement was up to 500 ms following the pump irradiation (defined as $t = 0$). After the 12 scans of the interferometer were complete, the stage was automatically repositioned to place the next well in the pump/IR beam path, which was controlled by a program in the spectrometer software (OPUS, Bruker). While repositioning the stage, the pump irradiation to the sample was cut with a shutter (shutter B) installed in the pump beam path. The timings of all TTL signals were confirmed using an oscilloscope.

3.8 Setting measurement conditions

The measurement range (1000–2500 cm^{-1}) and spectral resolution (8 cm^{-1}) were set in the spectrometer software (OPUS, Bruker). Next, we set the experimental parameters, such as the number of rapid scans after the TTL input to the interferometer trigger box (12) and the data accumulation number (500) for each well. The address data (x and y values) specifying the sample-containing wells were also loaded as an experimental parameter. After setting these parameters, the TR-IR measurements were performed. Interferograms were acquired in a double-sided, forward-backward data acquisition mode with a scan speed of 160 kHz.

3.9 Data analysis

The double-sided, forward-backward interferograms were split into single-sided interferograms and Fourier transformed to improve the time resolution, using a MATLAB program developed by Lórenz-Fonfría (Lórenz-Fonfría et al., 2014). Consequently, four sets of spectra were obtained at 10.0, 34.1, 10.0, and 23.3 ms intervals, despite the fact that a single scan of the interferometer required *ca.* 77 ms. Accordingly, the current time resolution can be quantified as < 35 ms. The "dark" spectrum was calculated as the average of all spectra recorded prior to pump irradiation, and it was then subtracted from each TR-IR spectrum (after pump irradiation) to obtain a series of TR-IR difference spectra.

4. Results and discussion
4.1 Observation of CO photolysis in CcO

Our TR-IR microspectroscopic system was first applied to measure CO photolysis in CcO. CcO is the terminal oxidase of cellular respiration, reducing O_2 to H_2O at a binuclear center composed of heme a_3 and Cu_B (Yoshikawa and Shimada, 2015). CO, instead of O_2, can also bind to heme a_3 as an inhibitor, and it is susceptible to photolysis (Einarsdóttir et al., 1993). The TR-IR difference spectra of CcO after CO photolysis in microcrystals are shown in Fig. 6A. A prominent band was observed at 1963 cm^{-1}, which was assigned to CO stretching of the heme-bound CO (ν_{Fe-CO}). Photodissociated CO is known to migrate to Cu_B before diffusing into the solvent space, (Einarsdóttir et al., 1993; Shimada et al., 2017) and this transient binding of CO to Cu_B can be characterized by the CO stretching peak at 2062 cm^{-1} (ν_{Cu-CO}). However, this signal was not observed here on account of the low time resolution (<35 ms). The time trace of the absorbance change (ΔAbs) at 1963 cm^{-1} is also shown in the figure. Here, the pump irradiation time was defined as $t = 0$ ms. The ν_{Fe-CO} band rapidly decreased by the first time point ($t = 9.9$ ms) and then gradually recovered. These intensity changes reflect the photodissociation of CO from heme a_3, followed by its recombination to the heme (Fig. 6C). The band intensity at $t = 20.0$ ms appeared slightly higher than that at $t = 9.9$ ms. Considering that the spectrum at time point n represents an average recorded over the time from time points $n - 1$ to n, the lower band intensity at $t = 9.9$ ms should be attributable to a partial contribution from the data prior to pump irradiation. The apparent rate (k_{app}) of CO recombination to heme a_3 in the CcO microcrystals was estimated to be 13.2 ± 0.8 s^{-1} by fitting ΔAbs at 1963 cm^{-1} after $t = 20.0$ ms with a single-exponential function.

We also performed TR-IR measurements of CO photolysis from CcO in solution, using the same experimental system and the same conditions as for the microcrystals (Fig. 6B). The concentration of CcO was 200 μM. ν_{Fe-CO} was again observed at 1963 cm^{-1}, as for the microcrystals. The k_{app} value in solution was estimated to be 8.7 ± 0.3 s^{-1}, which is essentially identical to that of the CcO microcrystals. For the solution data, we further estimated the population of dissociated CO at $t = 20.0$ ms, which was 85%, from the integrated intensity at 1963 cm^{-1} ($\varepsilon = 28$ mM^{-1} cm^{-2}) (Yoshikawa et al., 1977). This population was sufficiently high to yield the k_{app} value to a first approximation. The similar k_{app} values observed for CcO in microcrystals

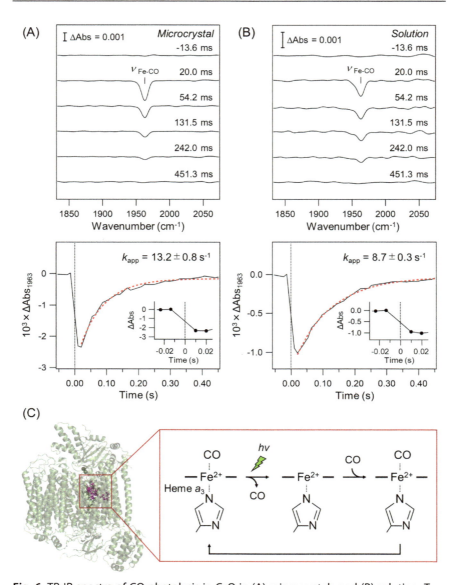

Fig. 6 TR-IR spectra of CO photolysis in C*c*O in (A) microcrystals and (B) solution. Top: Difference spectra after pump laser irradiation. Bottom: Time traces of ΔAbs at 1963 cm^{-1}. Single exponential fittings are indicated by red dashed lines. The data points around $t = 0$ are expanded in the inset. (C) Schematic representation of CO photolysis and recombination in C*c*O. The structure of C*c*O was taken from the PDB (ID: 3ABL) (Aoyama et al., 2009).

4.2 Observation of P450nor intermediate

and solution are considered reasonable, given that the binding of CO does not result in a global conformational change in CcO (Shimada et al., 2017).

We also utilized our TR-IR microspectroscopy system to investigate microcrystals of P450nor, a heme enzyme that binds NO and NADH and catalyzes the reduction of NO to N_2O during fungal denitrification (Shoun et al., 2012). The electronic and geometrical structures of the heme-bound NO after activation by hydride transfer from NADH have been a matter of considerable debate as this species is believed to be a key intermediate (Shiro et al., 1995). Thus, we employed TR-IR microspectroscopy in an effort to address this issue, using P450nor microcrystals pre-soaked with NADH and a caged substrate (caged NO) (Namiki et al., 1997; Tosha et al., 2017). It is worth mentioning that this approach, based on a caged substrate, enabled us to initiate the catalytic reaction by photo-triggering. Accordingly, the TR-IR spectra after caged NO photolysis with a pump pulse at 308 nm were measured using the same procedure established for the measurement of CO photolysis in CcO. It should be noted that the photo-decomposition of the caged NO meant that once the sample in a single well of the FTIR cell had been illuminated by a pump pulse, it could not be used again. Consequently, data accumulation for irreversible reactions such as this is normally more challenging than that for reversible reactions. However, we found that the intermediate of P450nor was stable with a lifetime of > 60 s in the microcrystals, which allowed for a prolonged period of data accumulation from a single well. Therefore, for each well, a "dark" spectrum was recorded for 60 s and an "intermediate" spectrum was then accumulated for 1 to 60 s following pump illumination, the rate of which was reduced to 1/120 Hz. The TR-IR difference spectrum ("intermediate" minus "dark") was averaged over 140 wells to increase the signal-to-noise ratio.

Additionally, the above TR-IR difference spectra were also measured using caged ^{15}NO, and the difference between ^{14}N and ^{15}N was calculated to extract the NO stretching band (Fig. 7). As a result, the NO stretching frequency of the intermediate was unambiguously determined to be $1298\ cm^{-1}$, which demonstrates that a nitroxyl radical anion species $(Fe^{3+}-NHO^{\bullet-})$ is present in the intermediate (Nomura et al., 2021). The NO stretching frequency of the intermediate in solution was $1290\ cm^{-1}$, indicating that the same species was generated both in microcrystals and in solution. Subsequent to these spectroscopic studies, an intermediate

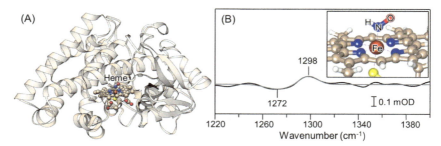

Fig. 7 TR-IR spectra of the P450nor intermediate in microcrystals. (A) Structure of P450nor (PDB ID: 7DVO)[15]. (B) $^{14}N-^{15}N$ difference spectrum of the P450nor intermediate. The spectral contribution from caged NO was subtracted. A Gaussian fitting curve for ν_{NO} is superimposed as a gray line. The inset depicts a simulated structure of the heme Fe^{3+}–NHO^- intermediate. *This figure was modified from Nomura et al. (2021).*

structural analysis was conducted using freeze-trap crystallography with XFEL, and the NO reduction mechanism by P450nor was determined by integrating the structural and spectroscopic data (Nomura et al., 2021).

5. Conclusion

A TR-IR microspectroscopic system with a fixed-target FTIR cell has been developed for measurements of protein microcrystals and was first tested to probe CO photolysis in C*c*O. This example application illustrates the versatility of our system. Next, the key intermediate of P450nor in microcrystals was analyzed by TR-IR microspectroscopy. The obtained findings demonstrate the potential for observing intermediates of enzymes *in crystallo* by using caged substrates with photo-triggering. Various caged compounds, such as caged ATP and photosensitive electron donors, are available (Ellis-Davies, 2007; Lübben and Gerwert, 1996). Engineered proteins covalently bound to photosensitive compounds have also been developed (Spradlin et al., 2016; Liu et al., 2022). Therefore, over the next decade, the application of caged compounds in TR X-ray crystallography and TR-IR microspectroscopy is expected to increase. The time resolution of the current TR-IR system may be insufficient for capturing protein intermediates of interest. Very recently, a novel technique using dual-comb quantum cascade lasers has emerged, offering microsecond time resolution for TR-IR spectroscopy (Norahan et al., 2021). This technique is promising for monitoring early intermediates of proteins' reaction. Therefore,

TR-IR microspectroscopy is anticipated to become a crucial addition to the suite of techniques employed in dynamic structural biology, in conjunction with TR X-ray crystallography.

Acknowledgments

We would like to thank Dr. Yuji Furutani and Dr. Victor A. Lórenz-Fonfría for their generosity in providing the MATLAB program to analyze the double-sided, forward-backward interferograms.

References

Aoyama, H., et al. (2009). A peroxide bridge between Fe and Cu ions in the O2 reduction site of fully oxidized cytochrome c oxidase could suppress the proton pump. *Proceedings of the National Academy of Sciences of the United States of America, 106,* 2165–2169.

Barth, A. (2007). Infrared spectroscopy of proteins. *Biochimica et Biophysica Acta, 1767,* 1073–1101.

Einarsdóttir, Ó., et al. (1993). Photodissociation and recombination of carbonmonoxy cytochrome oxidase: Dynamics from picoseconds to kiloseconds. *Biochemistry, 32,* 12013–12024.

Ellis-Davies, G. C. R. (2007). Caged compounds: Photorelease technology for control of cellular chemistry and physiology. *Nature Methods, 4,* 619–628.

Hosaka, T., et al. (2022). Conformational alterations in unidirectional ion transport of a light-driven chloride pump revealed using X-ray free electron lasers. *Proceedings of the National Academy of Sciences of the United States of America, 119,* e2117433119.

Kötting, D., & Gerwert, K. (2005). Monitoring protein-ligand interactions by time-resolved FTIR difference spectroscopy. *Methods in Molecular Biology, 305,* 261–286.

Liu, X., et al. (2022). Excited-state intermediates in a designer protein encoding a photo-trigger caught by an X-ray free-electron laser. *Nature Chemistry, 14,* 1054–1060.

Lórenz-Fonfría, V. A., Muders, V., Schlesinger, R., & Heberle, J. (2014). Changes in the hydrogen-bonding strength of internal water molecules and cysteine residues in the conductive state of channelrhodopsin-1. *The Journal of Chemical Physics, 141,* 22D507.

Lübben, M., & Gerwert, K. (1996). Redox FTIR difference spectroscopy using caged electrons reveals contributions of carboxyl groups to the catalytic mechanism of haem-copper oxidases. *FEBS Letters, 397,* 303–307.

Namiki, S., Arai, T., & Fujimori, K. (1997). High-performance caged nitric oxide: A new molecular design, synthesis, and photochemical reaction. *Journal of the American Chemical Society, 119,* 3840–3841.

Nango, E., et al. (2016). A three-dimensional movie of structural changes in bacter-iorhodopsin. *Science (New York, N. Y.), 354,* 1552–1557.

Nomura, T., et al. (2021). Short-lived intermediate in N_2O generation by P450 NO reductase captured by time-resolved IR spectroscopy and XFEL crystallography. *Proceedings of the National Academy of Sciences of the United States of America, 118,* e2101481118.

Norahan, M. J., et al. (2021). Microsecond-resolved infrared spectroscopy on nonrepetitive protein reactions by applying caged compounds and quantum cascade laser frequency combs. *Analytical Chemistry, 93,* 6779–6783.

Oda, K., et al. (2021). Time-resolved serial femtosecond crystallography reveals early structural changes in channelrhodopsin. *eLife, 10,* e62389.

Shimada, A., et al. (2017). A nanosecond time-resolved XFEL analysis of structural changes associated with CO release from cytochrome C oxidase. *Science Advances, 3,* e1603042.

Shiro, Y., et al. (1995). Spectroscopic and kinetic studies on reaction of cytochrome P450nor with nitric oxide. *The Journal of Biological Chemistry, 270*, 1617–1623.

Shoun, H., Fushinobu, S., Jiang, L., Kim, S. W., & Wakagi, T. (2012). Fungal denitrification and nitric oxide reductase cytochrome P450nor. *Philosophical Transactions of the Royal Society B: Biological Sciences, 367*, 1186–1194.

Spradlin, J., et al. (2016). Insights into an efficient light-driven hybrid P450 BM3 enzyme from crystallographic, spectroscopic and biochemical studies. *Biochimica et Biophysica Acta, 1864*, 1732–1738.

Tosha, T., et al. (2017). Capturing an initial intermediate during the P450nor enzymatic reaction using time-resolved XFEL crystallography and caged-substrate. *Nature Communications, 8*, 1585.

Yoshikawa, S., & Shimada, A. (2015). Reaction mechanism of cytochrome *c* oxidase. *Chemical Reviews, 115*, 1936–1989.

Yoshikawa, S., Choc, M. G., O'Toole, M. C., & Caughey, W. S. (1977). An infrared study of CO binding to heart cytochrome *c* oxidase and hemoglobin A. *The Journal of Biological Chemistry, 252*, 5498–5508.

CHAPTER SEVEN

Multiplexing methods in dynamic protein crystallography

Margaret A. Klureza[a], Yelyzaveta Pulnova[b], David von Stetten[c], Robin L. Owen[d], Godfrey S. Beddard[e,f], Arwen R. Pearson[a], and Briony A. Yorke[f,*]

[a]Institute for Nanostructure and Solid State Physics, University of Hamburg, HARBOR, Hamburg, Germany
[b]ELIbeamlines, Extreme Light Infrastructure, Dolni Brezany, Czechia
[c]European Molecular Biology, Laboratory (EMBL), Hamburg, Germany
[d]Diamond Light Source Ltd, Harwell Science and Innovation, Campus, Didcot, Oxfordshire, United Kingdom
[e]School of Chemistry, University of Edinburgh, David Brewster Road, United Kingdom
[f]School of Chemistry, University of Leeds, Woodhouse Lane, Leeds, United Kingdom
*Corresponding author. e-mail address: B.A.Yorke@leeds.ac.uk

Contents

1. Introduction	178
1.1 Laue and XFEL X-ray sources	179
2. Theory	180
2.1 Principles of multiplexing measurements	180
2.2 Error and the crystallographic multiplexing advantage	183
3. Experimental design	189
3.1 Combining HATRX with existing pump-probe techniques	190
3.2 S-matrix selection	192
3.3 Temporal encoding	193
3.4 Data collection	198
4. Data processing	199
5. Conclusion and outlook	201
Acknowledgments	201
References	202

Abstract

Time-resolved X-ray crystallography experiments were first performed in the 1980s, yet they remained a niche technique for decades. With the recent advent of X-ray free electron laser (XFEL) sources and serial crystallographic techniques, time-resolved crystallography has received renewed interest and has become more accessible to a wider user base. Despite this, time-resolved structures represent < 1 % of models deposited in the world-wide Protein Data Bank, indicating that the tools and techniques currently available require further development before such experiments can become truly routine. In this chapter, we demonstrate how applying data multiplexing to time-resolved crystallography can enhance the achievable time resolution

Methods in Enzymology, Volume 709
ISSN 0076-6879, https://doi.org/10.1016/bs.mie.2024.10.009
Copyright © 2024 Elsevier Inc. All rights reserved, including those for text and data mining, AI training, and similar technologies.

177

at moderately intense monochromatic X-ray sources, ranging from synchrotrons to bench-top sources. We discuss the principles of multiplexing, where this technique may be advantageous, potential pitfalls, and experimental design considerations.

1. Introduction

The dynamics of biological macromolecules are integral to their function; consequently, determining the nature of the molecular motions that occur during function is essential to unraveling underlying biological mechanisms. Such dynamics span a wide range of timescales. The fastest events are bond breaking/formation and side group rotations, which occur on femtosecond (fs) to picosecond (ps) timescales. These may be coupled to slower conformational changes, such as local rearrangements of the active site of an enzyme in response to substrate binding (nanoseconds to microseconds), allosteric binding (microseconds to seconds), or larger domain motions (milliseconds to seconds or even longer) (Levantino, Yorke, Monteiro, Cammarata, & Pearson, 2015). Not only are these individual processes important in their own right, but the interplay between events that occur across the full range of timescales also provides key insight into the structure-function relationship. X-ray crystallography is well established as a method to determine the static structure of biological macromolecules at atomic resolution. Since diffraction is an ensemble measurement averaged over the number of molecules in the crystal and the X-ray exposure time, τ_{exp}, it generally provides limited information about dynamics. However, when the dynamics are synchronized throughout all molecules within the X-ray illuminated volume, time-resolved crystallography can be used to capture the structures of transient species – but only if their lifetime is greater than the exposure time necessary to produce an interpretable diffraction image. The total number of photons diffracted by a crystal within a discrete exposure time is described by Darwin's formula (Darwin, 1914; Blundell & Johnson, 1976). Here, the intensity of a particular reflection is proportional to the number of photons incident on the crystal during the exposure, the properties of the crystalline lattice, the scattering atoms, and the volume of the crystal. Eq. 1 shows a simplified form of the formula (Holton & Frankel, 2010; Storm, Axford, & Owen, 2021), highlighting the relationship between the diffracted intensity I_{hkl} (photons), the incident X-ray intensity I_0 (photons), and the Bragg angle (θ).

$$I_{hkl} \propto I_0 \frac{(3 + \cos 4\theta)}{\sin \theta} \tag{1}$$

The incident intensity is the product of the X-ray flux, the area of the beam, and the exposure time. The intensity of diffraction decreases with increasing Bragg angle and the spatial resolution of X-ray diffraction data is proportional to the inverse of the exposure time. Consequently, the flux of the X-ray source imposes an inherent limit on the time resolution achievable while maintaining high spatial resolution. Monochromatic fourth generation synchrotron beamlines, with flux $\sim 10^{15}$ photons per second, require exposure times on the order of tens of microseconds (Grieco et al., 2024). Sub-microsecond time-resolved diffraction (TR-X) usually requires either polychromatic (Laue) synchrotron radiation or linear accelerator (LINAC) sources such as free electron lasers (FELs) (Caramello & Royant, 2024) or short pulse facilities (SPFs) (Jensen et al., 2021).

1.1 Laue and XFEL X-ray sources

Laue beamlines at synchrotrons have bandwidths of up to 5 %, resulting in photon fluxes that are 4–7 orders of magnitude greater than their monochromatic counterparts and enabling sub-nanosecond time resolution. However, Laue methods produce diffraction data that is more challenging to process than their monochromatic equivalents. Characterizing the exact wavelength of radiation contributing to a particular reflection requires normalization against an experimentally measured spectrum of the source. Additionally, Laue data show a moderate loss in resolution as compared to monochromatic diffraction on the same crystals. This may be attributed to increased background scattering, leading to a reduction in signal-to-noise ratio (Ren et al., 1999). The exposure time at Laue synchrotron and LINAC sources is only limited by the pulse duration of the instrument. Ultra-bright XFEL sources routinely produce 50–100 fs pulses with 10^{12} photons per pulse. Femtosecond time resolution allows sampling of the fastest nuclear motions and has the additional benefit that the diffraction may be produced before radiation damage can propagate through the crystal (Nass, 2019; Dickerson, McCubbin, & Garman, 2020). However, since each crystal is usually destroyed by a single pulse, data must be collected using a serial crystallographic approach. Serial data are collected from many thousands of randomly oriented crystals. This means that a different approach to indexing and integration is required. Since the crystal does not rotate during exposure, only partial reflections that contribute an unknown fraction of the complete intensity of a full reflection are recorded. The orientation of the crystals is random in all directions, and so images are not simply ordered according to a progressive rotation around a single axis, but instead must be individually indexed to determine their spatial relationship to each other. As a result, obtaining a complete dataset of sufficient quality requires on the order of 10,000 indexed diffraction images. While Laue and XFEL experiments are extremely powerful, limited access to beamlines and the complexity of data processing currently restrict mainstream use of these techniques. In this chapter, we discuss a multiplexing technique (Hadamard time-resolved crystallography, or HATRX) (Yorke, Beddard, Owen, & Pearson, 2014) that can access sub-microsecond time regimes using more widely available monochromatic synchrotron macromolecular beamlines. In essence, combining the diffraction signal from a series of discrete time points into a single image increases the intensities of the measured reflections in the image. After indexing and integration, the data is then mathematically transformed to return the original intensities for each time point. This results in processable intensities recorded from exposure times that would otherwise be only a fraction of those required to produce an interpretable diffraction pattern. This technique can be applied using both single-crystal and serial crystallographic approaches and is independent of the reaction initiation method used. Here, we introduce the principles of multiplexing, where this technique may be advantageous in improving temporal resolution, its limitations, experimental design considerations, and data processing.

2. Theory
2.1 Principles of multiplexing measurements

Data multiplexing refers to the simultaneous measurement of multiple signals prior to downstream deconvolution using a suitable mathematical transform. Multiplexing has been used in optical (James & Sternberg, 1969) and infrared spectroscopy (Golay, 1949; Jacquinot, 1954), NMR (Ernst & Anderson, 1966), and image processing (Oliver & Pike, 1974) techniques for decades. Perhaps the best known example of a multiplexing technique is Fourier transform infrared spectroscopy (FTIR) (Griffiths, 1983), in which a spectrum is produced from multiplexed frequencies using the Fourier transform. In these applications, a multiplexing approach is advantageous (in comparison to measuring a set of individual components), since the multiplexed signal will have a greater overall magnitude. While an equivalent number of measurements must still be made in order to eventually recover the true signal, multiplexing offers a means to exploit the increase in signal to improve the signal-to-noise ratio (SNR). In the case where errors are dominated by detector noise, the improvement is known as the Fellgett advantage (Fellgett, 1951; Hirschfeld, 1976) and results in greater sensitivity and resolution.

While multiplexing methods are well established, their application in time-resolved experiments is relatively new (Beddard, 2009; Yorke et al., 2014; Beddard & Yorke, 2016). Time-resolved crystallographic experiments are typically structured around a pump-probe scheme (Porter, 1950), where an excitation event (the "pump") initiates a chemical reaction or other structural perturbation at time $t = 0$. After a set time delay, Δt, the structure is probed at a discrete time point, t_i, by X-ray diffraction. The experiment is then repeated at different time delays $\Delta t_1, \Delta t_2, \ldots \Delta t_n$ until the whole reaction is mapped out in time. Such time-resolved experiments can be described by a measurement matrix, M, that represents the time, t, at which the probe pulse is measured with respect to the time of the pump, t_0. For a classical pump-probe experiment this is an $n \times n$ identity matrix, I_n, where n is the number of time points and repeats of the experiment (Fig. 1A). The measured signal at discrete time points, $f_{i.n}$, may then be represented as a signal vector, f:

$$\mathbf{f} = [f_1, f_2, \cdots f_{i=n}] \quad (2)$$

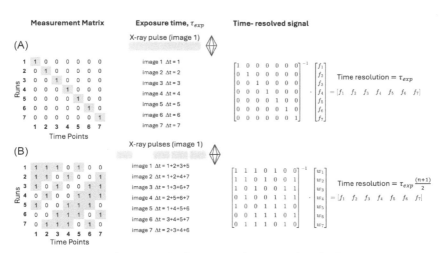

Fig. 1 Comparison of the standard Pump-probe (A) and multiplex experiments (B). The elements in the measurement matrix determine when the probe should be on = 1 and off = 0. For each run of the experiment the X-ray source is encoded with pulses according to the sequence in the respective row of measurement matrix. The signal from each pulse in a run is integrated onto the detector prior to transformation to return the individual time points.

In a time-resolved multiplexing experiment (Fig. 1B), the intensities at multiple discrete time points are summed or integrated on the detector during each run of the experiment. This means that the time resolution is equal to the total exposure time (τ_{exp}) divided by the number of measurements multiplexed during each run of the experiment. Every row of the measurement matrix is used to determine the sequence of time points to be measured in each diffraction image. The matrix is chosen such that it is an orthogonal matrix of 0's and 1's, and that each row, and each column of the matrix sums to $\frac{n+1}{2}$. If m_{ji} are the elements of the measurement matrix, then the measured signal w_j from one sequence is summed on the detector giving:

$$w_j = \sum_{i}^{n} f_i m_{ji}$$

(3)

The multiplexed intensities w_j are recorded in a multiplex domain signal vector, $\mathbf{w}$, of order n:

$$\mathbf{w} = \begin{bmatrix} w_1 \\ w_2 \\ \vdots \\ w_n \end{bmatrix}$$

(4)

The multiplexed data are then transformed to the time domain f by multiplying by the inverse of the measurement matrix $\mathbf{M}^{-1}$:

$$\mathbf{f} = \mathbf{w}\mathbf{M}^{-1}$$

(5)

A complete set of multiplexed diffraction data is collected by temporally encoding the probe and integrating the signal according to each row of the measurement matrix. As a result, the total number of diffraction images for the multiplexed experiment is the same as the standard pump-probe dataset. However, the multiplexed data differ from a set of individual time-resolved measurements in two important ways:

1. Each measured multiplexed intensity will be larger than the measured intensities for the individual components. This increases the flux-limited time resolution by a factor of $\frac{n+1}{2}$ total exposure is the sum of multiple individual time points recorded in each run of the experiment.

2. Since each multiplexed signal contains data from multiple time points, and each time point is measured multiple times the transformed time-resolved signals are an average of all measurements across the n datasets.

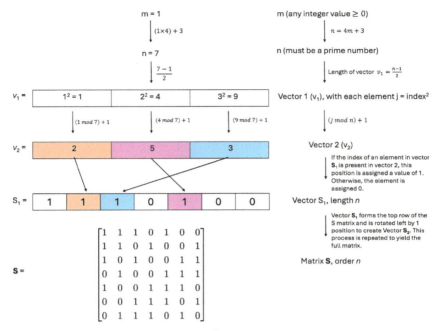

Fig. 2 Quadratic residues construction of a 7 × 7 **S**-matrix.

Hadamard time-resolved X-ray crystallography (HATRX) (Yorke et al., 2014) employs Hadamard **S**-matrices as the measurement matrix for the experimental encoding, as these provide an optimal sampling scheme when multiplexing on/off measurements (Drnovšek, 2013; Harwit & Sloane, 1979). There are numerous methods for constructing Hadamard and **S**-matrices that are described in detail elsewhere (Sloane, 1982; Seberry & Yamada, 2020). The matrices used throughout this chapter were constructed using the quadratic residues method shown in Fig. 2. In this construction, the order of the matrix, n, is a prime number which satisfies the equality $n = 4m + 3$, where m is any integer ≥ 0.

2.2 Error and the crystallographic multiplexing advantage

Structural determination of conformational changes in protein structures using time-resolved X-ray crystallography relies on the accurate measurement of small changes in the diffracted intensity over time. The key motivation for multiplexing in time-resolved crystallography is the increase in the number of measurements of equivalent reflections and the increase in recorded diffraction intensity that is achieved since the diffraction integrated by the detector is greater when the exposure time is longer or diffraction is integrated over a number of time points. The increase in intensity is exploited by performing indexing, integration, scaling and merging prior to transformation from the multiplex to time domain. In a crystallographic experiment the accuracy of the electron density map calculated from the set of intensities is affected by errors inherent in the measurement process; the indexing, integration and scaling procedures; sample imperfections (including non-isomorphism); and the phases used in the Fourier transform. The fundamental limit on the precision of the measured intensities is shot noise, a quantum effect that results in random fluctuations in the incident X-ray intensity. Shot noise follows a Poisson distribution, where the variance in the number of counts in any given reflection $varI = \langle I \rangle$ and the standard deviation, σ_p, is the square root of the mean intensity $\langle I \rangle$:

$$\sigma_P = \sqrt{\langle I \rangle} \tag{6}$$

Although counting errors due to shot noise increase with exposure time, they do so as the square·root of the signal and so the SNR, $\langle I \rangle / \sigma_p$, increases.

2.2.1 Experimental error .

Indexing algorithms (Gevorkov et al., 2019; Beyerlein et al., 2017; Ginn et al., 2016; Battye, Kontogiannis, Johnson, Powell, & Leslie, 2011; Kabsch, 2010; Duisenberg, 1992) rely on identifying the location of intense reflections to calculate the lattice basis vectors for accurate indexing and prediction of the location of all potential reflections prior to integration. Reflections are only identified during peak finding if the intensity is much greater than the background noise. Generally, the threshold of peak detection requires that the intensity is at least five times larger than the background noise for it to be reliably identified as a signal (Rose, 1948). The estimated positions and profiles of reflections are used during integration where the intensities are calculated using either summation of the pixel values within the predicted reflection area (strong reflections) or profile fitting (weak reflections). At this stage the background is calculated from pixels surrounding the reflection and subtracted from the intensity. The main source of the background noise present in diffraction images recorded using photon counting pixel array detectors is diffuse scattering, which can be classified as either incoherent or variational (Glover, Harris, Helliwell, & Moss, 1991). Incoherent scattering includes inelastic

(Compton) scattering from the sample itself, air between the sample and detector, buffer surrounding the crystal and the mounting material. Scattering from the internal solvent within the sample is not entirely incoherent but also contributes to the background, peaking in intensity at approximately 3.3 Å (Thüne & Badger, 1995). Variational scattering from intramolecular thermal atomic motions (protein scattering) and intermolecular lattice dynamics (phonon scattering) (Meisburger, Case, & Ando, 2023) within the crystal contributes to the background signal between and underneath Bragg peaks. Protein scattering peaks at around 3.8 Å and reduces the apparent intensity of measured reflections. Although variational diffusive scatter contains information about protein dynamics, in crystallographic data reduction this background signal adds to the overall error in Bragg peak intensity determination. The standard deviation of the background (σ_B) is the square root of the average background intensity under each reflection, $\langle I_B \rangle$.

$$\sigma_B = \sqrt{\langle I_B \rangle} \tag{7}$$

The integrated intensity is calculated from the pixels within the reflection minus the mean of the pixels in a defined region surrounding the reflection (White et al., 2013):

$$I = I_P - I_B \tag{8}$$

and the standard deviation of the intensity of a reflection after background subtraction is found by propagating the shot noise, errors in the intensity (σ_P^2), background (σ_B^2), and the number of pixels with the reflection integration area (N):

$$\sigma_{total} = \sqrt{\sigma_P^2 + N\sigma_B^2} \tag{9}$$

Absorption of X-rays by the crystal, buffer and mounting material systematically decreases the intensity of all reflections. While the relative errors associated with random noise decrease with increasing intensity, systematic errors result in an overall fractional reduction in the precision of the measured intensities for example, an increase in X-ray absorption of 5% would result in a 5% decrease in the intensity of all reflections (Holton, Classen, Frankel, & Tainer, 2014; Diederichs, 2016). A pervasive and unavoidable source of systematic error in X-ray crystallography is radiation damage due to absorption and inelastic scattering of X-rays. Radiation damage changes the structure of protein molecules, ultimately disrupting the crystalline lattice leading to an overall loss in diffracted intensity. It is especially relevant for time-resolved crystallography experiments, which by their nature cannot be performed at 100 K. HATRX multiplexing allows extremely low-dose structures to be calculated for radiation sensitive samples, such as those containing metal ions. A 10 μm protein crystal exposed for 1 ms on a beamline with flux 2×10^{12} photons per second in the focused beam can absorb an average diffraction weighted dose of approximately 3 kGy, sufficient to initiate the formation of photoelectrons (Dickerson, McCubbin, Brooks-Bartlett, & Garman, 2024). Using the HATRX method, the first 'time point' or more precisely dose point dataset can be calculated from data with a corresponding dose of $\frac{n+1}{2}$ of the total (where n is the dimension of the **S**-matrix). This may be particularly useful to fully characterize the effects of radiation damage to radiation sensitive samples at room temperature prior to time-resolved studies.

2.2.2 Instrument error

Further systematic errors are introduced by instrumental factors, including fluctuations in the X-ray intensity, shutter timing, incorrect para-meterization of the experiment geometry, movement of the sample due to vibration or thermal convection, and errors added by the detector. The time profile of X-ray intensity during each exposure is an important consideration since fluctuations in the incident intensity can obscure time-resolved changes in the diffracted signal. The time profile of synchrotron radiation is determined by the distribution of electron bunches around the storage ring, known as the fill pattern or operation mode. For time-resolved experiments, the length of a single pulse and the time between pulses is determined by the fill pattern. For Laue time-resolved experi-ments, the synchrotron is generally operated in hybrid mode, where the X-ray probe pulse originates from a single bunch (with duration ~50–200 ps) surrounded by more densely populated fill pattern. The single pulse is isolated from the remainder of the fill pattern using a rotating shutter (chopper). Other time-resolved operation modes, where a smaller number of electron bunches are evenly spaced around the ring, are also possible. In a standard operation mode, the bunch structure produces pulses with picosecond duration separated by nanosecond gaps (Schroer et al., 2022; Filling Modes - esrf.fr, n.d.; Imai & Hatsui, 2024; Storage Ring Operation Modes - ops.aps.anl.gov, n.d.). For monochromatic pump-probe experiments with time resolutions on the order of milliseconds the beam profile is considered to be continuous (at least on average) in standard operation mode. However, as the time resolution approaches the temporal spacing between electron bunches, synchronization between the electron bunches, shutters, sample positioning and detector becomes increasingly important. This is especially true for the HATRX experiment because fluctuations in intensity within a single run of the experiment or over multiple datasets results in fractional errors in the transformed time-resolved intensities. Increasing the exposure time by multiplexing can reduce the impact of random error,s but does not improve systematic relative errors; these are instead handled during scaling. During merging and scaling the intensities of multiple measurements of equivalent reflections are averaged and an estimate of the variance of each reflection, σ^2 is given by:

$$\sigma^2 = \sigma^2_{counting} + KI_{hkl}^2 \tag{10}$$

where K is a scale factor calculated by least squares optimization (Hamilton, Rollett, & Sparks, 1965) so that the observed spread of intensities matches their variance or by a Bayesian inference approach in which the Wilson distribution is used as the prior probability distribution (Dalton, Greisman, & Hekstra, 2022). Increasing the number of measurements for each equivalent reflection improves the stability of the scaling procedure, the error estimation and ultimately the quality of the data. The various models and corrections applied during data reduction mean that the data used to calculate electron density maps are not directly extracted from the pixel counts recorded in diffraction images. Estimating the effect of multiplexing on the SNR of the time-resolved transformed intensities, and hence potential multiplexing advantage, is therefore far more complicated for crystallographic experiments than the spectroscopic experiments for which Hadamard S-matrix multiplexing was first proposed (Sloane, Fine, Phillips, & Harwit, 1969).

2.2.3 Detector noise

Random errors can also be introduced by the detector due to dark noise, which is the electronic signal generated by thermal fluctuations even when no radiation is incident on the detector. Although cooling can reduce dark noise, it is typically corrected for after data collection. Additional electronic noise may be produced during the readout process. This readout noise occurs when the charge generated by a detected photon is amplified to produce a measurable voltage change and is an important factor to consider when using charge integrating detectors. The optimality of Hadamard S-matrices for experimental multiplexing is proven for the case where random noise added by the detector dominates the uncertainty in a measurement (Cheng, 1987; Drnovšek, 2013). In this case the multiplexed signal vector, w, is the product of the encoding matrix, $\mathbf{S}$, with the time domain signal, f, plus the random noise added by the detector, ϵ:

$$w = \mathbf{S}f + \epsilon \tag{11}$$

and the time domain signal is:

$$f = \mathbf{S}^{-1}(w + \epsilon) \tag{12}$$

Since the signal accumulates onto the detector, there is some resultant averaging on the noise. The best estimate of the data, $\hat{f}$, is the data itself, and so $\hat{f} - f = \mathbf{S}^{-1}\epsilon$. When the terms in this expression are expanded out, the mean square error (mse) can be calculated (Beddard & Yorke, 2016). If the noise is random and added onto the signal by the detector after multiplexing (detector noise) there may be an increase in the SNR for the HATRX experiment compared to taking the arithmetic mean of n individual measurements. For random and uncorrelated noise $\epsilon_k \epsilon_k = \sigma^2$, $\epsilon_j \epsilon_k = 0$ and the mean square error of the Hadamard transformed data is calculated according to the method described in (Beddard & Yorke, 2016):

$$mse = \frac{4\sigma^2 n}{(n + 1)^2} \tag{13}$$

In comparison, the traditional pump-probe data would be calculated as the mean of n repeated individual measurements with an mse of σ^2. The multiplexing advantage, Q, is then defined as the ratio of the mean square errors for the standard and multiplexing approaches:

$$Q = \frac{n + 1}{2\sqrt{n}} \tag{14}$$

Multiplexing may, therefore, provide an optimal SNR advantage for the transformed time-resolved data when images are recorded using charge-integrating detectors, which add noise to the signal due to the dark current and readout noise (Nitzsche & Riesenberg, 2003). Conversely, detector noise is largely absent when using photon counting detectors (Donath et al., 2023) in this case, background scattering, instrument and photon counting errors dominate.

Noise due to fluctuations in the source, η, are added to the signal before multiplexing and so the measured signal is then:

$$\mathbf{f} = \mathbf{S}^{-1}(\mathbf{w} + \eta) + \epsilon \tag{15}$$

If the source noise is also uncorrelated, $\eta_k \eta_k = \sigma'^2$ and $\eta_j \eta_k = 0$, then the mean square error is calculated from $\hat{f} - f = \eta + \mathbf{S}^{-1}\epsilon$ as

$$mse = \sigma'^2 + \frac{4\sigma^2 n}{(n+1)^2} \tag{16}$$

If the source noise is much larger than detector noise, then $Q \approx 1$ and there is no formal Felgett advantage in the transformed data. If the noise is correlated (e.g. shot noise) then $Q \approx 1/\sqrt{2}$ and there is a decrease in the SNR of the transformed data. However, this decrease rests on the assumption that it is possible repeat the measurement exactly at each value of Δt. In the case of HATRX, the time-resolved multiplexing advantage arises due to the increase in the relative intensity of reflections in the multiplex domain used during data reduction.

3. Experimental design

As described above, HATRX offers a means to increase the temporal resolution attainable at a monochromatic synchrotron beamline by multiplexing the diffraction signal over different timepoints. This approach can be coupled with myriad existing time-resolved pump-probe methods; as such, it provides a large degree of flexibility. However, making effective use of HATRX does introduce some additional experimental design considerations. In this section, we describe a workflow for designing and carrying out a HATRX experiment that will provide insight into a particular reaction or dynamic process within a biological macromolecule (Fig. 3): assessing crystalline properties, incorporating the particular pump-probe technique, selecting an appropriate S-matrix, encoding the temporal probe sequence within the X-ray beam, and finally collecting data. We will then discuss how best to process the resulting HATRX data, including the necessary transformation, in Section 5.

Any crystallographic experiment is necessarily constrained by certain properties of the sample: how many crystals are readily attainable and how large do they grow, to what resolution do they diffract, how well do they tolerate radiation, and so forth (Schulz, Yorke, Pearson, & Mehrabi, 2022). For a HATRX experiment, it is important to begin by evaluating these properties (including via preliminary diffraction data collection if needed), as

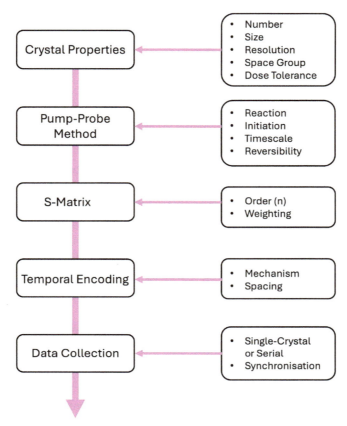

Fig. 3 Essential stages and decisions in the experimental design for a multiplexing approach to time-resolved crystallography.

the method provides enough flexibility to accommodate various approaches based on the system of interest. For instance, either a single-crystal or serial data collection strategy may be used for a HATRX experiment. Additionally, prior characterization of the reaction kinetics in the crystal, or at minimum the crystallization conditions, and its (ir)reversibility is essential for developing the time-resolved approach. This information will then guide the experimental design with regards to the selection of the **S**-matrix and the implementation of multiplexing through temporal encoding (Fig. 4).

3.1 Combining HATRX with existing pump-probe techniques

As described in Section 3, the temporal encoding of a HATRX experiment occurs by gating the X-ray beam or via gating the detector

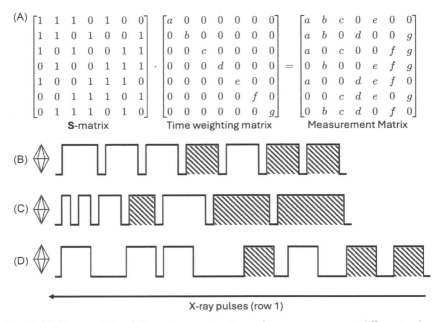

Fig. 4 (A) Time weighted **S**-matrix construction, where a–g represent different values of τ_{exp}. (B) X-ray pulses temporally encoded according to the first row of a standard **S**-matrix ($n = 7$) where "on" is represented by an open pulse and "off" is shaded. (C) X-ray pulses temporally encoded according to a time weighted **S**-matrix where τ_{exp} increases from a–h. (D) X-ray pulses temporally encoded according to the first row of a standard **S**-matrix with varying spacing between elements.

during acquisition of an image where the beam itself cannot be encoded with sufficient temporal resolution) – in other words, the probe of a pump-probe experiment. Consequently, the Hadamard modality is independent of any particular approach to data collection (single-crystal or serial) or reaction initiation (the pump) and so is compatible with a wide range of existing time-resolved techniques, including mixing (Mehrabi, Schulz, Agthe, et al., 2019; Schmidt, 2020; Olmos et al., 2018; Stubbs et al., 2024), light-induced reactions (whether inherently photoreactive (Caramello & Royant, 2024) or photocaged (Monteiro, Amoah, Rogers, & Pearson, 2021; Mehrabi, Schulz, Dsouza, et al., 2019)), temperature changes (Wolff et al., 2023), and electric field stimulation (Hekstra et al., 2016). Selecting among these is therefore a matter of identifying the reaction or phenomenon of interest, its timescale, its reversibility, and how it can be induced. Attention to the timescale is particularly key in guiding the selection of an **S**-matrix to encode relevant timepoints, while the reaction's induction mechanism and/or reversibility may inform the data collection strategy.

3.2 S-matrix selection

Selecting a suitable **S**-matrix with which to multiplex a HATRX experiment requires balancing several competing considerations, including not only the desired time resolution and overall experimental timescale, but also the mechanism for implementing the probe sequence within an X-ray beam, the sample consumption dictated by the choice of a serial or single-crystal data collection strategy, and the crystal's inherent radiation sensitivity. The most effective way to increase the time resolution would appear to be the use of a high-order matrix; for example, a Hadamard **S**-matrix of order $n = 103$ would theoretically increase the time resolution by a factor of 52 (i.e. $\frac{n+1}{2}$). However, since the experiment must be repeated for each row of the encoding matrix (representing a unique sequence of multiplexed time points), this would require a total of 103 complete datasets. This approach introduces several challenges, including increases in both the time needed for data collection and the computational expense of matrix inversion and processing such large volumes of images. In the context of serial crystallography, with possibly more than 10,000 (micro)crystals per dataset, using such a high order matrix would ultimately require over a million crystals. For a single-crystal experiment the requirement to collect each row of the matrix, over at least a wedge of reciprocal space, would result in the accumulation of dose far beyond the Garman limit (Owen, Rudiño-Piñera, & Garman, 2006). Clearly, then, increasing the order of the **S**-matrix indefinitely is not a viable way to achieve ever-higher time resolution via HATRX.

Utilizing an S-matrix of a more moderate order (e.g., $n = 19$) to achieve an improvement of around 1 order of magnitude is more realistic, and in many cases will provide sufficient resolution to determine transient states previously inaccessible by standard monochromatic serial crystallographic techniques. However, as biological macromolecular dynamics span such a broad range, there are still processes that would require even finer time resolution. Under such circumstances, it is possible to satisfy these competing demands and maximize the overall temporal resolution by selecting an **S**-matrix of moderate order and then dividing the exposure time, τ_{exp}, into intervals of varying length. Three different possible timing schemes are shown in Fig. 4, including the standard **S**-matrix scheme (Fig. 4B), a time weighted scheme (Fig. 4C) and a variable interval scheme (Fig. 4D) Time weighting constitutes multiplication of the Hadamard **S**-matrix by a time weighted diagonal matrix. The resulting measurement matrix maintains the "on"/"off" structure (Fig. 4A), but each "on" element is a number corresponding to the length of the time interval (Fig. 4C). In strict mathematical terms, the inclusion of entries other than 1 s and 0 s would mean that the encoding matrix no longer meets the defining criteria of a Hadamard **S**-matrix. However, since the timing weighted **S**-matrix remains invertible the multiplexed signal can still be transformed to deconvolve the time-resolved diffraction signal. The time resolution then varies along the reaction coordinate according to the values in the diagonal weighting matrix which can be chosen to reflect the timescales of various processes along a reaction pathway. Similarly, the timing of "off" elements can be modulated to ensure each element is appropriately placed along the reaction coordinate, as long the timing scheme is conserved throughout each run of the experiment (Fig. 4D).

3.3 Temporal encoding

The most significant technical challenge when performing a HATRX experiment is encoding the pulse sequence that defines the temporal multiplexing. The temporal profile of the pulse train and systematic errors in the rising and falling edge of each pulse introduce fractional errors in the measured intensities. Random or systematic errors in the synchronisation between the pump and probe pulses (phase jitter and frequency drift) are also important considerations; however, these are negligible if τ_{exp} is much longer than the period of the jitter or drift. The fill pattern is an important consideration when designing the HATRX experiment because the spacing of bunches will determine the minimum time between measurements and the best way to encode the Hadamard sequence. The maximum number of electron bunches that may be stored in the ring and the minimum spacing between them is determined by the operating frequency of the machine and the circumference of the storage ring. The distribution of the electron bunches determines the lifetime of the beam and is limited by 'electron evaporation' known as the Touschek effect (Carmignani, 2015); this is minimized when the total current is spread over a large number of bunches. Additionally, the density of the electron bunches determines the likelihood of ion trapping, an effect caused by electrons colliding with residual gas molecules in the ring. Ion trapping also reduces the lifetime of the beam and is minimized by introducing large gaps between bunches. The brilliance of the beam, however, is maximized by filling the maximum number of bunches. A compromise between brilliance and lifetime is made and the majority or even all of the ring is filled with evenly spaced bunches. For slower time resolutions (longer than 1 μs), in standard operation mode the average intensity is measured and so a dense fill pattern with nanosecond spacing between bunches is desirable.

The Hadamard encoding can be achieved by either temporally blocking or deflecting the X-ray beam or by electronically gating the detector. The Hadamard transform enhances errors caused by noise on the probe, and so the rising and falling edge of pulses in the sequence must be minimized or at least well characterized and corrected for. Gating the detector rather than using a chopper avoids problems arising from significant rise and fall times, frequency drift and phase jitter in a mechanical shutter. In an ideal multiplexing experiment, the total intensity of each pulse should be recorded and time-stamped, both to monitor the accuracy of the timings and so that the time-resolved intensities can be scaled according to the true incident intensity. The experimental timing should be set by the source clock, to ensure that the pump, experimental shutter and detector are synchronized with the temporal structure of the source. There are numerous methods to encode the X-ray beam, all of which have various advantages and disadvantages that must be taken into account during the early stages of experimental design.

3.3.1 Shutters

For synchrotron time-resolved experiments, both the length of a single pulse and the time between pulses are determined by the fill mode. In hybrid mode, the X-ray probe pulse originates from a single bunch with duration of approximately 50–200 ps. The single pulse is then isolated from the remainder of the fill pattern using a rotating shutter (chopper). Such chopper systems are phase locked to the synchrotron clock (Meents et al., 2009; Cammarata et al., 2009), which results in excellent phase stability with long term jitter < 5 ns. Air bearings allow operation at 500 Hz with the chopper disc rotating at 30,000 rpm. The actual pulse length is a convolution of the rotation speed, X-ray beam size, and slit size, such that a 50 μm slit chopping a 50 μm beam at 30,000 rpm can produce 350 ns pulses. Modification of such a chopper, so as to encode a standard **S**-matrix, is unfortunately challenging due to imbalance around the circumference leading to instability. Additionally, different chopper blades must be manufactured if alterations to the **S**-matrix are required. However, since **S**-matrices are cyclical, it is possible to encode each run of an $n \times n$ experiment with a single chopper blade, as the phase of the blade with

Multiplexing methods in dynamic protein crystallography 195

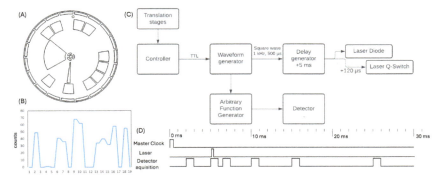

Fig. 5 Detector gated HATRX. (A) Chopper blade design (SciTech 300CD) tested at the PXS beamline to encode **S**-matrix pulse sequence (n = 19). The outer openings control active frequency stabilization, a laser diode is aligned to the small opening at the first position, when open the signal produced by an Si photodiode (Thorlabs SM1PD1B) triggers the detector, a delay generator is used to iterate through the sequences by delaying the start of data acquisition. Inner openings balance the blade to ensure stability. (Mai et al., 2023). (B) Encoded X-ray pulses measured using a pin diode X-ray detector (xPIN). (C) Schematic showing the temporal control of a laser initiated HATRX experiment performed at beamline I24 (Diamond Light Source). A similar configuration has also been implemented at the T-REXX end station (EMBL@PetraIII (von Stetten et al., 2019)). A shutterless fixed target serial crystallographic approach was used and the sample translation controller (Geobrick LV-IMS-II) acted as the master trigger. After sample positioning, the stages send a signal to the ZEBRA control system (Cobb, Chernousko, Uzun, & Light, 2013) which produces a TTL signal to arm the detector and trigger a waveform generator (Keysight 33500B) generating a square wave (50 Ohm, 1 kHz, 500 μs, 5 V). The wave was in turn used to trigger both the arbitrary function generator (Tektronix AFG1022) controlling detecting gating (Pilatus 6 M), and a delay generator (Standford research systems, DG535) controlling synchronisation of the laser trigger to the laser diode and Q-switch (Standa Q1TH, pulse length 900 ns, max power 35 μJ) (D) Example timing diagram for experimental configuration shown in (C).

reference to the synchrotron clock needs only to be shifted by one position when changing the run of the experiment. This approach is more suitable when using a laser driven plasma X-ray source (PXS) (Schick et al., 2012), such as the PXS beamline at ELI Beamlines (Mai, Pulnova, Nejdl, & Bulanov, 2023). A PXS is a pulsed source (1 KHz) which uses a high intensity (10^{16} W cm^{-2}) infrared laser focused on a moving target to produce pulses of radiation at a wavelength determined by the characteristic Kα energy of the target material. A chopper blade (Fig. 5A) with openings determined by a Hadamard **S**-matrix can encode the laser so that X-ray pulses are produced according to the chopper encoded sequence.

Encoding the laser pulses is far less challenging than directly encoding the X-ray beam; however, the PXS source is limited by the frequency of the laser and the flux and stability of the X-ray pulses produced.

An $n = 19$ chopper blade tested at ELI Beamlines was able to encode the X-ray pulse structure, although significant shot-to-shot variation resulted in the pulse at 0.04–0.05 s being effectively absent (Fig. 5B).

3.3.2 Bragg switches

Bragg switches are adaptive X-ray optics capable of temporal modulation of the path of an X-ray beam and perform a similar function to the acousto-optic modulators and digital mirror devices used in Hadamard spectroscopy (Genchi, Bucci, Laptenok, Giammona, & Liberale, 2021) and imaging techniques (Zhang, Wang, Zheng, & Zhong, 2017). Bragg switching results in transient modulation of the diffraction angle of X-rays incident on the surface of a Bragg mirror due to lattice distortions induced by a surface acoustic wave (SAW) (Tucoulou et al., 1997; Vadilonga et al., 2017). The lattice distortions act as a diffraction grating, which can be switched on and off by controlling the propagation of the SAW through the illuminated volume of the crystal. SAW Bragg switches controlled by radio frequency signal generation can temporally encode hard X-ray pulses for HATRX experiments on timescales ranging from 100 ns to 100 µs (Schmidt et al., 2024). Sub-picosecond pulses can be generated by exploiting the photo-acoustic effect which occurs when a Bragg crystal is excited by ultrafast optical pulses (Bucksbaum & Merlin, 1999; Loether et al., 2014; Sander et al., 2016). Excitation below the bandgap of the switching material results in the generation of coherent optical phonons, which induce periodic lattice distortions while avoiding electronic excitation. When the lattice distortions produce a superlattice, they displace the incident X-ray wave vector (K) diffraction according to the reciprocal lattice vector of the crystal (**G**) and the periodicity of the standing acoustic wave (q) (Bucksbaum & Merlin, 1999).

$$\Delta K = \mathbf{G} + q \tag{17}$$

SAW Bragg switches are currently reported to have a switching efficiency of 30% (Sander et al., 2016), and the maximum reported efficiency of a photoacoustic switch is 8% (Jarnac et al., 2017). Significant improvements in efficiency will therefore be required for a Bragg switch HATRX encoding, as currently the reduction in the intensity incident on the sample would counteract any gain in time resolution obtained by multiplexing.

3.3.3 Pseudo-single bunch kick and cancel

The Hadamard sequences could be introduced into the storage ring itself using, for example, a pseudo-single bunch kick and cancel (PSB-KAC) operational mode (Holldack et al., 2014; Hertlein et al., 2015; Sun, Robin, Steier, & Portmann, 2015). In this technique, electron bunches are selectively 'kicked' into a different orbit (known as the camshaft bunch) from the multi-bunch train using fast kicker magnets. The radiation from a camshaft bunch can then be selectively transmitted by a collimator in the experimental hutch. This method would have the added advantage of not requiring the synchrotron to run in a special time-resolved hybrid filling mode. Encoding the X-ray beam with a single bunch per time point would allow ~100 picosecond pulses with nanosecond spacing, and multiple bunches could be used per pulse. The main factor preventing the general use of this technique is the requirement for additional kicker magnets in the storage ring. However, multiple light sources have adopted this technology including the Advanced Light Source (Sun et al., 2015) and MAXIV (Olsson & Andersson, 2021), suggesting that it may become an increasingly accessible means to implement HATRX experiments in the future.

3.3.4 Detector gating

Finally, a simple and flexible approach to HATRX probe encoding is to use a (quasi-)continuous X-ray beam and instead only integrate the diffraction signal during discrete time intervals. This can be achieved by gating the X-ray detector prior to read-out of the integrated image. Shutterless data collection (Mueller, Wang, & Schulze-Briese, 2012) does not rely on the mechanical constraints of X-ray shuttering and therefore avoids problems arising from frequency drift and phase jitter. Photon counting detectors can be gated extremely rapidly; for example, Eiger2 detectors may achieve gating time resolutions finer than 60 ns (Donath et al., 2023). Consequently, the only additional equipment required on beamlines with these detectors are suitable arbitrary function and delay generators (Fig. 5C) The arbitrary function generator (AFG) is programmed with a waveform representing the Hadamard **S**-matrix pulse sequence which controls the acquisition timepoints on the detector. Delays between the acquisition and initiation trigger allow flexibility in the timing of initiation with respect to the data collection. Detector gating can be used to encode weighted **S**-matrix pulse sequences (Fig. 4) and to introduce variable delays between pulses (Figs. 4D and 5D). The major disadvantage of encoding the pulse sequences by detector gating is that the crystal is exposed to the X-ray beam even when data is not being recorded, which accounts for almost 50% of the pulse sequence acquisition time. It is therefore important to consider the dose limit of the system of interest and determine whether encoding the X-ray beam would be more appropriate.

3.4 Data collection

After implementing a suitable encoding method for the X-ray probe, the practical aspects of HATRX data collection do not significantly differ from any other time-resolved crystallographic experiment. From an experimental design standpoint, however, the temporal encoding of the probe and sample/dose considerations should be considered when selecting an appropriate the data collection strategy (i.e., single-crystal or serial) and the experimental synchronization.

3.4.1 Single-crystal and serial data collection strategies

HATRX experiments are compatible with both single-crystal and serial data collection and data from single-crystal experiments can be recorded as stills or with oscillation. The only constraint on these various approaches is the need to collect the full set of multiplexed images (e.g., seven images for a 7×7 encoding matrix) from the same crystal orientation and volume. Whereas traditional serial crystallographic experiments might collect only a single diffraction image from any one crystal, HATRX multiplexing requires that each sample remain stationary in the beam for long enough to collect multiple exposures - one for each row of the encoding **S**-matrix. This means that fixed targets are the most appropriate sample delivery system for serial HATRX experiments (Owen et al., 2023). For a single-crystal oscillation experiment, the set of multiplexed images must be collected over the same oscillation range, necessitating rotation back to the start angle after each image is recorded. Real time hit detection (Barty et al., 2014; Winter et al., 2022) gives an indication of how many repeats of each run of the experiment are required, but this does not confirm the completeness of each data set. Indexing and integration (White, 2019; Winter et al., 2022) should therefore be performed on-the-fly for both serial and single-crystal experiments. As with any time-resolved crystallographic experiment, the reversibility of the process of interest, the radiation sensitivity and the crystallization procedure all determine whether a single-crystal or serial data collection strategy would be most appropriate. Reversible reactions are compatible with either single-crystal or serial strategies, whereas irreversible reactions require a serial approach.

3.4.2 Synchronization

A general approach to synchronizing the sample position, reaction initiation, and detector can be used regardless of the precise methods used for encoding, initiation, and data collection. Standard beamline control software can be used to control the goniometer (single-crystal) or sample stages (serial), set the total exposure time, start data acquisition, and read out the detector. The data acquisition must be synchronized with the pulse sequence, as delays between the acquisition and initiation trigger allow flexibility in the timing of initiation with respect to the data collection. It is also advantageous to integrate signal from one time point prior to reaction initiation, so that the 'ground state' is recorded from the same samples as the various time points used for calculating time-resolved Fo-Fo isomorphous difference maps.

4. Data processing

While Hadamard encoding provides the potential to achieve a higher temporal resolution than would otherwise be possible at a monochromatic synchrotron beamline, it also introduces new challenges in processing and analysis of the resulting data, as the inverse Hadamard transform from multiplexed to time-resolved intensities must be included. Standard data reduction methods are used to produce a list of reflections with their associated Miller indices (H,K,L) merged intensities (I) and estimated errors (σI) for each run of the experiment (Kabsch, 2010; Battye et al., 2011; White, 2019; Winter et al., 2022). The inverse transform is then performed on scaled and merged intensities on a per-reflection basis. For each set of Miller indices, the corresponding intensity from every HATRX run dataset is used to construct a multiplexed signal column vector. Left multiplication of the signal vector by the inverse of the Hadamard matrix yields a vector corresponding to the intensity values for that reflection at every timepoint (Fig. 1B). The matrix multiplication is only possible when the order of the column vector is equal to the order of the measurement matrix. Thus any reflections not present across all datasets must be removed prior to Hadamard transformation. The intensities of each multiplexed dataset must also be on the same scale to ensure successful transformation and return time-resolved intensities. One approach is to determine an average multiplex scaling model by deriving a set of scale factors from the combined raw data across every run of the experiment. The multiplexed scaling model is then applied to each of the HATRX datasets individually during data reduction prior to the transform step. This approach is advantageous as it ensures that the datasets for all the Hadamard rows are on the same scale, as opposed to scaling each dataset separately during data reduction and then applying scale factors determined by cross-comparison of the merged multiplexed datasets afterwards.

Non-isomorphism represents a significant challenge when applying a multiplexing approach to time-resolved crystallography, as it can mask small but functionally significant changes in intensity. Furthermore, the isomorphism of a sample can change over the course of a reaction. Increases in the disorder of the crystalline lattice, the population of various intermediate states, and changes in the unit cell parameters can have a profound impact on the quality of multiplexed and transformed data. Analysis of the data quality both pre- and post-transform is therefore necessary. Standard data reduction statistics, including R_{split}, $\langle I/\sigma I \rangle$, and $CC_{1/2}$, each provide useful insight into the overall quality of the multiplexed data. Since the different multiplexed datasets represent the signal from different combinations of time points, comparison of the statistics across each set provides some indication about changes in data quality over the time series. Similarly, comparison of the mosaic spread of each set can indicate whether the crystal has rotated or translated to a position with a different thickness during the time taken to record an image. Such motion would preclude accurate measurement of changes in intensity arising from the process of interest over the timescale of the experiment. Splitting each multiplexed set into batches and independently processing each one allows assessment of data quality by calculation of Pearson correlation coefficients of the intensities before (CC_{Had}) and after (CC_{Time}) the data are transformed. Finally, it is important to assess the quality of the resulting electron density map and ensure that it does not simply reflect the phasing model. While any crystallographic electron density should be checked for model bias, this is a particularly crucial step in the context of a HATRX experiment, where demultiplexing the time-resolved signal inherently requires some additional mathematical manipulation of the integrated intensities. Given this, we recommend the use of the "von Stetten test" implemented on the T-REXX end station on beamline P14 at PETRAIII (Von Stetten et al., 2019), in which an individual well-ordered tyrosine (or other structurally distinctive) residue from the phasing model is removed mutated to a glycine residue (effectively deleting the side chain). Use of this mutated model for phasing, followed by a few rounds of restricted refinement, enables verification of the electron density map for positive difference density around the side chain of the underlying tyrosine residue. If this is indeed present, it indicates that the overall electron density map is not simply an artifact of the phasing model, but rather has predictive power and indicates that experimental signal with a reasonable SNR is present even after the inverse Hadamard transform has been performed.

5. Conclusion and outlook

The advent of fourth generation synchrotron sources offers the potential to investigate a wide range of dynamic processes relevant to biological function using X-ray diffraction. HATRX can provide a way to extend the time resolution of a variety of time-resolved synchrotron experiments without requiring the higher flux of a Laue or XFEL source. Additionally, multiplexing may also be used to determine low-dose structures that would otherwise be inaccessible using standard synchrotron techniques. As with all time-resolved experiments, careful characterization of the system of interest and planning of the experimental design are essential. Certain factors, including non-isomorphism and changes in the crystal orientation during data collection, could result in misleading results, and so we recommend collecting data with considerably higher multiplicity than would be strictly necessary for standard structural determination. Future development of encoding methods (including alternative X-ray sources such as PXS), optimization of weighted measurement matrices, and improved techniques to analyze multiplexed and transformed data will all serve to further expand the applicability of this technique for a wider user base.

Acknowledgments

We are grateful to Helen Ginn and Graeme Winter for helpful discussions regarding data processing and analysis. We thank Diamond Light Source, ELI beamlines and EMBL@ PetraIII for beamtime. ARP acknowledges funding from the Clusters of Excellence "CUI: Advanced Imaging of Matter" of the Deutsche Forschungsgemeinschaft (DFG) - EXC 2056 - project ID 390715994 and "The Hamburg Centre for Ultrafast Imaging" - EXC 1074 - project ID 194651731. Bundesministerium für Bildung und Forschung (Verbundforschungsprojekte 05K16GU1, 05K19GU1 & 05K22GU6). MAK acknowledges support from a Helmholtz Information and Data Science Academy (HIDA) short-term fellowship as well as a graduate fellowship from the NSF-Simons Center for Mathematical and Statistical Analysis of Biology at Harvard (award number #1764269).

References

Barty, A., Kirian, R. A., Maia, F. R., Hantke, M., Yoon, C. H., White, T. A., & Chapman, H. (2014). Cheetah: Software for high-throughput reduction and analysis of serial femtosecond X-ray diffraction data. *Journal of Applied Crystallography, 47*(3), 1118–1131.

Battye, T. G. G., Kontogiannis, L., Johnson, O., Powell, H. R., & Leslie, A. G. (2011). Imosflm: A new graphical interface for diffraction-image processing with mosflm. *Acta Crystallographica Section D: Biological Crystallography, 67*(4), 271–281.

Beddard, G. (2009). *Applying maths in the chemical and biomolecular sciences: An example-based approach.* USA: Oxford University Press.

Beddard, G. S., & Yorke, B. A. (2016). Pump–probe spectroscopy using the hadamard transform. *Applied Spectroscopy, 70*(8), 1292–1299.

Beyerlein, K. R., White, T. A., Yefanov, O., Gati, C., Kazantsev, I. G., Nielsen, N.-G., ... Schmidt, S. (2017). Felix: An algorithm for indexing multiple crystallites in X-ray free-electron laser snapshot diffraction images. *Journal of Applied Crystallography, 50*(4), 1075–1083.

Blundell, T. L., & Johnson, L. N. (1976). *Protein crystallography.* London: Academic Press.

Bucksbaum, P., & Merlin, R. (1999). The phonon bragg switch: A proposal to generate sub-picosecond X-ray pulses. *Solid State Communications, 111*(10), 535–539.

Cammarata, M., Eybert, L., Ewald, F., Reichenbach, W., Wulff, M., Anfinrud, P., ... Polachowski, S. (2009). Chopper system for time resolved experiments with synchrotron radiation. *Review of Scientific Instruments, 80*(1).

Caramello, N., & Royant, A. (2024). From femtoseconds to minutes: Time-resolved macromolecular crystallography at XFELs and synchrotrons. *Acta Crystallographica Section D: Structural Biology, 80*(2).

Carmignani, N. (2015). *Touschek lifetime studies and optimization of the european synchrotron radiation facility: Present and upgrade lattice. In Springer Thesis.* Springer Nature.

Cheng, C.-S. (1987). An application of the kiefer-wolfowitz equivalence theorem to a problem in hadamard transform optics. *The Annals of Statistics,* 1593–1603.

Cobb, T., Chernousko, Y., Uzun, I., & Light, D. (2013). Zebra: A flexible solution for controlling scanning experiments. *Proceedings of Icalepcs,* 739–763.

Dalton, K. M., Greisman, J. B., & Hekstra, D. R. (2022). A unifying Bayesian framework for merging X-ray diffraction data. *Nature Communications, 13*(1), 7764.

Darwin, C. G. (1914). XXXIV. The theory of X-ray reflexion. *The London, Edinburgh, and Dublin Philosophical Magazine and Journal of Science, 27*(158), 315–333.

Dickerson, J. L., McCubbin, P. T., & Garman, E. F. (2020). Raddose-XFEL: Femtosecond time-resolved dose estimates for macromolecular x-ray free-electron laser experiments. *Journal of Applied Crystallography, 53*(2), 549–560.

Dickerson, J. L., McCubbin, P. T., Brooks-Bartlett, J. C., & Garman, E. F. (2024). Doses for X-ray and electron diffraction: New features in raddose-3d including intensity decay models. *Protein Science, 33*(7), e5005.

Diederichs, K. (2016). Crystallographic data and model quality. *Nucleic Acid Crystallography: Methods and Protocols,* 147–173.

Donath, T., Šišak Jung, D., Burian, M., Radicci, V., Zambon, P., Fitch, A. N., ... Schulz-Briese, C. (2023). Eiger2 hybrid-photon-counting x-ray detectors for advanced synchrotron diffraction experiments. *Journal of Synchrotron Radiation, 30*(4).

Drnovšek, R. (2013). On the s-matrix conjecture. *Linear Algebra and Its Applications, 439*(11), 3555–3560.

Duisenberg, A. J. (1992). Indexing in single-crystal diffractometry with an obstinate list of reflections. *Journal of Applied Crystallography, 25*(2), 92–96.

Ernst, R. R., & Anderson, W. A. (1966). Application of fourier transform spectroscopy to magnetic resonance. *Review of Scientific Instruments, 37*(1), 93–102.

Fellgett, P. B. (1951). *Theory of infrared sensitivities and its application to investigations of stellar radiation in the near infrared* (Ph.D. thesis).

Filling Modes - esrf.fr. (n.d.). https://www.esrf.fr/Accelerators/Operation/Modes. Accessed 19-07-2024.

Genchi, L., Bucci, A., Laptenok, S. P., Giammona, A., & Liberale, C. (2021). Hadamard-transform spectral acquisition with an acousto-optic tunable filter in a broadband stimulated raman scattering microscope. *Optics Express, 29*(2), 2378–2386.

Gevorkov, Y., Yefanov, O., Barty, A., White, T. A., Mariani, V., Brehm, W., ... Chapman, H. N. (2019). Xgandalf–extended gradient descent algorithm for lattice finding. *Acta Crystallographica Section A: Foundations and Advances, 75*(5), 694–704.

Ginn, H. M., Roedig, P., Kuo, A., Evans, G., Sauter, N. K., Ernst, O. P., ... Stuart, D. I. (2016). Taketwo: An indexing algorithm suited to still images with known crystal parameters. *Acta Crystallographica Section D: Structural Biology, 72*(8), 956–965.

Glover, I., Harris, G., Helliwell, J., & Moss, D. (1991). The variety of X-ray diffuse scattering from macromolecular crystals and its respective components. *Acta Crystallographica Section B: Structural Science, 47*(6), 960–968.

Golay, M. J. (1949). Multi-slit spectrometry. *JOSA, 39*(6), 437–444.

Grieco, A., Boneta, S., Gavira, J. A., Pey, A. L., Basu, S., Orlans, J., ... Martin-Garcia, J. M. (2024). Structural dynamics and functional cooperativity of human nqo1 by ambient temperature serial crystallography and simulations. *Protein Science, 33*(4), e4957.

Griffiths, P. R. (1983). Fourier transform infrared spectrometry. *Science (New York, N. Y.), 222*(4621), 297–302.

Hamilton, W., Rollett, J. T., & Sparks, R. (1965). On the relative scaling of X-ray photographs. *Acta Crystallographica, 18*(1), 129–130.

Harwit, M., & Sloane, N. J. (1979). *Hadamard transform optics.* New York: Academic Press.

Hekstra, D. R., White, K. I., Socolich, M. A., Henning, R. W., Srajer, V., & Ranganathan, R. (2016). Electric-field-stimulated protein mechanics. *Nature, 540*(7633), 400–405. https://doi.org/10.1038/nature20571.

Hertlein, M., Scholl, A., Cordones, A., Lee, J., Engelhorn, K., Glover, T., ... Robin, D.S. (2015). X-rays only when you want them: Optimized pump–probe experiments using pseudo-single-bunch operation. *Journal of Synchrotron Radiation, 22*(3), 729–735.

Hirschfeld, T. (1976). Fellgett's advantage in UV–vis multiplex spectroscopy. *Applied Spectroscopy, 30*(1), 68–69.

Holldack, K., Ovsyannikov, R., Kuske, P., Müller, R., Schälicke, A., Scheer, M., ... Föhlisch, A. (2014). Single bunch X-ray pulses on demand from a multi-bunch synchrotron radiation source. *Nature Communications, 5*(1), 4010.

Holton, J. M., Classen, S., Frankel, K. A., & Tainer, J. A. (2014). The R-factor gap in macromolecular crystallography: An untapped potential for insights on accurate structures. *The FEBS Journal, 281*(18), 4046–4060.

Holton, J. M., & Frankel, K. A. (2010). The minimum crystal size needed for a complete diffraction data set. *Acta Crystallographica Section D: Biological Crystallography, 66*(4), 393–408.

Imai, Y., & Hatsui, T. (2024). Quantifying bunch-mode influence on photon-counting detectors at spring-8. *Journal of Synchrotron Radiation, 31*(2).

Jacquinot, P. (1954). The luminosity of spectrometers with prisms, gratings, or Fabry-Pérot etalons. *JOSA, 44*(10), 761–765.

James, J. F., & Sternberg, R. S. (1969). *The design of optical spectrometers.* London: Chapman and Hall.

Jarnac, A., Wang, X., Bengtsson, A., Ekstrom, J., Enquist, H., Jurgilaitis, A., ... Larsson, J. (2017). Communication: Demonstration of a 20 ps X-ray switch based on a photo-acoustic transducer. *Structural Dynamics, 4*(5).

Jensen, M., Ahlberg Gagnér, V., Cabello Sánchez, J., Bengtsson, A., Ekstrom, J. C., Björg Úlfarsdóttir, T., ... Katona, G. (2021). High-resolution macromolecular crystallography at the femtomax beamline with time-over-threshold photon detection. *Journal of Synchrotron Radiation, 28*(1), 64–70.

Kabsch, W. (2010). XDS. *Acta Crystallographica Section D: Biological Crystallography, 66*(2), 125–132.

Levantino, M., Yorke, B. A., Monteiro, D. C., Cammarata, M., & Pearson, A. R. (2015). Using synchrotrons and XFELS for time-resolved X-ray crystallography and solution scattering experiments on biomolecules. *Current Opinion in Structural Biology, 35*, 41–48.

Loether, A., Gao, Y., Chen, Z., DeCamp, M., Dufresne, E., Walko, D., & Wen, H. (2014). Transient crystalline superlattice generated by a photoacoustic transducer. *Structural Dynamics, 1*(2).

Mai, D.-D., Pulnova, Y., Nejdl, J., & Bulanov, S. (2023). Laser driven Cu-tape-based PXS for pump-probe spectroscopy and imaging. *Compact Radiation Sources from EUV to Gamma-Rays: Development and Applications*, PC125820I.

Meents, A., Reime, B., Kaiser, M., Wang, X.-Y., Abela, R., Weckert, E., & Schulze-Briese, C. (2009). A fast X-ray chopper for single-bunch extraction at synchrotron sources. *Journal of Applied Crystallography, 42*(5), 901–905.

Mehrabi, P., Schulz, E. C., Agthe, M., Horrell, S., Bourenkov, G., Von Stetten, D., ... Miller, R. J. D. (2019). Liquid application method for time-resolved analyses by serial synchrotron crystallography. *Nature Methods, 16*(10), 979–982.

Mehrabi, P., Schulz, E. C., Dsouza, R., Mu¨ller-Werkmeister, H. M., Tellkamp, F., Dwayne Miller, R. J., & Pai, E. F. (2019). Time-resolved crystallography reveals allosteric communication aligned with molecular breathing. *Science, 365*(6458), 1167–1170.

Meisburger, S. P., Case, D. A., & Ando, N. (2023). Robust total X-ray scattering workflow to study correlated motion of proteins in crystals. *Nature Communications, 14*(1), 1228.

Monteiro, D. C. F., Amoah, E., Rogers, C., & Pearson, A. R. (2021). Using photocaging for fast time-resolved structural biology studies. *Acta Crystallographica Section D: Structural Biology, 77*(10), 1218–1232.

Mueller, M., Wang, M., & Schulze-Briese, C. (2012). Optimal fine φ-slicing for single-photon-counting pixel detectors. *Acta Crystallographica Section D: Biological Crystallography, 68*(1), 42–56.

Nass, K. (2019). Radiation damage in protein crystallography at X-ray free-electron lasers. *Acta Crystallographica Section D: Structural Biology, 75*(2), 211–218.

Nitzsche, G., & Riesenberg, R. (2003). Noise, fluctuation, and HADAMARD-transform spectrometry. *Proceedings of SPIE, 5111*, 273–282. https://doi.org/10.1117/12.510052.

Oliver, C., & Pike, E. (1974). Multiplex advantage in the detection of optical images in the photon noise limit. *Applied Optics, 13*(1), 158–161.

Olmos, J. L., Pandey, S., Martin-Garcia, J. M., Calvey, G., Katz, A., Knoska, J., ... others, Schmidt, M. (2018). Enzyme intermediates captured "on the fly" by mix-and-inject serial crystallography. *BMC Biology, 16*, 1–15.

Olsson, D. K., & Andersson, A. (2021). Studies on transverse resonance island buckets in third and fourth generation synchrotron light sources. *Nuclear Instruments and Methods in Physics Research Section A: Accelerators, Spectrometers, Detectors and Associated Equipment, 1017*, 165802.

Owen, R. L., De Sanctis, D., Pearson, A. R., & Beale, J. H. (2023). A standard descriptor for fixed-target serial crystallography. *Acta Crystallographica Section D: Structural Biology, 79*(8), 668–672.

Owen, R. L., Rudiño-Piñera, E., & Garman, E. F. (2006). Experimental determination of the radiation dose limit for cryocooled protein crystals. *Proceedings of the National Academy of Sciences, 103*(13), 4912–4917.

Porter, G.-N. (1950). Flash photolysis and spectroscopy. A new method for the study of free radical reactions. *Proceedings of the Royal Society of London. Series A: Mathematical and Physical Sciences, 200*(1061), 284–300.

Ren, Z., Bourgeois, D., Helliwell, J. R., Moffat, K., Srajer, V., & Stoddard, B. L. (1999). Laue crystallography: Coming of age. *Journal of Synchrotron Radiation, 6*, 891–917.

Rose, A. (1948). In L. Marton (Vol. Ed.), *Advances in electronics and electron physics: Vol. 1*. New York: Academic Press.

Sander, M., Koc, A., Kwamen, C., Michaels, H., Pudell, J., Zamponi, F., ... Gaal, P. (2016). Characterization of an ultrafast bragg-switch for shortening hard X-ray pulses. *Journal of Applied Physics, 120*(19).

Schick, D., Bojahr, A., Herzog, M., Schmising, C., Shayduk, R., Leitenberger, W., ... Bargheer, M. (2012). Normalization schemes for ultrafast X-ray diffraction using a table-top laser-driven plasma source. *Review of Scientific Instruments, 83*(2).

Schmidt, M. (2020). Reaction initiation in enzyme crystals by diffusion of substrate. *Crystals, 10*(2), 116.

Schmidt, D., Hensel, D., Petev, M., Khosla, M., Brede, M., Vadilonga, S., & Gaal, P. (2024). Wavegate: A versatile tool for temporal shaping of synchrotron beams. *Optics Express, 32*(5), 7473–7483.

Schroer, C. G., Wille, H.-C., Seeck, O. H., Bagschik, K., Schulte-Schrepping, H., Tischer, M., ... Weckert, E. (2022). The synchrotron radiation source PETRA III and its future ultra-low-emittance upgrade PETRA IV. *The European Physical Journal Plus, 137*(12), 1312.

Schulz, E. C., Yorke, B. A., Pearson, A. R., & Mehrabi, P. (2022). Best practices for time-resolved serial synchrotron crystallography. *Acta Crystallographica Section D: Structural Biology, 78*(1), 14–29.

Seberry, J., & Yamada, M. (2020). *Hadamard matrices: Constructions using number theory and linear algebra*. John Wiley & Sons.

Sloane, N. J. A. (1982). *Hadamard and other discrete transforms in spectroscopy. Fourier, hadamard, and hilbert transforms in chemistry*. Springer, 45–67.

Sloane, N., Fine, T., Phillips, P., & Harwit, M. (1969). Codes for multiplex spectrometry. *Applied Optics, 8*(10), 2103–2106.

Storage Ring Operation Modes. ops.aps.anl.gov. (n.d.). https://ops.aps.anl.gov/SRparameters/node5.html. Accessed 19-07-2024.

Storm, S. L., Axford, D., & Owen, R. L. (2021). Experimental evidence for the benefits of higher X-ray energies for macromolecular crystallography. *IUCrJ, 8*(6), 896–904.

Stubbs, J., Hornsey, T., Hanrahan, N., Esteban, L. B., Bolton, R., Malý, M., ... West, J. (2024). Droplet microfluidics for time-resolved serial crystallography. *IUCrJ, 11*(2), 237–248.

Sun, C., Robin, D., Steier, C., & Portmann, G. (2015). Characterization of pseudo-single bunch kick and cancel operational mode. *Physical Review Special Topics—Accelerators and Beams, 18*(12), 120702.

Thüne, T., & Badger, J. (1995). Thermal diffuse X-ray scattering and its contribution to understanding protein dynamics. *Progress in Biophysics and Molecular Biology, 63*(3), 251–276.

Tucoulou, R., Roshchupkin, D., Schelokov, I., Brunel, M., Ortega, L., Ziegler, E., ... Douillet, S. (1997). High frequency electro-acoustic chopper for synchrotron radiation. *Nuclear Instruments and Methods in Physics Research Section B: Beam Interactions with Materials and Atoms, 132*(1), 207–213.

Vadilonga, S., Zizak, I., Roshchupkin, D., Petsiuk, A., Dolbnya, I., Sawhney, K., & Erko, A. (2017). Pulse picker for synchrotron radiation driven by a surface acoustic wave. *Optics Letters, 42*(10), 1915–1918.

Von Stetten, D., Agthe, M., Bourenkov, G., Polikarpov, M., Horrell, S., Yorke, B., ... Schneider, T.R. (2019). TREXX: A new endstation for serial time-resolved crystallography at PETRA III. *Acta Crystallographica Section A: Foundations and Advances, 75*, e26.

White, T. A. (2019). Processing serial crystallography data with crystfel: A step-by-step guide. *Acta Crystallographica Section D: Structural Biology, 75*(2), 219–233.

White, T. A., Barty, A., Stellato, F., Holton, J. M., Kirian, R. A., Zatsepin, N. A., & Chapman, H. N. (2013). Crystallographic data processing for free-electron laser sources. *Acta Crystallographica Section D: Biological Crystallography, 69*(7), 1231–1240.

Winter, G., Beilsten-Edmands, J., Devenish, N., Gerstel, M., Gildea, R. J., McDonagh, D., ... Evans, G. (2022). DIALS as a toolkit. *Protein Science, 31*(1), 232–250. https://doi.org/10.1002/PRO.4224.

Wolff, A. M., Nango, E., Young, I. D., Brewster, A. S., Kubo, M., Nomura, T., ... Thompson, M.C. (2023). Mapping protein dynamics at high spatial resolution with temperature-jump X-ray crystallography. *Nature Chemistry, 15*(11), 1549–1558.

Yorke, B. A., Beddard, G. S., Owen, R. L., & Pearson, A. R. (2014). Time-resolved crystallography using the Hadamard transform. *Nature Methods, 11*(11), 1131–1134. https://doi.org/10.1038/nmeth.3139.

Zhang, Z., Wang, X., Zheng, G., & Zhong, J. (2017). Hadamard single-pixel imaging versus Fourier single-pixel imaging. *Optics Express, 25*(16), 19619–19639.

CHAPTER EIGHT

Processing serial synchrotron crystallography diffraction data with DIALS

James Beilsten-Edmands[a,*,1], James M. Parkhurst[a,b,1],
Graeme Winter[a], and Gwyndaf Evans[a,b]

[a]Diamond Light Source Ltd, Harwell Science and Innovation Campus, Didcot, Oxfordshire, United Kingdom
[b]Rosalind Franklin Institute, Harwell Science and Innovation Campus, Didcot, Oxfordshire, United Kingdom
*Corresponding author. e-mail address: james.beilsten-edmands@diamond.ac.uk

Contents

1. Introduction	208
2. Methods	211
2.1 Model parameterisation using a multivariate-normal distribution	212
2.2 General impact of the RLP distribution	214
2.3 Parameter estimation via maximum likelihood methods	215
2.4 Integration and partiality estimation	217
3. DIALS command-line tools for processing SSX data	218
3.1 Indexing with dials.ssx_index	219
3.2 Integration with dials.ssx_integrate	219
3.3 Data reduction with dials.cosym and dials.scale	220
3.4 Associated tools from cctbx.xfel	221
4. Xia2.ssx: an automated processing pipeline	221
4.1 Integration and reduction of SSX data with xia2.ssx	222
4.2 Standalone data reduction with xia2.ssx_reduce	225
4.3 Merging in groups based on metadata	225
5. Demonstration and evaluation on example data	227
5.1 Processing anomalous SSX data: AcNiR above the K-edge	227
5.2 Processing SSX dose-series: photoreduction of Fe(III) FutA	233
6. Discussion	239
Acknowledgments	240
Appendix A. Block matrix inversion	240
Appendix B. Derivatives of the log-likelihood and Fisher Information matrix	241
References	242

Abstract

This chapter describes additions to the DIALS software package for processing serial still-shot crystallographic data, and the implementation of a pipeline, *xia2.ssx*, for

---

[1] The two authors contributed equally to this work.

Methods in Enzymology, Volume 709
ISSN 0076-6879, https://doi.org/10.1016/bs.mie.2024.10.004
Copyright © 2024 Elsevier Inc. All rights are reserved, including those for text and data mining, AI training, and similar technologies.

processing and merging serial crystallography data using DIALS programs. To integrate partial still-shot diffraction data, a 3D gaussian profile model was developed that can describe anisotropic spot shapes. This model is optimised by maximum likelihood methods using the pixel-intensity distributions of strong diffraction spots, enabling simultaneous refinement of the profile model and Ewald-sphere offsets. We demonstrate the processing of an example SSX dataset where the improved partiality estimates lead to better model statistics compared with post-refined isotropic models. We also demonstrate some of the workflows available for merging SSX data, including processing time/dose resolved data series, where data can be separated at the point of merging after scaling and discuss the program outputs used to investigate the data throughout the pipeline.

1. Introduction

Macromolecular X-ray crystallography has been a very successful technique for determining biological molecular structures, contributing to over 180,000 entries in the protein data bank (PDB) to date. The typical practice of determining structures at cryogenic temperatures, while justified in order to reduce radiation damage and enable complete data collection from a single crystal by rotation, is less comparable to native biological conditions than data collection at room temperature. This is of particular importance for the study of dynamics in structural biology by crystallography, a field that has had a renaissance over the past decade, enabled by several experimental developments in X-ray crystallography (Orville, 2018). This includes the development of sample delivery methods such as fixed-targets, liquid and viscous jets to rapidly present thousands of microcrystals for measurement (Grünbein & Nass Kovacs, 2019), in combination with the development of time-resolved serial femtosecond crystallography at X-ray free-electron lasers (XFELs) (Pandey et al., 2020), and synchrotron advancements including increased photon flux density and the development of micro-focus beamlines (Axford et al., 2012). The use of both XFEL and synchrotron facilities therefore enables biological processes to be investigated on a wide range of timescales at room temperature, down to the femtosecond timescale at XFELs or typically on the milli- or micro-second timescale at synchrotrons. At XFELs, the impact of radiation damage is mitigated by the extremely fast timescales and high X-ray intensities, which produce strong diffraction on a single image before sample destruction (Neutze et al., 2000) or site-specific radiation damage, if a suitably short pulse is used (Hough & Owen, 2021). At synchrotrons, collecting only a few images (or a single still image) from any single crystal limits the radiation dose, which is calculable and controllable. In both cases, a complete dataset of sufficient data quality can only be achieved by combining data from hundreds or thousands of crystals.

From a data processing perspective, these experimental developments present new challenges. These techniques are capable of measuring diffraction from over hundreds of crystals a second, however the number of images that need to be collected cannot be specified ahead of the experiment due to a number of factors. Firstly, not every image will contain diffraction from a crystal, due to the nature of the rapid sample delivery techniques. Secondly, the overall merged data quality (resolution, completeness, internal consistency) from a large number of crystals will determine whether biological insight can be attained, therefore enough data must be collected, but over-collecting will reduce the number of different experiments that can be performed in the allocated experiment time. This means that fast feedback is critical to a successful experiment and making efficient use of available beamtime; in its simplest form this can be live hit-finding analysis to give an indication of the number of crystals exposed, but as each step of data analysis provides more information, ideally the complete merged data would be returned as soon as possible (i.e. within a matter of minutes). Examples of software that provide online data analysis at XFELs include Cheetah (Barty et al., 2014), OnDA (Mariani et al., 2016), IOTA (Lyubimov et al., 2016), which themselves run analysis algorithms contained in packages such as CrystFEL (White et al., 2012), DIALS (Winter et al., 2018), cctbx.xfel.merge (Brewster et al., 2019a) and XDS (Kabsch, 2010b). In addition, live analysis may be performed as part of specialised hardware-accelerated processing, for example in the Jungfraujoch system (Leonarski et al., 2023).

A second challenge is that processing still-shot diffraction data presents algorithmic challenges compared with rotation data. For rotation data, each reflection is rotated through the diffraction condition, so the measured intensity is the integral of the complete diffraction spot and the profile (shape) of the spot can be empirically determined for suitably fine-sliced data. For still-shot data, each diffraction pattern only measures a slice through the volume of the diffraction spots. In addition, most reflections will not be measured at the exact diffraction condition for the ideal lattice, therefore the diffraction profile must be estimated to model these offsets in order to determine an accurate partiality estimate (sometimes referred to as the Ewald offset correction) that puts the measured intensities on the same scale within an image. One method to address this for serial crystallography is the post-refinement method, where following integration of reflection intensities on the images, the crystal parameters such as unit cell, orientation and mosaic spread, which together determine the partiality, are refined alongside the scaling corrections to give the most internally consistent merged intensities (Uervirojnangkoorn et al., 2015). However, post-refinement approaches are constrained by the fidelity of the models employed, which typically apply isotropic models of mosaic block size and spread (Kabsch, 2014; Sauter et al., 2014), plus a polychromatic beam-model when modelling XFEL data (White, 2019; Ginn et al., 2016). Work by Mendez et al. (2020) showed how a full experiment model can be refined against individual pixel data using maximum likelihood methods, leading to improved data quality but relying on significant computational resources that would be impractical for processing during experimental data collection. Recent work by Brehm et al. (2023) describe a more general gaussian model for describing anisotropic peak shapes that leads to improved data quality, a similar approach to the integration algorithm that we describe here, first described in Parkhurst (2020). Another algorithmic challenge when combining thousands of measurements is the indexing ambiguity problem for polar space groups, whereby all images must be consistently indexed prior to scaling and merging, a particular challenge in serial crystallography due to the sparsity of common reflections between still images. Algorithms for indexing ambiguity resolution of rotation data have been described for rotation data (Brehm & Diederichs, 2014; Gildea & Winter, 2018) and have implementations in XDS, DIALS and CrystFEL.

In this work we describe and evaluate recent additions and developments in the DIALS software package for the processing of still-shot serial synchrotron crystallography (SSX) data. This includes a description of an enhanced profile modelling algorithm for integration of still images, using a 3D gaussian description to model anisotropic spot profiles and reflection offsets based on pixel intensity distributions (Parkhurst, 2020) now implemented in DIALS for a monochromatic beam model. Using SSX data collected on beamline I24 at Diamond Light Source, we show how this leads to improved partiality estimates compared with post-refined isotropic partiality models, with improved structure quality for a given number of crystals, or requiring significantly fewer crystals to reach a given overall R-factor. In addition, we describe an automated processing pipeline for serial data processing, *xia2.ssx*, running stepwise processes from DIALS that enables full processing from images to merged data, fulfilling an equivalent role to the *xia2* (Winter, 2010) pipeline for processing crystallographic rotation data. The *xia2.ssx* pipeline includes additional workflows required for serial crystallography, such as separation of data by experimental criteria, for example in dose or times series experiments, as well as quick merging of integrated data during data accumulation in an SSX experiment. Given the diverse nature of serial crystallography experiments, there is no single workflow for all experiments, and therefore the focus here is on demonstrating the features that are available for processing and the program outputs that can be used for evaluating the data processing. These developments in DIALS and *xia2* provide a consistent data processing experience at synchrotrons from rotation to serial crystallography, lowering the barrier to entry for experimentalists experienced in rotation crystallography processing methods.

2. Methods

Crystalline diffraction produces observed diffraction spots that have a spatial spread on the detector surface, since a crystal does not form a perfect infinite lattice and the X-ray beam has a finite spatial divergence and wavelength distribution. In the reciprocal space formalism, there is a distribution of intensity about the ideal reciprocal lattice points (RLPs), while the diffraction condition as described by the Ewald sphere is broadened by the divergence and wavelength distribution of the beam.

During a rotation experiment, the entire reflection is rotated through the diffracting condition, such that the intensity weighted centre of mass of the observed reflection on the detector image corresponds to the predicted position of the reflection, for a suitably well modelled crystal. For still-shot diffraction data, the intensity distribution of an observed spot on the detector is the result of the convolution of the Ewald Sphere distribution with the broadened reciprocal lattice point. As such, the weighted centre of mass of the observed reflection, when mapped into reciprocal space, will not lie on the centre of the reciprocal lattice point, with the deviation depending on the shape of the reciprocal lattice points and the orientation of the crystal, thus varying from reflection to reflection. To correct for this effect, a per-reflection scale factor is required, which has been termed an Ewald offset correction or partiality in previous literature (Kabsch, 2014; White et al., 2013). The term 'partiality' will be used in this work as a shorthand for this geometry dependent scale factor, however this is distinct from the definition of partiality of a reflection in a rotation experiment, as the full reflection intensity is not measured in a still-shot image.

Within DIALS, the existing stills profile model consists of two components (Sauter et al., 2014); the mosaic block size which results in spherical broadening of RLPs, and the angular spread of mosaic blocks, which results in a 'spherical cap' around the reciprocal lattice vector (Nave, 1998). For these spherical (isotropic) models, transformations can be applied to determine the predicted centres of mass of the spots on the detector from the models (Sauter et al., 2014).

In this section, we describe an approach using a multivariate normal distribution (MVN) to describe anisotropic RLPs. For non–isotropic RLPs, the centre of mass offset is highly dependent on the shape of the profile. The use of a MVN description combined with a maximum likelihood optimisation leads to mathematical simplifications and allows simultaneous modelling of the Ewald offsets and RLP shapes with a per-image model of unit cell, crystal orientation and anisotropic mosaic spread.

2.1 Model parameterisation using a multivariate-normal distribution

The distribution of RLPs in reciprocal space, can be modelled as a multivariate normal distribution convolved with the reciprocal lattice and described by a 3D covariance matrix $\mathbf{M}$. A valid covariance matrix is required to be positive semi-definitive, which is enforced by using a Cholesky decomposition, i.e. $\mathbf{M} = \mathbf{L}\mathbf{L}^*$, where $\mathbf{L}$ is a lower triangular matrix with positive diagonal elements. Therefore, the RLP covariance matrix is fully described by 6 parameters which are the non–zero elements of $\mathbf{L}$, where

$$\mathbf{L} = \begin{pmatrix} m_1 & 0 & 0 \\ m_2 & m_3 & 0 \\ m_4 & m_5 & m_6 \end{pmatrix} \tag{1}$$

In the reciprocal-space coordinate frame, this describes an arbitrarily orientated ellipsoidal distribution. Spherical spot shapes can be modelled with a constrained form of $\mathbf{L}$ with $m_1 = m_3 = m_6 > 0$, $m_2 = m_4 = m_5 = 0$. In these models, which will be referred to as $\mathbf{M_E}$ (ellipsoidal) and $\mathbf{M_S}$ (spherical), all RLPs have an identical covariance matrix. Fig. 1A shows example RLP distributions for an ellipsoidal covariance matrix.

To model an angular mosaic spread, a radial dependence must be introduced to the per-reflection RLP distribution to describe RLPs that subtend a constant angle with the reciprocal space origin, which can model

Processing serial synchrotron crystallography diffraction data with DIALS

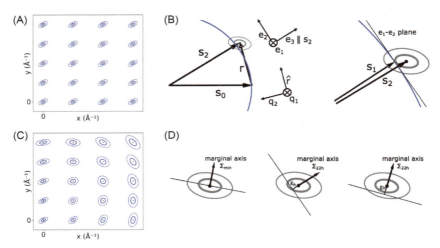

Fig. 1 (A) Example reciprocal lattice point (RLP) distributions described by a reflection independent ellipsoidal covariance matrix $\mathbf{M_E}$, plotted for contours of fixed probability density. (B) Schematic of vector and coordinate frame definitions. $\mathbf{s_2}$ is the vector to the RLP centre and $\mathbf{r} = \mathbf{s_2} - \mathbf{s_0}$. In general, $|\mathbf{s_2}| \neq |\mathbf{s_0}|$ and the Ewald sphere does not intersect at the RLP centre. The $\mathbf{e}$ vectors define the orthonormal local reflection-specific coordinate frame, which is used to separate the multivariate gaussian into a conditional distribution in the $\mathbf{e_1}$-$\mathbf{e_2}$ plane and a marginal distribution along $\mathbf{e_3} \parallel \mathbf{s_2}$. The vectors $\mathbf{q_1}$, $\mathbf{q_2}$ and $\hat{\mathbf{r}}$ form an orthonormal set that is used to rotate an angular component of mosaic spread into the reciprocal space coordinate frame for a given reflection, with the angular component in the $\mathbf{q_1}$-$\mathbf{q_2}$ plane. $\mathbf{s_1}$, the observed centre of mass of the diffraction spot, is approximated by transforming the mean of the conditional distribution to the laboratory frame (see text). (C) Example RLP distributions described by covariance matrices $\mathbf{M^h}$ adding an angular mosaic spread to $\mathbf{M_E}$, plotted for contours of fixed probability density (relative to the probability density at the centre of each RLP). (D) Illustration of quantities used in calculation of the partiality of a spot, which depends on the distance from the Ewald sphere, ϵ, along the marginal axis, and on the variance along this axis Σ_{22h}. The highest integrated probability density occurs in the first case, for an intersection at the RLP centre perpendicular to the direction of minimum variance (Σ_{min}).

an angular distribution of mosaic domain orientations. First, a decomposition matrix $\mathbf{L_A}$ is defined, where

$$\mathbf{L_A} = \begin{pmatrix} m_1 & 0 & 0 \\ m_2 & m_3 & 0 \\ 0 & 0 & 0 \end{pmatrix} \qquad (2)$$

which parameterises a 2D MVN in the xy plane. The angular covariance $\mathbf{M_A^h}$ of reflection $\mathbf{h}$ is scaled in proportion to $|\mathbf{r}|$, with $\mathbf{M_A^h} = (|\mathbf{r^h}|\mathbf{L_A})(|\mathbf{r^h}|\mathbf{L_A})^*$.

Finally, a transformation matrix $\mathbf{Q^h}$ is defined to rotate the covariance into the reciprocal space coordinate frame for each reflection, where

$$\mathbf{Q^h} = \begin{pmatrix} \hat{\mathbf{q}}_1 \\ \hat{\mathbf{q}}_2 \\ \hat{\mathbf{r}} \end{pmatrix} \tag{3}$$

$\mathbf{q}_1 = \hat{r} \times \hat{s}_0$ and $\mathbf{q}_2 = \hat{r} \times \hat{q}_1$ (illustrated in Fig. 1B). Therefore, the angular covariance for a reflection $\mathbf{h}$ is described by the $|\mathbf{r}|$-dependent covariance matrix $\mathbf{M^h} = (\mathbf{Q^h})^T \mathbf{M}_A^h \mathbf{Q^h}$, a 2D MVN in the $\mathbf{q}_1$-$\mathbf{q}_2$ plane perpendicular to $\mathbf{r}$.

The coordinate frames are shown in Fig. 1A. For this angular component, the 2D MVN can be constrained to be isotropic, i.e. circular, rather than elliptical, with the constraints $m_1 = m_3 > 0$, $m_2 = 0$. Purely angular mosaic spread models are known to inadequately describe observed low resolution reflections, as the angular component of covariance tends to zero as $|\mathbf{r}| \to 0$ (Sauter et al., 2014). Therefore, an additional $\mathbf{r}$-invariant mosaicity component should be included, to account for the finite size of the RLPs at low resolution, which can be achieved by including the $|\mathbf{r}|$-invariant ellipsoidal/spherical mosaicity model. An overall model then defines the reflection covariance most generally as $\mathbf{M^h} = (\mathbf{Q^h})^T \mathbf{M}_A^h \mathbf{Q^h} + \mathbf{M_E}$: this describes a reflection-independent 3D MVN distribution of the RLPs due to domain size effects, with 2D angular broadening from orientational mosaic spread. Example RLP distributions of $\mathbf{M^h}$ are shown in Fig. 1C.

2.2 General impact of the RLP distribution

To optimise the model, the relationship between the mosaicity model parameterisation and the expected distribution of diffracted beam vectors on the Ewald sphere (referred to as the general impact) must be known. This distribution can be described simply, whereas the projection onto a flat detector plane is more complex. Under the assumption that the Ewald sphere is flat on the scale of a spot, the general impact can be approximated by the shape of the spot on a plane tangential to the Ewald sphere surface. To work in this reference frame, it is convenient to define a reflection specific coordinate frame with basis vectors $(\hat{e}_1, \hat{e}_2, \hat{e}_3)$, where $\mathbf{e}_1 = \hat{s}_2 \times \hat{s}_0$, $\mathbf{e}_2 = \hat{s}_2 \times \hat{e}_1$, $\hat{e}_3 = \hat{s}_2$, i.e. $\hat{e}_1$ and $\hat{e}_2$ define a tangent plane to the Ewald sphere (Fig. 1B). The definitions of $\mathbf{e}_1$ and $\mathbf{e}_2$ are the same as in the coordinate system defined in Kabsch (2010a), but the $\hat{e}_3$ definition differs. These basis vectors can be used to form the rows of a rotation matrix $\mathbf{R_e} = (\hat{e}_1, \hat{e}_2, \hat{e}_3)$, which transforms the laboratory space coordinate system for a given reflection such that the $\mathbf{s_2}$ vector lies along the z-axis of the local reflection coordinate system. Therefore, given a covariance matrix $\mathbf{M}$, the RLP distribution in the local coordinate system is a MVN $P(\hat{e}_1, \hat{e}_2, \hat{e}_3)$ with mean $\mu = \mathbf{R_e s_2}$ and a covariance $\Sigma = \mathbf{R_e M R_e^T}$. To compute the general impact requires calculating the conditional distribution of the MVN on the tangent plane, i.e. it requires the decomposition of P into the 1D marginal distribution $P(\hat{e}_3)$ and the 2D conditional distribution $P(\hat{e}_1, \hat{e}_2 | \hat{e}_3)$, with $P(\hat{e}_1, \hat{e}_2, \hat{e}_3) = P(\hat{e}_1, \hat{e}_2 | \hat{e}_3) \, P(\hat{e}_3)$. A convenient property of the MVN distribution is that the marginal and conditional distributions are also normal distributions. First, we define the partition of the mean vector and covariance matrix as follows:

$$\mu = \begin{pmatrix} \mu_1 \\ \mu_2 \end{pmatrix}, \quad \Sigma = \begin{pmatrix} \Sigma_{11} & \Sigma_{12} \\ \Sigma_{21} & \Sigma_{22} \end{pmatrix}. \tag{4}$$

Here μ_1 and Σ_{11} are the 2D vector mean and covariance of the conditional distribution, and μ_2 and Σ_{22} are the scalar mean and variance of the marginal distribution. Using the properties of block matrix inversions (Appendix A), the conditional distribution at the intersection with the Ewald sphere can be shown to have a mean and variance

$$\bar{\mu} = \mu_1 + (\Sigma_{12})(\Sigma_{22})^{-1}(|s_0| - \mu_2), \quad \bar{\Sigma} = \Sigma_{11} - (\Sigma_{12})(\Sigma_{22})^{-1}(\Sigma_{21}) \tag{5}$$

The mean of the conditional distribution describes the central impact of the spot on the Ewald sphere, and the variance describes the general impact of the spot. The centre of mass of the reflection recorded on the Ewald sphere can be approximated by transforming the mean of the conditional distribution back into the laboratory frame:

$$\mathbf{s_1} = \mathbf{R_e^T} \begin{pmatrix} \bar{\mu} \\ |s_0| \end{pmatrix} \tag{6}$$

2.3 Parameter estimation via maximum likelihood methods

The starting point for model optimisation is the pixel intensity values of a set of indexed spots and an experimental model (beam, detector, crystal). The pixel intensity values constitute the general impact of a spot on the detector, which are mapped into the local reflection-specific coordinate frame. In the local coordinate frame, this distribution can be approximated as a bivariate normal distribution, whereas the projection onto the detector surface is in general not normal. This transformation and approximation significantly simplify the estimation of the model parameters and allows the parameters to be estimated via a maximum likelihood estimator. The probability of observing a count at a position x, y, z in the reflection-specific coordinate system is given by $P(x, y, z) = P(x, y | z)P(z)$. Each RLP centre is predicted to be a distance $\epsilon = |\mathbf{s_2}| - |\mathbf{s_0}|$ from the.

Ewald sphere along $\mathbf{e_3}$, and the centre of each pixel in the observed spot maps to a point $\mathbf{x} = (x, y)$ in the $(\mathbf{e_1}, \mathbf{e_2})$ plane. Given the expected mean and covariance from the model, the probability density of observing a count at $(\mathbf{x}, \epsilon)$ is

$$P(\mathbf{x} \mid \epsilon) = \frac{1}{\sqrt{(2\pi)^2 |\bar{\mathbf{\Sigma}}|}} \exp\left(-\frac{1}{2}(\mathbf{x} - \bar{\mu})^\mathbf{T}\bar{\mathbf{\Sigma}}^{-1}(\mathbf{x} - \bar{\mu})\right) \tag{7}$$

$$P(\epsilon) = \frac{1}{\sqrt{(2\pi)\Sigma_{22}}} \exp\left(-\frac{1}{2}\epsilon^2\Sigma_{22}^{-1}\right) \tag{8}$$

Therefore, given N observed strong spots, where a spot h has contributions from O_h detector pixels, j, and each pixel has observed counts of $c_{h,j}$, the log likelihood can be written as:

$$L = -\frac{1}{2}\sum_h^N \sum_j^{O_h} c_{h,j}[\ln(\Sigma_{22h}) + \epsilon_h^2/\Sigma_{22h} + \ln(|\bar{\mathbf{\Sigma}}_h|) \tag{9}$$
$$+ (\mathbf{x}_{h,j} - \bar{\mu}_h)^T\bar{\mathbf{\Sigma}}_h^{-1}(\mathbf{x}_{h,j} - \bar{\mu}_h)]$$

For each spot, the total observed counts, c_{tot}, mean, $\bar{x}$ and covariance $\mathbf{S}$ of each spot are given by

$$c_{\text{tot}} = \sum_j^O c_j, \quad \bar{\mathbf{x}} = \frac{1}{c_{\text{tot}}}\sum_j^O c_j\mathbf{x}_j, \quad \mathbf{S} = \frac{1}{c_{\text{tot}}}\sum_j^O c_j(\mathbf{x}_j - \bar{\mathbf{x}})(\mathbf{x}_j - \bar{\mathbf{x}})^T \tag{10}$$

Using the cyclic permutation property of the matrix trace operator, the log likelihood can be written in terms of these statistics as

$$L = -\frac{1}{2}\sum_h^N c_{\text{tot},h}[\ln(\Sigma_{22h}) + \epsilon_h^2/\Sigma_{22h} + \ln(|\bar{\mathbf{\Sigma}}_h|) + \text{tr}(\bar{\mathbf{\Sigma}}_h^{-1}\mathbf{S}_h) \tag{11}$$
$$+ (\bar{\mathbf{x}}_h - \bar{\mu}_h)^T\bar{\mathbf{\Sigma}}_h^{-1}(\bar{\mathbf{x}}_h - \bar{\mu}_h)].$$

This removes the inner sum over the spot pixels, and since the statistics c_{tot}, $\bar{x}$ and $\mathbf{S}$ do not depend on the model parameters, they only need to be calculated once which therefore allows quick evaluation of the log-likelihood at each step of optimisation. To estimate the parameters, the Fisher scoring algorithm (Fisher, 1922) is used, which only requires calculation of the first derivatives of L.

Further details of the calculations of the derivatives and Fisher information matrix can be found in Appendix B. The expected general impact depends on the mosaicity parameterisation in addition to the crystal model ($\mathbf{U}$ and $\mathbf{B}$ matrices), as the crystal model determines the predicted $\mathbf{s_2}$ and both must be optimised. The starting point for the optimisation is the crystal model from indexing and an isotropic $\mathbf{M}$ with diagonal elements σ_D^2, where σ_D is the standard deviation of the beam divergence (Kabsch, 2010b), calculated on the indexed spots. Maximum likelihood optimisation is performed on the mosaicity parameterisation, followed by the crystal model, after which the indexed spots are re-predicted and the spot statistics recalculated. By default, three of these macrocycles are performed for each indexed crystal.

2.4 Integration and partiality estimation

Following optimisation of the profile model, unit cell and orientation of the crystal using the strong spots, all reflections are predicted on the image and integrated using the procedure described in Winter et al. (2018). For each spot, a shoebox is generated and separated into foreground and background pixels. Foreground pixels are defined as those within the χ^2 quantile with probability 0.9973 (univariate 3 σ level) in the reflection-specific coordinate system. The integrated intensity is determined by summation of the foreground pixels minus an estimate of the background. To calculate the reflection dependent scale factor, i.e. the partiality, it is assumed that the Ewald sphere is flat on the scale of a RLP. For a monochromatic beam, i.e. a δ-function wavelength distribution, the Ewald sphere intersects the RLP as a plane, and the probability density of the RLP in this plane is given by the conditional distribution. As the conditional distribution is itself a multivariate normal distribution, the relative integrated probability density is therefore the probability density of the marginal distribution

$$\rho_h \propto \frac{1}{\sqrt{\Sigma_{22h}}} \exp\left(-\tfrac{1}{2}\epsilon_h^2/\Sigma_{22h}\right). \tag{12}$$

This depends on the shortest distance between the RLP centre and the intersection of the Ewald sphere and also on Σ_{22h}, the variance of the marginal distribution perpendicular to the Ewald sphere. Therefore, ρ_h is largest when Σ_{22h} is the variance along the shortest elliptical axis and $\epsilon_h = 0$, i.e. the Ewald sphere intersects perpendicular to the shortest axis, which we refer to as Σ_{min}, as illustrated in Fig. 1D. For an intersection off the centre of the RLP at an arbitrary orientation, the relative partiality is therefore given by

$$p_h = \frac{\rho_h}{\rho_{min}} = \frac{\frac{1}{\sqrt{\Sigma_{22h}}} \exp\left(-\frac{1}{2}\epsilon_h^2 / \Sigma_{22h}\right)}{\frac{1}{\sqrt{\Sigma_{min}}}} = \sqrt{\frac{\Sigma_{min}}{\Sigma_{22h}}} \exp\left(-\frac{1}{2}\epsilon_h^2 / \Sigma_{22h}\right) \quad (13)$$

Importantly, for anisotropic RLPs, the partiality of each reflection depends on both the displacement of the RLP centre from the Ewald sphere and the orientation of the RLP covariance relative to the Ewald sphere for a given reflection. For mosaicity models with a reflection-independent covariance matrix, all reflection covariances have the same Σ_{min}. For radially dependent models Σ_{min} is the minimum covariance of the reflection-independent part of the model. The relative uncertainty in $1/p$ can also be estimated as

$$\frac{\sigma(1/p_h)}{1/p_h} = p_h \frac{\partial(1/p_h)}{\partial \epsilon_h} \sigma(\epsilon_h) \approx \frac{\epsilon_h}{\Sigma_{22h}} \sqrt{\frac{\Sigma_{22h}}{n_{obs}}}, \quad (14)$$

where the estimate of the uncertainty in ϵ_h is given by the uncertainty of the mean of a normal distribution from n_{obs} observations of spots upon which the mosaicity model is optimised.

3. DIALS command-line tools for processing SSX data

The DIALS package contains a set of command line programs to enable sequential processing of rotation data from raw images to merged intensity files. In addition, the individual programs contain a range of algorithms and options suitable to different use cases including rotation and still-shot data processing. To facilitate the processing workflow for processing blocks of serial still-shot data, two new programs were implemented, *dials.ssx_index* and *dials.ssx_integrate*. These programs perform indexing and integration on blocks of still-shot images, using appropriate parameters for still-shot data and aggregating summary statistics into HTML reports and log outputs. This section summarises which aspects of the existing DIALS algorithms are used in these programs, and the new additions to data reduction programs to enable still-shot processing. These components then form the basis of processing workflows for still-shot data processing, as implemented in the *xia2.ssx* pipeline which is detailed in Section 4.

3.1 Indexing with dials.ssx_index

The *dials.ssx_index* program runs the same underlying algorithms as *dials.index* on the output of *dials.find_spots*, a table of the strong spots per image. Indexing is attempted on each image that contains at least 10 strong spots, using the indexing option *stills.indexer=stills*. Typically, the *unit_cell* and *space_group* are set if known, which greatly increases the indexing success rate for still-shot data. Multiple lattices are searched for on each image, using the *FFT1D* indexing algorithm. If the *FFT1D* algorithm is unsuccessful and a unit cell is provided, indexing is also attempted with the *real_space_grid_search* algorithm (Gildea et al., 2014). Following refinement of the crystal model, with a fixed beam and detector model, the stills indexer refines two parameters describing the crystal mosaicity as a combination of a mosaic block size parameter and mosaic spread parameter, as described in Sauter et al. (2014). A HTML report is generated, containing plots of per-image quantities such as the percentage of strong spots indexed, and distributions of cell parameters. The output data files can be viewed in the *dials.image_viewer* and *dials.reciprocal_lattice_viewer*.

3.2 Integration with dials.ssx_integrate

The *dials.ssx_integrate* program runs the ellipsoid profile modelling and integration algorithm, as described in Section 2 on each indexed crystal. The default parametrisation used is the six-parameter ellipsoidal covariance matrix $\mathbf{M_E}$, with no angular component. This profile model was found to best-describe typical well-diffracting SSX datasets. Following max-likelihood optimisation of the profile and crystal model, the full set of intensities are predicted, background-subtracted intensities are determined by summation integration and partiality values are estimated, as described in Section 2.4. In *dials.ssx_integrate* this routine is selected by default with the option *algorithm=ellipsoid*. Alternatively, the two-parameter mosaicity model determined by the *stills* indexer can be used to predict the location and spatial extent of spots on the image, followed by summation integration as described in Winter et al. (2018), which is selected with the option *algorithm=stills*. A HTML report is generated, containing plots of per-image metrics including $I/\sigma(I)$ and the mosaicity model parameters, and the output can be viewed as before.

3.3 Data reduction with dials.cosym and dials.scale

For still-shot data, existing symmetry-determination algorithms are typically unreliable for determination of an unknown point group symmetry. Therefore, after integration, a particular space group must have been chosen to continue processing. However, only the point group affects indexing ambiguity resolution and scaling. For space groups where the point group symmetry is lower than the lattice symmetry, there is more than one valid way of indexing the diffraction pattern of a given crystal, however a consistent choice must be achieved across all crystals. The *dials.cosym* program can be used to resolve this indexing ambiguity for still-shot data, with a higher chance of success as the signal-to-noise ratio and resolution range of each image increases. An option was added to *dials.cosym* to provide a reference file, containing intensity information (MTZ or CIF file format), or a structural model (PDB or CIF file format) from which estimated intensities can be calculated. If a reference file is provided, *dials.cosym* selects the indexing solution which best correlates with the reference.

For scaling still-shot data, the kB scaling model in the *dials.scale* program is used, which consists of a scale factor (k) and B-factor for each crystal (Beilsten-Edmands et al., 2020). There are two different approaches for scaling which are fundamentally different in how they scale with the number of crystals. If a reference model is provided, each image can be scaled independently against the reference intensities. The reference intensity values are used as fixed merged intensity values $\langle I_h \rangle$ in the scaling objective function and outlier rejection formula. In this case, the scale factor determination for a given crystal is a linear least-squares problem with two free parameters, which can be quickly solved separately for each crystal and is trivially parallelizable. If no reference model is used, the best-estimate set of merged intensities $\langle I_h \rangle$ is calculated at each step of the optimisation by merging data from all datasets, and changes during the minimisation process. In this case, scaling is a non-linear least-squares problem with $2n$ free parameters, where n is the total number of crystals. This has a much higher computational complexity, higher memory requirements and becomes slow for many thousands of crystals. Another important aspect of data reduction is the adjustment of intensity uncertainties. In *dials.scale*, a two-parameter error model is optimised to adjust the uncertainty estimates, $(\sigma')^2 = a^2(\sigma^2 + (bI)^2)$, where a and b are the error model parameters to optimise. Importantly, for still data integrated with the ellipsoid profile model, the uncertainties of the partiality estimates are calculated from the profile model and propagated to σ before the error model optimisation is performed. The inclusion of a partiality uncertainty is a key difference to rotation data and increases the σ of intensity observations with low partiality.

3.4 Associated tools from cctbx.xfel

The DIALS package depends on the cctbx package (Grosse-Kunstleve et al., 2002) for its core crystallographic calculations. Within cctbx there is a program for data reduction of XFEL data, *cctbx.xfel.merge*, which takes as input DIALS integrated data files integrated with the *stills* integration algorithm. As described in previous publications (Sauter, 2015; Brewster et al., 2019b), *cctbx.xfel.merge* implements a post-refinement algorithm that optimises the two-parameter mosaicity model of the *stills* indexer, the crystal orientation, a scale and B-factor as part of data reduction. Additionally, a three-parameter error model $(\sigma')^2 = \text{SdFac}^2(\sigma^2 + (\text{SdB} * I) + (\text{SdAdd} * I)^2)$ is optimised, while indexing ambiguity resolution is performed using the underlying methods from *dials.cosym*.

4. Xia2.ssx: an automated processing pipeline

The individual DIALS programs described above form the building blocks of a processing pipeline for still-shot data. The pipeline for processing rotation data, as well as the analogous pipeline for still-shot data, is illustrated in Fig. 2. For still-shot processing, an accurate parameterisation of the detector is essential for reliable indexing and integration, whereas for

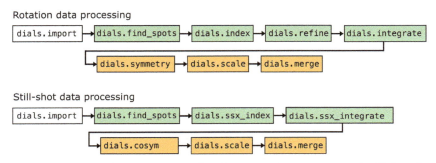

Fig. 2 The sequential processing pipeline for processing rotation and still-shot data with DIALS programs.

rotation data, this can be refined between indexing and integration. Non-isomorphism must be considered as part of the data reduction process, and consideration must be given to careful processing and separation of mixed-state serial data, such as time or dose series data. To handle these challenges, an automated processing pipeline was developed in the *xia2* suite, to provide a simple tool designed to handle these challenges and provide aggregated feedback to the user. For specialised analyses that do not fit into this model pipeline, the individual component programs in DIALS can be combined into alternative workflows.

4.1 Integration and reduction of SSX data with xia2.ssx

The *xia2.ssx* pipeline combines the individual DIALS programs into a workflow as illustrated in Fig. 3, which enables processing from raw images to merged data. The minimal required inputs for full processing are the image files, plus the expected unit cell and space group. The requirement for specifying a unit cell and space group is a practical consideration: in contrast to rotational crystallography, where the unit cell can reliably be determined through indexing the

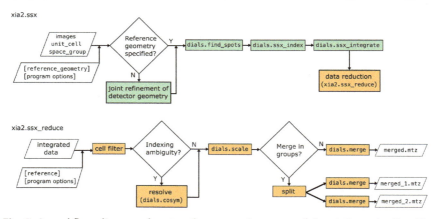

Fig. 3 A workflow diagram showing the processing steps of the *xia2.ssx* pipeline. The minimal required input are the image data, expected unit cell and space group. If no reference geometry is specified, a joint refinement of the detector geometry is performed, used indexed data from a subset of the images. Then, spot-finding, indexing and integration are performed in batches of 1000 images, before data reduction, which can be run as the standalone program *xia2.ssx_reduce*. In data reduction, a unit cell filter removes non-isomorphous crystals, before indexing ambiguity resolution using *dials.cosym*, if appropriate based on the crystal symmetry or cell parameters. Then each crystal is scaled, optionally in parallel batches against a reference set of intensities if specified. Finally, data can be split into groups based on experimental condition before each group is merged, producing a merged MTZ output file.

3D reciprocal lattice, and the space group symmetry can be discovered by testing the existence of rotational and screw axes, the existing algorithms are less reliable for still-image analysis due to the sparsity of reflections measured per crystal. In addition, serial crystallography experiments are typically performed to study particular aspects of well-characterised systems, so the unit cell, space group and reference intensity data are typically known at the start of data processing. If these parameters are not specified as input, the *xia2.ssx* pipeline will stop after spot-finding and indexing, outputting the most populous cell clusters and potential lattice symmetries for the user to consider and provide as input for a subsequent run.

Given this input, the first step is to determine an accurate model of the detector geometry. It has previously been shown that having a well-characterised experimental geometry improves the signal strength from serial crystallography data (Brewster et al., 2018), which can be determined from a joint refinement of the detector and crystal models, a feature that was already implemented within the *dials.refine* program and described in Brewster et al. (2018). The procedure used in *xia2.ssx* is to perform spot-finding (*dials.find_spots*) and indexing (*dials.ssx_index*) in batches of 1000 images, until a threshold number of crystals are indexed. Then a joint refinement of the crystal models and a single detector model are performed using *dials.refine* to determine the best-estimate of the detector geometry. This improved detector model is then fixed for indexing and integration of the full dataset. Alternatively, an external reference geometry can be given, either from previous DIALS processing of serial data or from DIALS processing of a rotation dataset. This can be used as a starting point for further joint refinement with the *starting_geometry* option, or as a fixed reference with no further joint refinement with the *reference_geometry* option. Following optimisation of the experimental geometry, spot-finding (*dials.find_spots*), indexing (*dials.ssx_index*) and integration (*dials.ssx_inte-grate*) are performed on batches of 1000 images. A summary of the processing statistics is logged at the completion of each process for each batch, with log files, HTML reports and output data from the individual programs output to batch subdirectories. Integration can be performed with the *ellipsoid* profile model (default) or *stills* integrator.

Following integration of all images, the first step of data reduction is to perform unit cell filtering, intended to remove a small fraction of the most non-isomorphous cells. For quick filtering, a filter based on the median unit cell values and absolute length (1.0 Å) and angle (1.0°) tolerances is applied by default. Alternatively, cells within a certain tolerance of a specified cell can be selected, useful for selecting a subset of multi-modal cell distributions, or full clustering analysis can be triggered using *dials.cluster_unit_cell*, with the largest cluster being retained. Before scaling, an assessment is performed to determine whether indexing ambiguity resolution is required, by comparing the space group symmetry (used in indexing and integration) to the lattice symmetry. The lattice symmetry is determined from the unit cell values including a tolerance factor (the parameter *lattice_symmetry_max_delta* (Le Page, 1982)), which allows for the case of 'accidental' indexing ambiguities due to approximately equal cell parameters in addition to the 27 space groups where the symmetry of the Bravais lattice is higher than the space group symmetry. If indexing ambiguities are possible, then the *dials.cosym* program, run with the space group specified, is used to resolve the indexing ambiguities. The current implementation of the cosym algorithm has an n^2 memory dependence, where n is the number of crystals, therefore a modified procedure is used, where the indexing ambiguity is solved for batches of crystals at a time (default batch size 1000). Then, the crystals within each batch are scaled, before a final run of *dials.cosym* where each batch is considered as a single dataset, to consistently index all batches. If a reference file has been provided for scaling, the indexing solution chosen is that which best correlates with the reference. For scaling, when a reference file is provided, *dials.scale* is run on batches of crystals using the reference intensities in scaling (see Section 3.3), which places all crystals on a common scale across all batches. If no reference is provided, all crystals are scaled together in a single run of *dials.scale*, however this becomes computationally expensive for many thousands of crystals. Before the final merging step, the data can optionally be split into groups based on metadata, such as time-point or dose, as described below. Finally, *dials.merge* is used to merge the data for each group and produced a merged MTZ file. Throughout the data reduction pipeline, a *partiality_threshold* of 0.25 is applied.

The *xia2.ssx* program provides the most important program options as directly settable command line parameters, which are a limited subset of the full parameters available in the individual DIALS programs. To configure more advanced parameters for individual programs, *xia2.ssx* makes use of a feature of the cctbx ecosystem called 'phil' files; plain-text files that specify program parameter values, which can be specified for the individual DIALS programs (an example is given in Section 5.2). As such, the *xia2.ssx* program defines a processing workflow and acts as a runner for the individual DIALS programs, setting suitable default program parameters for SSX datasets while allowing full user specification of program parameters.

4.2 Standalone data reduction with xia2.ssx_reduce

To enable quick reduction and aggregation of integrated data, the data reduction part of the *xia2.ssx* pipeline can be run as the standalone command *xia2.ssx_reduce*, which takes DIALS-integrated experiment and reflection files as inputs. If several datasets have been reduced separately but with the same reference set of intensities, then all crystals will be on the same scale and the scaled data can quickly be merged without repeating data reduction. To achieve this, the *xia2.ssx_reduce* program can be run with the option *steps=merge*. Fig. 4 shows an example workflow for processing and aggregating blocks of images in an efficient manner. With a reference geometry predetermined from a well-characterized sample, images can be integrated with a consistent detector geometry and scaled against a reference set of intensities. The output DIALS scaled files can then be used as input to *xia2.ssx_reduce*, with the option *steps=merge* triggering solely the merging step, as the scaled data are already on a common scale fixed by the reference. If no reference intensities were available for processing, the alternative workflow would be to use the DIALS integrated files as input to *xia2.ssx_reduce*, which would perform the full data reduction process.

4.3 Merging in groups based on metadata

Serial crystallography enables a new kind of experiment, where repeated measurements are taken on a given crystal, to measure the evolution with time after X-ray exposure and/or laser-excitation, giving a dose or time series with a defined repeat interval. For cases where the resultant change in diffraction intensities is expected to be modest, one workflow is to process all data together, to get data processed in a consistent manner and placed on a common scale, before separately merging based on experimental

Fig. 4 Schematic of typical workflow for processing experimental data, allowing quick merging of sets of images after initial processing using a reference set of intensities. The final processing step (*xia2.ssx_reduce*) only needs to merge the scaled and consistently indexed data.

condition. *xia2.ssx* contains a mechanism to allow splitting of scaled data before merging in separate groups. In *xia2.ssx/ xia2.ssx_reduce*, the option *dose_series_repeat=n* can be used to define the repeat interval *n* that should be used to split the data, i.e. every *n*th image is part of the same experimental condition. For generalised merging of images into specified groups in an automated manner based on experimental metadata, a YAML configuration file (as illustrated in Listing 1 below) can be used to define the location of metadata for grouping images for merging. For example, consider a HDF5 image file containing a mixture of pump and probe diffraction data, pre-filtered for empty images, where the pump status (on/off) is written into the file as a data array.

When supplied with a correctly specified YAML file, *xia2.ssx/ xia2.ssx_reduce* will split the scaled data into groups based on the pump status label and merge each group separately. Importantly, as the splitting is performed by inspecting the image index via the experiment list data structure, it does not rely on the scaled data being ordered and can therefore be used when merging multiple datasets together, such as in the workflow shown in Fig. 4.

Listing 1: Example definition of a *xia2.ssx* grouping YAML file, to trigger merging in groups based on the 'pump status' label.

```
---
metadata:
  pump:
    path/to/image1.h5: path/to/image1.h5:/
    entry/pump_status
    path/to/image2.h5: path/to/image2.h5:/
    entry/pump_status
grouping:
  merge_by:
    values:
      - pump
---
```

This provides a mechanism by which data acquisition software can write metadata into an image or metadata file, which can automatically be used by *xia2.ssx* to produce grouped output data. Alternatively, postprocessing can be used to determine a classifier, which when stored as a HDF5 array can be used by *xia2.ssx* to divide images into groups for merging.

5. Demonstration and evaluation on example data

5.1 Processing anomalous SSX data: AcNiR above the K-edge

A dataset consisting of 19,200 images was collected on a fixed-target chip at beamline I24, Diamond Light Source, containing microcrystals of copper nitrite reductase from *Achromobacter cycloclastes* (AcNiR). The collection wavelength was 1.376 Å, just above the Cu K-edge, and therefore a high level of anomalous signal was expected. AcNiR is known to crystallise into the $P2_13$ cubic space group with a cell length 96.4 Å, with a polymorph with $a = b = c = 97.8$ Å (Ebrahim et al., 2019).

In this demonstration, we first process the data with the *xia2.ssx* pipeline and discuss the processing steps. We then evaluate the ellipsoid profile model by comparison with the stills profile model, by reducing integrated data with the *cctbx.xfel.merge* program, so that post-refinement can be applied to the stills profile model as is standard practice. Structure refinement is performed with *REFMAC5* (Murshudov et al., 2011) via *DIMPLE* v2.6.2 (https://ccp4.github.io/dimple/). A reference model from PDB entry 2BW4 (Antonyuk et al., 2005) was used, enabling a consistent free set of reflections, and the *–anode* option was used to run *ANODE* (Thorn & Sheldrick, 2011) for calculating anomalous difference maps and anomalous peak heights.

5.1.1 Standard processing with xia2.ssx

The raw data consists of 19,200 images in CBF format collected on a Pilatus 6 M, named sequentially from image_00000.cbf to image_19199.cbf. The full command to process the data with *xia2.ssx* was.

```
xia2.ssx template=images/image_#####.cbf \.
  mask=pixels.mask \.
  geometry_refinement.n_crystals=2000 \.
  space_group=P213 \.
  unit_cell=96.4,96.4,96.4,90,90,90 \.
  anomalous=True \.
  d_min=1.8 \.
  reference=2bw4.pdb \.
  absolute_length_tolerance=0.3.
```

A mask was generated manually in the *dials.image_viewer* to mask the beam-stop shadow. A minimum of 2000 indexed crystals were required for the initial geometry refinement, and a reference model from PDB entry 2BW4 was used to generate a reference intensity set for data reduction. During the initial geometry refinement routine, spot-finding and indexing were performed on the first four batches of 1000 images, at which point 2215 crystals had been indexed, with a median cell constant of 96.66 ± 0.14 Å. During the geometry refinement, the spots RMSDs reduced from (0.309 mm, 0.163 mm, 0.126 °) to (0.141 mm, 0.085 mm, 0.061 °). Subsequently, the processing of the first four batches with the refined geometry gave 2403 indexed crystals, with a median cell constant of 96.39 ± 0.15 Å. During indexing and integration, 56 % of the images were indexed with at least one lattice, with multiple lattices per image being common, resulting in 16,440 integrated crystals. The average sigmas of the ellipsoid mosaicity model were (161, 378, 497) $\mu\text{Å}^{-1}$, i.e. an average ellipsoidal aspect ratio of 3.1, indicative of significant diffraction spot anisotropy as could be seen on the raw images.

A tighter unit cell filter tolerance was applied (*absolute_length_tolerance=0.3*) for the purposes of data reduction, to avoid inclusion of the polymorph cell, which filtered out 274 crystals. As the $P2_13$ space group point group symmetry is lower than the $P432$ lattice symmetry, the indexing ambiguity must be resolved by the resolve aspect of the pipeline, the details of which were discussed in Section 4. Fig. 5A shows the *dials.cosym* output from a representative batch of 1000 crystals, demonstrating clear separation of the two indexing modes for most crystals. The output from reindexing all batches with *dials.cosym* is shown in Fig. 5B, again showing clear clustering indicating resolution of the indexing modes. Another means to confirm the success of indexing ambiguity resolution is the Padilla–Yeates L-test (Padilla & Yeates, 2003), which is performed as part of the merging procedure in *dials.merge* and shown in Fig. 5C. In this case the L-statistic more closely matches the theoretical untwinned curve than the perfect twin curve, suggesting the data is not perfectly untwinned, consistent with the small fraction of crystals lying between the two main cosym clusters in Fig. 5A.

The overall merging statistics showed high internal consistency, with $CC_{1/2} = 0.997$, $R_{split} = 0.055$, $I/\sigma = 20.6$, with an anomalous multiplicity of 125.0. The final R/R_{free} values were 0.112/0.147 and the Cu anomalous peak heights were 50.3 σ & 44.2 σ. As a comparison, the data were also reduced without the use of a reference, which resulted in lower internal consistency, with $CC_{1/2} = 0.982$, $R_{split} = 0.083$, $I/\sigma = 17.1$, with an anomalous multiplicity of 137.9. The final R/R_{free} values were 0.165/0.210

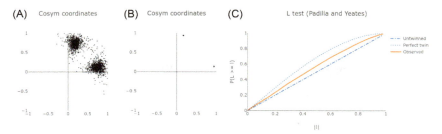

Fig. 5 Indexing ambiguity resolution of SSX data with *dials.cosym*. (A) output cosym coordinates plot for a representative batch of 1000 crystals, showing separation into two main clusters for most crystals. (B) indexing ambiguity resolution of all 16 batches, which separate into two tight clusters. (C) L-test metric plot for twinning; for this dataset the statistic does not exactly match the theoretical untwinned curve, suggesting that the data is not perfectly untwinned.

and the Cu anomalous peak heights were 34.6 σ & 29.1 σ. Aside from the reduced runtime, the main benefit of the reference model is to aid in outlier rejection, with the highest fraction of outliers occurring at lower resolutions, leading to stronger anomalous signal and significantly improved model *R*-factors.

5.1.2 Comparison with existing methods

To evaluate the ellipsoid partiality model, the data were processed using both the *ellipsoid* and *stills* integration algorithm. In addition to the different partiality estimates of the two algorithms, the set of reflections predicted also differs. To test the differences in partiality estimates, we first performed a comparison using a common set of reflections predicted by both integration algorithms. As the *stills* integration algorithm has been developed for use with post-refinement methods available in *cctbx.xfel.merge*, the base comparison case is to perform data reduction with *cctbx.xfel.merge*, enabling post-refinement for stills-integrated data and disabling post-refinement for ellipsoid-integrated data. Next, the differences in prediction were assessed by processing the full ellipsoid-integrated dataset with *cctbx.xfel.merge*. Finally, we compared processing of the full ellipsoid-integrated dataset reduced with *xia2.ssx_reduce* with the full stills-integrated dataset reduced with *cctbx.xfel.merge*.

A common set of stills and ellipsoid integrated reflections were chosen by matching the miller indices for each crystal in each image, for those reflections with an ellipsoid-integrated partiality estimate $p > 0.25$. For data reduction with *cctbx.xfel.merge*, the ellipsoid-integrated intensities were corrected for partiality and the partiality uncertainty was propagated to the intensity uncertainties before the data reduction. A reference model from PDB entry 2BW4 was used, anomalous pairs were separated and a resolution limit of 1.8 Å was applied. Fig. 6 shows the REFMAC R-factors and ANODE anomalous peaks heights for the data processed cumulatively in batches of 1000 crystals, up to a total of 16,166 crystals. The ellipsoid data R_{free} values are consistently lower than the stills-integrated data by around one percentage point. Alternatively, a given R/R_{free} is reached with many fewer crystals for the ellipsoid-integrated data. The Cu anomalous peak heights also show consistent improvement for the ellipsoid integrated data, indicating a more accurate anomalous signal. Processing statistics for the full 16,166 crystals are shown in Table 1. For this dataset, the ellipsoid model gives a clearly improved partiality estimate compared to post-refining a spherical model. We attribute this to the fact that the spot shapes have significant anisotropy, therefore there is limited improvement that can be gained from post-refining an isotropic mosaicity model.

Next, the full ellipsoid-integrated dataset ($p > 0.25$) was reduced with *cctbx.xfel.merge*, to investigate the effect of the additional predictions of the ellipsoid model, which increased the anomalous multiplicity from 98.8 to 141.4 for the full dataset (16,166 crystals). Overall, as listed in Table 1, the full ellipsoid-integrated dataset processed with *cctbx.xfel.merge* had similar data quality to the common subset dataset. The overall I/σ increased slightly from 8.65 to 8.82, with R_{split} lowered from 5.8 % to 5.5 %. R_{free} increased slightly

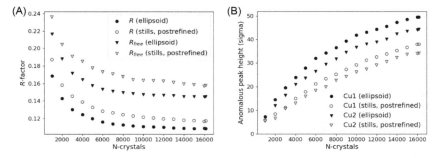

Fig. 6 A comparison of processing the AcNiR SSX dataset with the two integration algorithms, ellipsoid and stills, for a common set of reflections reduced with *cctbx.xfel.merge*. (A) REFMAC R/R_{free} and (B) ANODE Cu anomalous peak heights, for the dataset processed cumulatively in batches of 1000 crystals. For this dataset, the ellipsoid partiality model results in consistently improved R-factors and anomalous peak heights, which therefore enables a structure of a particular quality to be achieved with significantly fewer measured crystals.

Processing serial synchrotron crystallography diffraction data with DIALS

Table 1 Comparative statistics for processing the AcNiR dataset with the stills and ellipsoid integration algorithms, reduced with cctbx.xfel.merge.

Profile/integration algorithm	Stills	Ellipsoid	Ellipsoid
Reduction program	cctbx.xfel.merge	cctbx.xfel.merge	cctbx.xfel.merge
Post-refinement enabled	Yes	No	No
Common subset	Yes	Yes	No
$CC_{1/2}$	0.997	0.996	0.996
Rsplit	0.066	0.058	0.055
I/σ	6.60	8.65	8.82
Anom. multiplicity	60.5	98.8	141.4
R/R_{free}	0.117/0.158	0.109/0.145	0.111/0.147
wR/wR_{free}	0.122/0.167	0.116/0.157	0.119/0.158
Anom. peak heights	38.0, 34.1	49.6, 44.4	52.1, 45.5
Error model (SdFac, SdB, SdAdd)	1.31, 0.33, 0.40	1.01, 0.10, 0.64	0.97, 0.14, 0.79

The first two columns are statistics for a common subset of reflections integrated by both algorithms; the final column is the full ellipsoid-integrated dataset. This shows that the improved data quality for the ellipsoid model comes primarily from improved partiality estimates, rather than additional predicted reflections.

from 0.145 to 0.147 and the peak heights increased slightly from (49.6, 44.4) to (52.1, 45.5). The trends for the data processed cumulatively (not shown) showed similar behaviour, with R_{free} and anomalous heights marginally higher. Looking at the overall trends, it is clear that the improvement in data quality is coming primarily from the improved partiality estimates rather than the additional predicted reflections.

The statistics for the full ellipsoid-integrated dataset (16,166 crystals, $p > 0.25$), reduced with *xia2.ssx_reduce* are shown in Table 2, alongside statistics for the full stills-integrated dataset reduced with *cctbx.xfel.merge*. We cannot directly compare with the ellipsoid-integrated dataset reduced with *cctbx.xfel.merge*, as the disabling of post-refinement means that outliers cannot be removed during data reduction. In *xia2.ssx_reduce*, outlier rejection is performed in the scaling step, which reduces the anomalous multiplicity from 141.4 to 125.0, whereas in *cctbx.xfel.merge*, reflections can only be removed as

Table 2 Comparative statistics for processing the AcNiR dataset for the full stills-integrated dataset reduced with cctbx.xfel.merge and the full ellipsoid-integrated dataset reduced with xia2.ssx_reduce.

Profile/integration algorithm	Stills	Ellipsoid
Reduction program	cctbx.xfel.merge	xia2.ssx_reduce
Post-refinement enabled	Yes	No
$CC_{1/2}$	0.996	0.997
Rsplit	0.068	0.055
I/σ	6.5	20.6
Anom. multiplicity	63.3	125.0
R/R_{free}	0.117/0.160	0.112/0.147
wR/wR_{free}	0.121/0.168	0.100/0.133
Anom. peak heights	37.4, 33.9	50.3, 44.2
Error model (SdFac, SdB, SdAdd)	1.36, 0.31, 0.39	1.17, N.A., 0.08[a]

[a]average over data reduction batches.

part of re-prediction in post-refinement. The ellipsoid-integrated dataset reduced with *xia2.ssx_reduce* results in lower *REFMAC* R-factors and higher anomalous peak heights compared to the stills-integrated dataset reduced with *cctbx.xfel.merge*, consistent with the previous comparison on the common subset of reflections. Another notable difference is the difference in I/σ values, primarily due to the different error model corrections optimised during data reduction. In *xia2.ssx_reduce*, the *dials.scale* error model parameters refined to much lower values, i.e. the σ values are not increased to the same extent, giving higher relative I/σ. The ellipsoid-integrated σ values are additionally increased by the partiality uncertainty.

To verify the reliability of the σ estimates from *xia2.ssx_reduce*, it is instructive to look at the σ-weighted R-factors (wR/wR_{free}) from *REFMAC* (see Table 2). For the stills-integrated data post refined with *cctbx.xfel.merge*, $wR/wR_{\text{free}} = 0.121/0.168$, whereas for the ellipsoid-integrated reduced data with *xia2.ssx_reduce*, $wR/wR_{\text{free}} = 0.100/0.133$. The differences between the weighted R-factors are much larger than for the unweighted R-factors, indicating more reliable σ estimates from the new methods. Fig. 7 shows wR/wR_{free} by resolution for the stills-integrated

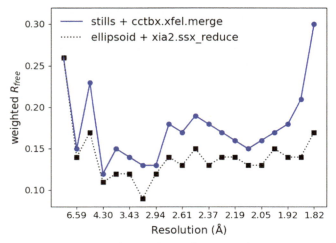

Fig. 7 REFMAC σ-weighted R_{free} by resolution for AcNiR data processing, for the full stills-integrated dataset reduced with *cctbx.xfel.merge* (blue, circles) and the full ellipsoid-integrated dataset reduced with *xia2.ssx_reduce* (black, squares). The *xia2.ssx_reduce* processing of the ellipsoid-integrated data gives significantly lower weighted R_{free} values, except in the lowest resolution shells. Both methods give a high R_{free} at the lowest resolution shell, indicating a general deficiency in modelling the lowest resolution reflections.

data, reduced with *cctbx.xfel.merge*, and the ellipsoid integrated data reduced with *xia2.ssx_reduce*. The wR_{free} values for processing with *xia2.ssx_reduce* are significantly lower, except in the lowest resolution shells. The high R-factors at the lowest resolution indicate a general deficiency in modelling the lowest resolution reflections and should be a key focus of further algorithm development.

5.2 Processing SSX dose-series: photoreduction of Fe(III) FutA

A room-temperature SSX dose-series dataset consisting of 256,000 diffraction images of the FutA iron-binding protein was analysed. This was collected as 10 measurements of repeated exposure equivalent to 22 kGy dose at each location on a fixed-target chip at beamline I24, Diamond Light Source, and was previously published in Bolton et al. (2024). This previous study confirmed the presence of an Fe3+ state, with subsequent photo-reduction of the Fe3+ site accompanied by a change in conformation of the Arg203 residue, which was observed to be visible at 22 kGy and strongest at 88 kGy. In this work, we demonstrate the features available in *xia2.ssx*/DIALS to process images collected under different experimental conditions. Although this example is a dose-series dataset, the methods are applicable in general to other types of serial diffraction measurements investigating dynamics such as laser-excitation series, X-ray pump-probe or ligand-binding studies.

The raw data consists of 256,000 images in CBF format, collected on a Pilatus 6 M, were sequentially named from image_000000.cbf to image_255999.cbf. First, a reference geometry was refined with the command:

```
xia2.ssx template=images/image_######.cbf \.
  dials_import.phil=import.phil \.
  mask=pixels.mask \.
  space_group=P21 \.
  unit_cell=39.7,78.7,48.4,90,97.8,90 \.
  geometry_refinement.n_crystals=2000 \.
  steps=None.
```

In the raw image headers, the beam centre was incorrectly specified by around 30 pixels along the fast axis, therefore this was overwritten on import by providing a 'phil' file for the *dials.import* program, containing the line *geometry.detector.fast_slow_beam_centre=1292,1310*. A mask was generated manually in the *dials.image_viewer* to mask the beam-stop shadow, while *steps=None* instructs no full processing to be done after geometry refinement. Then, due to a significant tilt in the detector orientation, another round of geometry refinement was performed at the start of full processing, starting with the geometry refined from the first process (contained in the *refined.expt* file), with the command:

```
xia2.ssx template=images/image_######.cbf \.
  mask=pixels.mask \.
  space_group=P21 \.
  unit_cell=39.7,78.7,48.4,90,97.8,90 \.
  starting_geometry=geometry_refinement/refined.expt \.
  geometry_refinement.n_crystals=2000 \.
  steps=find_spots+index+integrate.
```

The change in detector geometry, from initial to final refined geometry, is shown in Fig. 8A, which shows a relative displacement and tilt. During indexing and integration, 25.1 % of images were indexed with at least one lattice. In total, 96,612 lattices were independently integrated. The repeated measurements are currently treated independently in indexing and

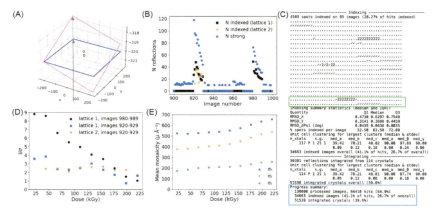

Fig. 8 Sample output for FutA SSX dose series data processing. (A) Comparison of detector models in the image headers (dark blue) and after joint geometry refinement (pink), showing a relative displacement and tilt. (B) Example indexing plot from a *dials.ssx_index.html* report, showing trends in the number of strong spots and indexing success for diffraction at images 920–929 and 980–989 as 10 measurements are made at each location on the chip. (C) A subsection of the *xia2.ssx* log output for indexing and integration of this batch of 1000 images, with the 100-image range from (B) highlighted by the added green box. Over the 1000 images, some physical crystals are successfully indexed on each image, but this is not always the case: '.' indicates an unindexed image, a tick is a singe lattice indexed and '2' is two lattices indexed on an image. A cumulative summary is also presented after each batch (added blue box). (D) Mean I/σ by dose, based on data in the *dials.ssx_integrate.html* report, for the indexed crystals from (B). A clear reduction in I/σ is seen for the crystal in images 980–989, whereas lattice 2 in images 920–929 has no evident change in I/σ with dose. (E) The average ellipsoid mosaicity components per dose point, calculated over the whole dataset after data reduction. The mosaicity increases gradually from 22 to 110 kGy, after which the increase is more pronounced.

integration resulting in up to 10 integrated lattices for each physical crystal. A selection of example output is show in Fig. 8B–D.

To scale all integrated data together before splitting and merging each dose point, the *xia2.ssx_reduce* command was used:

```
xia2.ssx_reduce batch_*/integrated* d_min=2.1 dose_series_repeat=10
```

In this case, we took advantage of the fact that there is a repeating pattern of experimental conditions by using the option *dose_series_repeat=10* to instruct merging into 10 groups based on the image index. A *d_min* value of 2.1 Å was used for data reduction to be consistent with the previously published data. No reference was used for data reduction in this example. 96,080 lattices remained after data reduction, which gave between 11,981 (66 kGy) and 6297 (220 kGy) lattices per dose point. For this dataset, there is no indexing ambiguity due to symmetry or accidental ambiguity due to similar lattice parameters. The merged MTZ file for each dose point was used as input for structure determination with *REFMAC5* via *DIMPLE*, with a starting model of PDB entry 8OGG (Bolton et al., 2024).

Table 3 shows a selection of data quality metrics for this processing, where all dose points are scaled together. Above 110 kGy, there is a reduction in I/σ that is indicative of radiation damage and cannot be accounted for solely by the reduced number of integrated crystals or reduction in multiplicity. In addition, the relative Wilson B-factor increases steadily up until the 176 kGy dose point. As a relative B-factor is applied to each image in scaling, data with a small amount of global radiation damage can in theory be corrected, such that the Wilson B-factors, when calculated from data placed on the same relative scale, would be approximately constant. Therefore, the increase in relative B-factor indicates that even at low doses the effect of radiation damage is causing a reduction of intensity with resolution to a greater extent than that which can be corrected for by a simple global B-factor correction. The effect of radiation damage can also be seen in the average mosaicity model parameters (Fig. 8E), which increase as a function of dose, particularly towards the end of the dose series.

To investigate the lower dose points, the data from 22 kGy to 88 kGy were scaled together. To achieve this, first the integrated data were split into separate files by dose using the *dials.split_still_data* command, then *xia2.ssx_reduce* was run on files only containing the first four doses.

```
dials.split_still_data batch_*/integrated*
series_repeat=10.
xia2.ssx_reduce group_{0,1,2,3}* d_min=2.1
dose_series_repeat=10.
```

Table 4 shows the processing statistics for this processing. The merging statistics for the four dose points are overall similar to when all ten dose points were processed together, with I/σ and multiplicity increased due to fewer reflections being rejected as outliers. Fig. 9 shows $2F_o - F_c$ electron density maps for the 22 kGy and 88 kGy datasets, as well as $F_o - F_c$ difference density maps. The density maps indicate a partial conformational change of the Arg203 residue, which is more pronounced at 88 kGy, as shown in the previous study (Bolton et al., 2024). Furthermore, at 88 kGy there is loss of density at the water site above the Fe and loss of density at the Tyr13 residue, i.e. site-specific changes have occurred.

Table 3 Summary merging statistics for ten-point FutA SSX dose series processed with xia2.ssx_reduce.

Dose (kGy)	22	44	66	88	110	132	154	176	198	220
n–crystals	10,149	11,395	11,981	11,623	10,960	9832	8170	7980	7153	6297
Multiplicity	34.9	41.6	46.9	47.2	45.2	41.2	38.1	37.2	35.4	32.2
I/sigma	12.3	13.7	14.0	13.1	11.7	10.1	8.5	7.3	6.2	5.3
$CC_{1/2}$	0.89	0.91	0.92	0.92	0.92	0.93	0.93	0.92	0.93	0.93
R_{split}	0.19	0.17	0.16	0.16	0.16	0.17	0.17	0.17	0.18	0.18
Wilson B (Å^2)	11.7	13.8	16.4	19.2	22.0	24.7	26.8	27.8	28.1	28.0
R_{work}	0.192	0.193	0.193	0.194	0.192	0.189	0.185	0.180	0.175	0.173
R_{free}	0.231	0.229	0.232	0.234	0.238	0.231	0.228	0.220	0.222	0.218

The resolution range for all datasets is 47.6 Å to 2.1 Å.

Table 4 Summary statistics for combined reduction of the first four dose points of the FutA SSX dose series processed with xia2.ssx reduce (resolution range 47.6 Å to 2.1 Å).

Dose (kGy)	22	44	66	88
n-crystals	10,149	11,396	11,981	11,621
Multiplicity	36.7	43.2	48.2	48.2
I/sigma	13.3	14.6	14.7	13.6
$CC_{1/2}$	0.90	0.92	0.93	0.93
R_{split}	0.19	0.17	0.16	0.16
Wilson B (Å^2)	11.4	13.7	16.6	19.6
R_{work}	0.197	0.197	0.196	0.195
R_{free}	0.236	0.233	0.234	0.233

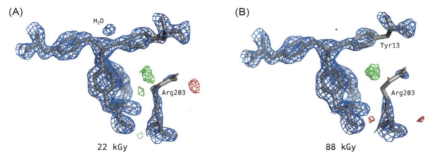

Fig. 9 Electron density maps of the Fe environment ($2F_o - F_c$, blue, 1.5σ) and difference density ($F_o - F_c$, green, 3σ) for FutA after accumulation of (A) 22 kGy and (B) 88 kGy. In both structures, the difference densities indicate partial conformational change of the Arg203 residue, which is greater at 88 kGy, consistent with results in Bolton et al. (2024). At 22 kGy, density is still present at the water above the Fe site, which is absent at 88 kGy, while loss of density at the Tyr13 residue is evident at 88 kGy dose. Images were generated with PyMOL (DeLano et al., 2002).

In this section, we demonstrated how mixed-state data can be scaled together with *xia2.ssx* before merging, and the tools that are available to split mixed-state data before scaling and merging. The purpose was to demonstrate the available tools rather than provide a recommendation for one correct approach to handle all possible experimental cases. The correct approach will depend on the level of isomorphism between the different experimental conditions and the scientific question being addressed by the experiment. Further algorithm developments could improve processing for repeated measurements on the same crystals, such as performing joint indexing on serial images from the same physical crystal to improve indexing success across the series of images. In the cases of indexing ambiguity resolution, this data could then be considered a single entity for reindexing.

6. Discussion

In this work, we have demonstrated the methods available in DIALS and *xia2* for processing serial still-shot diffraction data. This includes a generalised anisotropic model to describe the distribution of diffraction intensity about reciprocal lattice points, an empirical approach which can describe anisotropic mosaic block sizes and orientational distributions. This *ellipsoid* model is optimised using a maximum-likelihood algorithm that simultaneously considers the pixel intensity distribution, reflection offsets, and spot shapes. In addition to providing more accurate partiality estimates, this approach also offers the advantage of estimating the standard error of the partiality. This allows for more reliable weighting of integrated intensities during data reduction and merging. Currently only a mono-chromatic beam model has been implemented in DIALS, which is suitable for SSX but will have lower fidelity for pink-beam SSX or SFX. The inclusion of wavelength spread as part of the profile model optimisation and integration (Parkhurst, 2020), could bring further data quality improvements for those experiments.

We described and demonstrated some of the workflows available in the *xia2.ssx* and *xia2.ssx_reduce* programs for processing serial still-shot diffraction data. This includes automated detector geometry refinement, workflows for handling scaling and merging of data as well as the means to manage mixed-state datasets such as dose or time series datasets. As the right processing approach will depend on the dataset and experiment in question, the aim was to demonstrate the features available and the use of the program outputs to evaluate the data processing steps, which we encourage any users of the programs to become familiar with. We demonstrated an example dataset where the ellipsoid integration algorithm gives improvements upon isotropic models, however there can still be a significant fraction of outliers after integration. In these cases, the use of a reference intensity set during scaling is beneficial for outlier rejection. For typical

SSX experiments, where the purpose of the study is to investigate dynamics of a known structure, suitable reference intensity sets are typically available. Currently we do not include a means to post-refine the ellipsoid profile model, which could further improve the final data quality. To post-refine the ellipsoid model would require rewriting the scaling objective function in a likelihood formalism, which could then be maximised in combination with the likelihood function of the ellipsoid model (Eq. 9). This contrasts with post-refinement methods which minimise the deviation between predicted and observed spot positions. However, post-refinement of the ellipsoid model would have a much higher requirement on computational resources than the integration step, as it would require log–likelihood optimisation not limited to only the strong reflections, and must be done for all crystals together, compared to integration which is done individually per crystal. Another important aspect of SSX processing is the handling of non–isomorphism. *xia2.ssx* contains methods to handle cell non–isomorphism and indexing non–isomorphism, while DIALS includes further tools to investigate intensity non–isomorphism, such as $\Delta CC_{1/2}$ analysis (Winter et al., 2022). The ability to use the output of the *xia2.ssx* pipeline for further analysis provides flexibility in the analysis that can be performed with DIALS and *xia2*, enabling the user to tailor the analysis to tackle technique-specific or dataset-specific challenges in serial synchrotron crystallography.

Acknowledgments

DIALS/xia2 processing was performed with DIALS software version 3.21.0 (August 2024), *cctbx.xfel.merge* processing was performed with cctbx version at commit 773e9a2. We are grateful to beamline I24, Diamond Light Source, for providing the AcNiR dataset. We are grateful to Ivo Tews and beamline I24, Diamond Light Source, for providing the FutA dose-series dataset.

The development of DIALS has been or is supported by Diamond Light Source, STFC via CCP4,

Biostruct-X project No. 283570 of the EU FP7, the Wellcome Trust (grant nos. 202933/Z/16/Z and 218270/Z/19/Z), and US National Institutes of Health grants GM095887 and GM117126.

Appendix A. Block matrix inversion

If a matrix, A, can be partitioned as:

$$A = \begin{pmatrix} A_{11} & A_{12} \\ A_{21} & A_{22} \end{pmatrix} \tag{15}$$

then, as shown in Petersen & Pedersen (2012), its inverse is

$$A = \begin{pmatrix} A^{11} & A^{12} \\ A^{21} & A^{22} \end{pmatrix} \tag{16}$$

where

$$
\begin{aligned}
A^{11} &= (A_{11} - A_{12}A_{22}^{-1}A_{21})^{-1} = A_{11}^{-1} + A_{11}^{-1}A_{12}(A_{22} - A_{21}A_{11}^{-1}A_{12})^{-1}A_{21}A_{11}^{-1} \\
A^{12} &= -(A_{11} - A_{12}A_{22}^{-1}A_{21})^{-1}A_{12}A_{22}^{-1} = (A^{21})^T \\
A^{21} &= -A_{22}^{-1}A_{21}(A_{11} - A_{12}A_{22}^{-1}A_{21})^{-1} = (A^{12})^T \\
A^{22} &= A_{22}^{-1} + A_{22}^{-1}A_{21}(A_{11} - A_{12}A_{22}^{-1}A_{21})^{-1}A_{12}A_{22}^{-1} = (A_{22} - A_{21}A_{11}^{-1}A_{12})^{-1}
\end{aligned} \tag{17}
$$

Appendix B. Derivatives of the log-likelihood and Fisher Information matrix

The Fisher scoring algorithm uses the Fisher Information matrix $\mathbf{I}$, which is the negative of the expected value of the Hessian Matrix. The Fisher Information matrix is guaranteed to be positive semi-definite, leading to more robust convergence. Additionally, only the first derivatives need to be calculated, resulting in a less computationally intensive implementation (Osborne, 1992). In the Fisher method, at each iteration, $(t + 1)$, the parameters β are updated according to

$$\beta(t + 1) = \beta(t) + I(\beta(t))^{-1}V(\beta(t)), \tag{18}$$

where $V(\beta_{(k)})$ is known as the score function and the elements of the Fisher Information matrix $\mathbf{I}$ are given by

$$I_{kl} = -E\left[\frac{\partial^2 L}{\partial \beta_k \partial \beta_l}\right] \tag{19}$$

where L is the log-likelihood.

As shown in (Petersen & Pedersen, 2012), the derivative of the determinant, trace and inverse of a matrix, A, are given by:

$$\frac{\partial}{\partial x}|A| = |A|\,\mathrm{tr}\left(A^{-1}\frac{\partial A}{\partial x}\right),\ \frac{\partial}{\partial x}\mathrm{tr}(A) = \mathrm{tr}\left(\frac{\partial A}{\partial x}\right),\ \frac{\partial}{\partial x}A^{-1} = -A^{-1}\frac{\partial A}{\partial x}A^{-1}. \tag{20}$$

Using these identities, the first derivatives of the log-likelihood function in Eq. 11 with respect to a parameter β_k are given by

$$
\begin{aligned}
\frac{\partial L}{\partial \beta_k} = &-\frac{1}{2}\sum_i^N c_{\mathrm{tot},i}\left[\Sigma_{22i}^{-1}\frac{\partial \Sigma_{22i}}{\partial \beta_k} - \Sigma_{22i}^{-2}\frac{\partial \Sigma_{22i}}{\partial \beta_k}\epsilon_i^2 + 2\Sigma_{22i}^{-1}\epsilon_i\frac{\partial \epsilon_i}{\partial \beta_k}\right] \\
&-\frac{1}{2}\sum_i^N c_{\mathrm{tot},i}\left[\mathrm{tr}\left(\boldsymbol{\Sigma}_i^{-1}\frac{\partial \boldsymbol{\Sigma}_i}{\partial \beta_k} - \boldsymbol{\Sigma}_i^{-1}\frac{\partial \boldsymbol{\Sigma}_i}{\partial \beta_k}\boldsymbol{\Sigma}_i^{-1}\left(\mathbf{S}_i + (\bar{\mathbf{x}}_i - \bar{\mu}_i)(\bar{\mathbf{x}}_i - \bar{\mu}_i)^T\right)\right)\right] \\
&-\frac{1}{2}\sum_i^N c_{\mathrm{tot},i}\left[2\mathrm{tr}\left(\boldsymbol{\Sigma}_i^{-1}\left((\bar{\mathbf{x}}_i - \bar{\mu}_i)\frac{\partial \bar{\mu}_i^T}{\partial \beta_k}\right)\right)\right]
\end{aligned} \tag{21}
$$

The derivatives of $\overline{\Sigma}$, $\overline{\mu}$, Σ_{22} and ϵ are given by

$$\frac{\partial \Sigma}{\partial \beta_k} = \frac{\partial \Sigma_{11}}{\partial \beta_k} - \frac{\partial \Sigma_{12}}{\partial \beta_k}\Sigma_{22}^{-1}\Sigma_{21} + \Sigma_{12}\Sigma_{21}\Sigma_{22}^{-2}\frac{\partial \Sigma_{22}}{\partial \beta_k} - \Sigma_{12}\Sigma_{22}^{-1}\frac{\partial \Sigma_{21}}{\partial \beta_k}$$

$$\frac{\partial \mu}{\partial \beta_k} = \frac{\partial \mu_1}{\partial \beta_k} + \frac{\partial \Sigma_{12}}{\partial \beta_k}\Sigma_{22}^{-1}\varepsilon - \Sigma_{12}\Sigma_{22}^{-2}\frac{\partial \Sigma_{22}}{\partial \beta_k}\varepsilon - \Sigma_{12}\Sigma_{22}^{-1}\frac{\partial \varepsilon}{\partial \beta_k} \qquad (22)$$

$$\frac{\partial \varepsilon}{\partial \beta_k} = -\frac{\partial \mu_2}{\partial \beta_k}$$

As $\Sigma = \mathbf{R_e M R_e^T}$, the derivative of sigma in terms of the mosaicity model parameters is given by

$$\frac{\partial \Sigma}{\partial \beta_k} = \mathbf{R_e}\frac{\partial M}{\partial \beta_k}\mathbf{R^T}, \qquad (23)$$

where the local reflection-specific coordinate system is approximated to be fixed during optimisation of crystal cell and orientation parameters.

The Fisher information matrix requires evaluation of the expected value of the second derivatives.

The expected values of the first and second moments of the distribution of the spots about the Ewald Sphere are $E[\epsilon_i] = 0$ and $E[e_i^2] = \Sigma_{22i}$. The first and second moments of the general impact of the spot in the local coordinate system are $E[(\overline{x}_i - \overline{\mu}_i)] = 0$ and $E[S_i] = \overline{\Sigma}_i$. The elements of the Fisher information matrix only require the first derivatives and are calculated as

$$\begin{aligned} I_{kl} &= \frac{1}{2}\sum_i^N c_{\text{tot},i}\left[\Sigma_{22i}^{-2}\frac{\partial \Sigma_{22i}}{\partial \beta_k}\frac{\partial \Sigma_{22i}}{\partial \beta_l} + 2\Sigma_{22i}^{-1}\frac{\partial \epsilon_i}{\partial \beta_k}\frac{\partial \epsilon_i}{\partial \beta_l}\right] \\ &+ \frac{1}{2}\sum_i^N c_{\text{tot},i}\left[\text{tr}\left(\overline{\Sigma}_i^{-1}\frac{\partial \overline{\Sigma}_i}{\partial \beta_k}\overline{\Sigma}_i^{-1}\frac{\partial \overline{\Sigma}_i}{\partial \beta_l} + 2\overline{\Sigma}_i^{-1}\left(\frac{\partial \overline{\mu}_i}{\partial \beta_k}\frac{\partial \overline{\mu}_i}{\partial \beta_l}\right)\right)\right] \end{aligned} \qquad (24)$$

References

Antonyuk, S. V., Strange, R. W., Sawers, G., Eady, R. R., & Hasnain, S. S. (2005). *Proceedings of the National Academy of Sciences, 102*(34), 12041–12046.

Axford, D., Owen, R. L., Aishima, J., Foadi, J., Morgan, A. W., Robinson, J. I., ... Fry, E. E. (2012). *Acta Crystallographica Section D: Biological Crystallography, 68*(5), 592–600.

Barty, A., Kirian, R. A., Maia, F. R., Hantke, M., Yoon, C. H., White, T. A., & Chapman, H. (2014). *Journal of applied crystallography, 47*(3), 1118–1131.

Beilsten-Edmands, J., Winter, G., Gildea, R., Parkhurst, J., Waterman, D., & Evans, G. (2020). *Acta Crystallographica Section D: Structural Biology, 76*(4), 385–399.

Bolton, R., Machelett, M. M., Stubbs, J., Axford, D., Caramello, N., Catapano, L., ... Tews, I. (2024). *Proceedings of the National Academy of Sciences, 121*(12), e2308478121. https://www.pnas.org/doi/abs/10.1073/pnas.2308478121.

Brehm, W., & Diederichs, K. (2014). *Acta Crystallographica Section D: Biological Crystallography, 70*(1), 101–109.

Brehm, W., White, T., & Chapman, H. N. (2023). *Acta Crystallographica Section A: Foundations and Advances, 79*(2), 145–162.

Brewster, A., Young, I., Lyubimov, A., Bhowmick, A., & Sauter, N. (2019a). *Computational Crystallography Newsletter, 10*, 22–39.

Brewster, A. S., Bhowmick, A., Bolotovsky, R., Mendez, D., Zwart, P. H., & Sauter, N. K. (2019b). *Acta Crystallographica Section D: Structural Biology, 75*(11), 959–968.

Brewster, A. S., Waterman, D. G., Parkhurst, J. M., Gildea, R. J., Young, I. D., O'Riordan, L. J., ... Sauter, N. K. (2018). *Acta Crystallographica Section D, 74*(9), 877–894. https://doi.org/10.1107/S2059798318009191.

DeLano, W. L., et al. (2002). CCP4 newsl. *Protein Crystallography, 40*(1), 82–92.

Ebrahim, A., Appleby, M. V., Axford, D., Beale, J., Moreno-Chicano, T., Sherrell, D. A., ... Owen, R. L. (2019). *Acta Crystallographica Section D: Structural Biology, 75*(2), 151–159.

Fisher, R. A. (1922). Philosophical transactions of the Royal Society of London. *Series A, Containing Papers of a Mathematical or Physical Character, 222*(594–604), 309–368.

Gildea, R. J., Waterman, D. G., Parkhurst, J. M., Axford, D., Sutton, G., Stuart, D. I., ... Winter, G. (2014). *Acta Crystallographica Section D: Biological Crystallography, 70*(10), 2652–2666.

Gildea, R. J., & Winter, G. (2018). *Acta Crystallographica Section D: Structural Biology, 74*(5), 405–410.

Ginn, H. M., Evans, G., Sauter, N. K., & Stuart, D. I. (2016). *Journal of Applied Crystallography, 49*(3), 1065–1072.

Grosse-Kunstleve, R. W., Sauter, N. K., Moriarty, N. W., & Adams, P. D. (2002). *Journal of Applied Crystallography, 35*(1), 126–136.

Grünbein, M. L., & Nass Kovacs, G. (2019). *Acta Crystallographica Section D: Structural Biology, 75*(2), 178–191.

Hough, M. A., & Owen, R. L. (2021). *Current Opinion in Structural Biology, 71*, 232–238.

Kabsch, W. (2010a). *Acta Crystallographica Section D, 66*(2), 133–144. https://doi.org/10.1107/S0907444909047374.

Kabsch, W. (2010b). *Acta Crystallographica Section D: Biological Crystallography, 66*(2), 125–132.

Kabsch, W. (2014). *Acta Crystallographica Section D: Biological Crystallography, 70*(8), 2204–2216.

Le Page, Y. (1982). *Journal of Applied Crystallography, 15*(3), 255–259.

Leonarski, F., Bru¨ckner, M., Lopez-Cuenca, C., Mozzanica, A., Stadler, H.-C., Matˇej, Z., ... Wang, M. (2023). *Journal of Synchrotron Radiation, 30*(1), 227–234. https://doi.org/10.1107/S1600577522010268.

Lyubimov, A. Y., Uervirojnangkoorn, M., Zeldin, O. B., Brewster, A. S., Murray, T. D., Sauter, N. K., ... Brunger, A. T. (2016). *Journal of Applied Crystallography, 49*(3), 1057–1064.

Mariani, V., Morgan, A., Yoon, C. H., Lane, T. J., White, T. A., O'Grady, C., ... Barty, A. (2016). *Journal of Applied Crystallography, 49*(3), 1073–1080.

Mendez, D., Bolotovsky, R., Bhowmick, A., Brewster, A. S., Kern, J., Yano, J., ... Sauter, N. K. (2020). *IUCrJ, 7*(6), 1151–1167.

Murshudov, G. N., Skub´ak, P., Lebedev, A. A., Pannu, N. S., Steiner, R. A., Nicholls, R. A., ... Vagin, A. A. (2011). *Acta Crystallographica Section D: Biological Crystallography, 67*(4), 355–367.

Nave, C. (1998). *Acta Crystallographica Section D: Biological Crystallography, 54*(5), 848–853.

Neutze, R., Wouts, R., Van der Spoel, D., Weckert, E., & Hajdu, J. (2000). *Nature, 406*(6797), 752–757.

Orville, A. M. (2018). *BMC Biology, 16*(55).

Osborne, M. R. (1992). *International Statistical Review/Revue Internationale de Statistique*, 99–117.

Padilla, J. E., & Yeates, T. O. (2003). *Acta Crystallographica Section D: Biological Crystallography, 59*(7), 1124–1130.

Pandey, S., Bean, R., Sato, T., Poudyal, I., Bielecki, J., Cruz Villarreal, J., ... Kupitz, C. (2020). *Nature Methods, 17*(1), 73–78.

Parkhurst, J. M. (2020). *Statistically robust methods for the integration and analysis of X-ray diffraction data from pixel array detectors* (Ph.D. thesis).

Petersen, K. B., & Pedersen, M. S. (2012). *Technical University of Denmark.* www. math. uwaterloo. ca/~ hwolkowi~....

Sauter, N. K. (2015). *Journal of Synchrotron Radiation, 22*(2), 239–248.

Sauter, N. K., Hattne, J., Brewster, A. S., Echols, N., Zwart, P. H., & Adams, P. D. (2014). *Acta Crystallographica Section D: Biological Crystallography, 70*(12), 3299–3309.

Thorn, A., & Sheldrick, G. M. (2011). *Journal of Applied Crystallography, 44*(6), 1285–1287.

Uervirojnangkoorn, M., Zeldin, O. B., Lyubimov, A. Y., Hattne, J., Brewster, A. S., Sauter, N. K., ... Weis, W. I. (2015). *Elife, 4,* e05421.

White, T. A. (2019). *Acta Crystallographica Section D: Structural Biology, 75*(2), 219–233.

White, T. A., Barty, A., Stellato, F., Holton, J. M., Kirian, R. A., Zatsepin, N. A., & Chapman, H. N. (2013). *Acta Crystallographica Section D: Biological Crystallography, 69*(7), 1231–1240.

White, T. A., Kirian, R. A., Martin, A. V., Aquila, A., Nass, K., Barty, A., & Chapman, H. N. (2012). *Journal of Applied Crystallography, 45*(2), 335–341.

Winter, G. (2010). *Journal of Applied Crystallography, 43*(1), 186–190.

Winter, G., Beilsten-Edmands, J., Devenish, N., Gerstel, M., Gildea, R. J., McDonagh, D., ... Evans, G. (2022). *Protein Science, 31*(1), 232–250.

Winter, G., Waterman, D. G., Parkhurst, J. M., Brewster, A. S., Gildea, R. J., Gerstel, M., ... Young, I. D. (2018). *Acta Crystallographica Section D: Structural Biology, 74*(2), 85–97.

CHAPTER NINE

Time-resolved scattering methods for biological samples at the CoSAXS beamline, MAX IV Laboratory

Fátima Herranz-Trillo[a], Henrik Vinther Sørensen[a,b], Cedric Dicko[c], Javier Pérez[d], Samuel Lenton[e], Vito Foderà[e], Anna Fornell[f], Marie Skepö[b], Tomás S. Plivelic[a], Oskar Berntsson[a], Magnus Andersson[g], Konstantinos Magkakis[g], Fredrik Orädd[g], Byungnam Ahn[a], Roberto Appio[a], Jackson Da Silva[a], Vanessa Da Silva[a], Marco Lerato[a], and Ann E. Terry[a,*]

[a]MAX IV Laboratory, Lund University, Lund, Sweden
[b]Department of Chemistry, Division of Computational Chemistry, Lund University, Lund, Sweden
[c]Division of Pure and Applied Biochemistry, Lund University, Lund, Sweden
[d]Synchrotron SOLEIL, Saint-Aubin - BP, Gif sur Yvette Cedex, France
[e]Department of Pharmacy, Faculty of Health and Medical Sciences, University of Copenhagen, Universitetsparken, Copenhagen, Denmark
[f]Division of Biomedical Engineering, Department of Materials Science and Engineering, Science for Life Laboratory, Uppsala University, Uppsala, Sweden
[g]Department of Chemistry, Umeå University, Umeå, Sweden
*Corresponding author. e-mail address: ann.terry@maxiv.lu.se

Contents

1. Introduction	246
2. Technical design of CoSAXS	248
2.1 Optical design	249
2.2 Experimental hutch	250
2.3 Vacuum vessel	252
2.4 Data acquisition and processing	252
3. SUrF—combined SAXS with UV–vis/Raman/fluorescence	254
3.1 Design of the multiprobe platform SUrF	255
3.2 Time-dependent acid-induced unfolding of BSA with SUrF	259
4. Time-resolved SAXS studies using microfluidic chips	265
4.1 Microfluidic chip design	267
4.2 Following surfactant-induced structural changes in a microfluidic chip	269
5. Structural kinetics investigated with stopped-flow	275
5.1 Structural changes induced by addition of Ca^{2+} ions	276
6. Time-resolved X-ray solution scattering, TR-XSS	282
6.1 Design of the TR-XSS experimental setup	283
6.2 Thermal response of lysozyme to laser induced temperature jumps	285

Methods in Enzymology, Volume 709
ISSN 0076-6879, https://doi.org/10.1016/bs.mie.2024.10.019
Copyright © 2024 Elsevier Inc. All rights are reserved, including those for text and data mining, AI training, and similar technologies.

245

6.3 Laser-induced activation of ATP binding in adenylate kinase 289
7. Future perspectives 290
Acknowledgments 292
References 292

Abstract

CoSAXS is a state-of-the-art SAXS/WAXS beamline exploiting the high brilliance of the MAX IV 3 GeV synchrotron. By coupling advances in sample environment control with fast X-ray detectors, millisecond time-resolved scattering methods can follow structural dynamics of proteins in solution. In the present work, four sample environments are discussed. A sample environment for combined SAXS with UV–vis and fluorescence spectroscopy (SUrF) enables a comprehensive understanding of the time evolution of conformation in a model protein upon acid-driven denaturation. The use of microfluidic chips with SAXS allows the mapping of concentration with very small sample volumes. For highly reproducible sequences of mixing of components, it is possible using stopped-flow and SAXS to access the initial effects of mixing at 2 millisecond timescales with good signal to noise to allow structural interpretation. The intermediate structures in a protein are explored under light and temperature perturbations by using lasers to "pump" the protein and SAXS as the "probe". The methods described demonstrate that features at low q, corresponding to cooperative motions of the atoms in a protein, could be extracted at millisecond timescales, which results from CoSAXS being a highly-stable, low background, dedicated SAXS beamline.

1. Introduction

In this chapter we introduce the capabilities for time resolved scattering methods for biologically relevant samples, which are available, mid 2024, at the CoSAXS beamline, MAX IV. These methods provide structural information from liquid samples based on the interpretation of small and wide angle X-ray scattering. The chapter does not represent an exhaustive list of all that CoSAXS offers to its scientifically diverse user community nor does it seek to review the time–resolved techniques offered at other large-scale facilities.

The time resolution achievable is very broad, from millisecond experiments to those over several seconds. This temporal scale covers a range of important biological processes, such as protein folding, membrane transport, and enzyme catalysis (for example, Deng, Zhadin & Callender, 2001; Khodadadi & Sokolov, 2015; Le Ferrand, Duchamp, Gabryelczyk, Cai & Miserez, 2019). There are methods that can be employed to study longer timescales, for example, regularly examining an aliquot of a reactant at different time intervals, which we can perform at CoSAXS. Advances in laboratory-based X-ray sources, for example, the metal jet technology, developed by *Excillum*, have improved the time resolution available at home sources to the order of seconds. Indeed, many of the measurements at CoSAXS are supported by static structural data collected on such sources. Instead, this chapter concentrates on examining the millisecond time-resolved changes where we take advantage of the very high brilliance offered by the 4th generation MAX IV synchrotron and state of the art X-ray detectors.

Small angle X-ray scattering (SAXS) is a relatively simple technique where the forward scattering at small angles, typically less than 4°, provides information about the size and shape of ensembles of atoms in solution, and possibly about interactions between these molecules. In this chapter we use model proteins to demonstrate the information we can extract from our methods. In comparison to macromolecular crystallography, where a protein is constrained within an artificially imposed crystalline form, SAXS data from, for example, proteins, amides, lipids, vesicles, are often gathered in as near in vivo conditions as possible. It should be noted that we still often must impose some limits on our samples to extract the structural information, for example, by suppressing intermolecular interactions by diluting the protein in its buffer. We then examine the biomolecules during imposed external stimuli, for example, whilst changing the buffer conditions, temperature, light excitation, pH, and monitoring how these molecules deform, aggregate, diffuse, etc. However, the interpretation of the SAXS signal is not straightforward; BioSAXS is the term commonly used to describe the analysis methods to extract information and 3D reconstructions from biologically relevant molecules and ensembles. Time-resolved X-ray solution scattering compares the difference in the scattering from perturbated to non-perturbated conditions and usually is strongly supported by Molecular Dynamics analysis. It is not the aim of this chapter to review these methods, instead we refer the interested reader to excellent reviews in the literature (Graewert & Svergun, 2013; Bizien, Durand, Roblina, Thureau, Vachette & Pérez, 2016; Brosey & Tainer, 2019; Kirby & Cowieson, 2014; Levantino, Yorke, Monteiro, Cammarata & Pearson, 2015).

As proteins and biological molecules do not have a large proportion of high atomic number elements in their composition, they will have very low electron density contrast when compared to the buffer or solution in which they are measured and thus the SAXS signal is very weak; typically, only 1 in 10^6 X-ray photons are scattered by such molecules. Very great care is therefore expended in the design and optimization of beamlines and equipment to remove any contributions from noise and background scattering in the measurements, utilizing very thin walled, in-vacuum sample environments for these methods. Working in vacuum limits the types and ranges of external stimuli which can be imposed on the samples and for this reason many of our sample environments necessarily operate in air. We will describe in this chapter how the use of these often commercially available equipment, for example, microfluidic chips and stopped-flow devices, can be coupled into the beamline as well as the developments at CoSAXS of combining scattering with spectroscopy and with laser-induced time-resolved methods.

CoSAXS has been in general user operation at the MAX IV 3 GeV synchrotron since December 2020, whilst also continually developing new capabilities. We routinely offer simultaneous SAXS/WAXS and USAXS measurements which we complement with a suite of sample environments, including automated BioSAXS measurements. In this way we support many scientific research areas from soft matter and life science to colloidal science. In the following, we describe how the CoSAXS optical and mechanical design achieves a highly stable, high-flux beamline. Coupled with fast photon counting X-ray detectors, this allows us to exploit the very high brilliance of the MAX IV 3 GeV ring for time-resolved methods. As CoSAXS serves many different types of experiments, its optical design and operating parameters are more flexible compared to beamlines offering a single experimental method. We can utilize a wide range of X-ray energies, beam size, focusing lengths and sample-to-detector distances. The following section describes how we have achieved the desired stability of the X-ray beam and detector system without compromising the range of experiments feasible at CoSAXS.

2. Technical design of CoSAXS

The design parameters for CoSAXS were chosen to offer a simultaneous q-range of $6 \times 10^{-4} < q < 6$ Å^{-1} with an X-ray photon flux of $>10^{13}$ photons/s at 12 keV. The beamline should operate at 4 to 20 keV with an energy discrimination of 2×10^{-4} $\Delta E/E$. It is also intended to be able to focus the beam from the sample position to fully defocused beam whilst still achieving high signal to noise and low background experiments. X-ray tracing, uniquely designed optics and finite elements of the mechanical stability of all the elements in the beamline have helped achieve this (Plivelic et al., 2019). Fig. 1 shows the layout of CoSAXS with the key components marked and their position along the beamline.

CoSAXS is equipped with a 2 m long, in-vacuum undulator (*Hitachi Metals Ltd*) (101 periods with 19.3 mm period length) with 4.2 mm minimum magnetic gap (K = 1.92 max). This provides a continuous energy range from 4 to 20 keV. Typically, we use the 5th to 11th harmonics for energies from 7 to 18 keV.

2.1 Optical design

The Si (111), horizontally deflecting, double crystal monochromator (hDCM) (*RI Instruments GmbH*) allows an energy range of 4 to 20 keV, with a narrow bandpass of $\Delta E/E$ of 2×10^{-4}. The horizontal geometry was chosen to preserve the vertical coherence inherent to the 3 GeV ring at MAX IV. The first crystal, positioned at 25 m from the source, is actively cryo-cooled (*Axilon AG*), the second crystal indirectly cooled through copper braids. The hDCM incorporates an inclined crystal geometry where the diffracting plane is Si (111), but the surfaces of the crystals are the (100) plane. This increases the footprint of the incident X-ray beam, thereby reducing the thermal load on the monochromator and increasing the

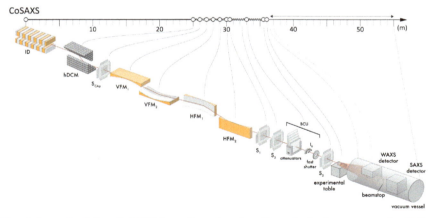

Fig. 1 Layout of the CoSAXS beamline showing the principal components of the beamline and their distance from the source. The diagnostic units, vacuum components, radiation protection shutters are not shown. Equally only 4 sets of slits are shown for clarity. ID: Insertion Device; hDCM: monochromator; VFM$_1$, VFM$_2$, HFM$_1$, HFM$_2$: mirrors; S$_1$, S$_2$, S$_3$: collimation slits; S$_{CAp}$: coherent aperture slit; BCU: beam conditioning unit; IM: incident intensity monitor, I$_0$. The components upstream of S1 are in the front end and optics hutch, components after are in the experimental hutch. The SAXS detector can move along the length of the vacuum vessel.

thermal stability of the monochromatic beam. There is no significant need to wait to achieve thermalization of the monochromator crystals following an electron beam dump, for example.

Downstream of the monochromator are paired Kirkpatrick-Baez (KB) mirrors (VFM_1, VFM_2, HFM_1, HFM_2), in a four-bounce geometry, for higher harmonic rejection and focusing. To cover the energy range of CoSAXS, the mirrors have two separate stripes of Si or Rh (*InSync Inc.*). Independent vertical and horizontal focusing is achieved by controlled bending with mechanical benders and strain gauges for monitoring the applied forces (*IRELEC*). This configuration allows selectable foci from the shortest distance at the nominal sample position, to any position downstream thereof; at the sample position an X-ray beam size of $30 \times 19 \, \mu m$ (width x height, FWHM) is achieved. The predicted performance of the beamline and mirrors is detailed in Plivelic et al. (2019).

Within the optics hutch are a white beamstop, bremsstrahlung collimator and slits, which are all necessary components in a synchrotron beamline to reduce the high energy background and improve signal to noise. Diamond screens coupled with CCD cameras can be inserted into the monochromatic beam to aid in beamline alignment and optimization. All the optical components are housed within a lead/steel shielded optics hutch and remotely controlled with feedback systems, including vacuum pumps, gauges and temperature monitoring. A downstream hutch, the experimental hutch, is where the users conduct their experiments. There is no access to any hutch during an experiment so all components and experimental conditions must be controlled remotely.

2.2 Experimental hutch

The components in the experimental hutch are also shown in Fig. 1. At the entrance to the experimental hutch, the beam passes through a double set of slits, S_1 and S_2, (*JJ X-ray A/S*) to suppress parasitic X-ray scatter, especially from the mirror surfaces. Two different blades are within this slit assembly, either so-called scatterless hybrid slits (Li, Beck, Huang, Choi & Divinagracia, 2008) with single crystal InP wafers mounted on tungsten carbide blades or conventional tungsten carbide blades; typically, the former are used for most beamline configurations.

Immediately after the slits is an in-vacuum beam conditioning unit (BCU), which includes a 3-axis filter unit in which piezo stages (*Micro Control Systems, Inc.*) translate different thickness foils of aluminum, titanium and silicon into the beam path, the latter two materials typically being used during beamline alignment and not during experimental measurements. For some experiments, the incident X-ray flux is too high and radiation damage occurs within the sample, this effect can be reduced by attenuating the beam with the aluminum foils, if no other mitigation strategy is possible. Within the BCU is also the incident beam intensity monitor (I_0). This consists of a 6 µm thick high purity aluminum foil (*Goodfellow Cambridge Ltd*) mounted at 45° to the X-ray beam axis with a silicon photodiode facing the foil to measure linearly the incident beam intensity. The BCU also houses a fast operating piezo shutter (<8 ms to fully open) (*Cedrat Technologies*), which is synchronized with the data acquisition to reduce radiation damage within the sample.

Between the BCU and the sample position is a telescopic unit (not shown in Fig. 1), motorized along the beam path, to minimize the X-ray path in air for the differing experimental requirements to achieve the desired low backgrounds. There is a final set of in-vacuum scatterless hybrid slits (S_3) (*JJ X-ray A/S*) again fitted with InP single crystal wafers, which move with the telescopic unit. These are the final guard slits before the sample. The telescopic unit can either be connected directly to the sample environment, i.e., "through-vacuum", lowest background configuration, where there are no vacuum windows in the X-ray beam path and continuous vacuum from the electron ring to the vacuum vessel housing the X-ray detectors, or with a 100 mm long final collimation tube mounted with 6 to 10 mm diameter, 1 µm thick Si_3N_4 windows (*Norcada Inc.*) to allow the installation of sample environments in air. An in-line beam viewer based on a 45° mounted mirror and a perpendicularly mounted CCD camera and focusing lens assembly (Da Silva, Appio, Mota-Santiago, Plivelic, Terry & Herranz, 2024) can be installed downstream of the slits within this telescopic unit.

A large granite experimental table was designed within MAX IV to allow mounting of sample environments of up to 50 kg on an optical breadboard 1 m x 1 m, the center of which (the nominal sample position) is located 36 m downstream of the source. This provides good thermal and mechanical stability to the sample environments. Large heavy sample environments are intended to be mounted directly onto the experimental table. Smaller sample environments can be mounted on additional vertical and horizontal translation stages. All axes for the experimental table and sample stages have absolute encoders (*Renishaw plc*) allowing accurate repositioning of the samples.

2.3 Vacuum vessel

The SAXS and WAXS detectors are installed within a 1 m diameter, 16 m long vacuum vessel (*Added Value Industrial Engineering Solutions*), the front of which is within the experimental hutch but extends into the MAX IV experimental hall outside of the hutch. The vacuum vessel is required to reduce the air scattering by the transmitted X-ray beam and to allow the detection of the weak SAXS signals which are at least 10^{-6} lower intensity than the X-ray beam transmitted through the sample. The vessel is constructed from 6 mm thick steel, satisfying radiation safety and mechanical stability requirements. A tapered nose cone is mounted on a gate valve at the front of the vessel and sample environments can be either connected directly to the nose cone or a 2 μm thick, 20 mm diameter Si_3N_4 window (*Norcada Inc.*) can be mounted.

The q-range which can be covered by the SAXS detector depends on the sample-to-detector distance (SDD), the lateral offset of the detector in the scattering plane, X-ray beam energy and the diameter of the X-ray beam stop. The in-vacuum SAXS detector is mounted within a carriage on rails to allow SDD from 0.75 m to 14.5 m and lateral translation of the SAXS detector across the entire detection face and vertical translation of approximately 2/3 of the height of the detector face. The minimum q, at 12.4 keV, is 8×10^{-4} $Å^{-1}$ at 14.5 m SDD. The beamstop (*ANSTO*) in front of the SAXS detector allows monitoring of the transmitted X-ray beam intensity (I_t).

The WAXS detector is mounted at the entrance to the vacuum vessel at a SDD of 0.5 m. This detector has limited horizontal translation of < 10 mm in the scattering plane. The maximum q, at 12.4 keV, is currently $\sim 2 Å^{-1}$.

2.4 Data acquisition and processing

In the standard SAXS/WAXS configuration, CoSAXS uses two mega-pixel, in-vacuum, hybrid photon-counting detectors for data acquisition. The SAXS detector is an EIGER®2 X 4 M (*Dectris AG*). The WAXS detector is a custom "L-shaped" PILATUS®3 X 2 M (*Dectris AG*) in which the bottom, right hand side quadrant is removed to allow the SAXS signal to pass unobstructed to the downstream SAXS detector. The detector configuration ensures good overlapping q-range at 3.5 m SDD and minimal overlap for 10 m SDD. Typically, more than 3 orders of q-range are achieved, $0.001 < q < 2 Å^{-1}$. When wider q is needed, a MYTHEN®2 1 K strip detector (*Dectris AG*) can be mounted in air on the side of the nose cone. Table 1 provides technical details of the CoSAXS detectors.

The control system at CoSAXS uses the Tango protocol (https://tango-controls.org) to communicate with all the beamline components. On top of that is Sardana (Coutinho et al., 2011) which provides a software environment for orchestrating the motion of IcePAP-controlled (Janvier, Clement, Fajardo & Cuni, 2013) stepper motors, signal acquisition, and macros to control the experimental conditions. The time synchronization of the experiment, sending and receiving of TTL triggers, motor movements, operation of the fast shutter, data acquisition from the SAXS and WAXS detectors, and signals and amplification of I_o and I_t is via two PandABoxes (Position and Acquisition Box) (Zhang, Abiven, Bisou, Renaud, Thibaux & Ta, 2017); one orchestrates the basic beamline

Table 1 Technical specifications of the SAXS and WAXS detectors available at CoSAXS.

	EIGER®2 X 4 M	PILATUS®3 X 2 M	MYTHEN®2 X 1 K
Number of pixels	4 M	2 M	Single strip detector
Sensor size (w × h) in pixels	2068 × 2162	1679 × 1475 (bottom right hand side quadrant removed)	
Pixel size	$75 \times 75\,\mu m^2$	$172 \times 172\,\mu m^2$	
Sensor material and thickness	Si, 450 μm	Si, 450 μm	Si, 450 μm
Energy range	6–40 keV	5–36 keV	6.6–40 keV
Maximum frame rate	500 Hz @ 16 bit	250 Hz @ 32 bit	1000 Hz @ 24 bit
Operating conditions	In vacuum $<10^{-2}$ mbar	In vacuum $<10^{-2}$ mbar	In air

For most experiments, the EIGER®2 X 4 M is used for SAXS experiments simultaneously with the PILATUS®3 X 2 M, both detectors are installed in a 16 m long, 1 m diameter vacuum vessel at CoSAXS. The SAXS detector is on a carriage and can be positioned along the length of the vacuum vessel, such that sample-to-detector distances from 0.75 m to 14.5 m can be achieved. The MYTHEN®2 1 K is mounted in air on the nose cone of the vacuum vessel and is used if an extended wider q-range is desired, for example, for time-resolved X-ray solution scattering experiments. No flexibility of the SDD for the PILATUS®3 X 2 M and MYTHEN®2 1 K detectors is possible. Please note the X in the Dectris detector name refers to detectors for synchrotron use.

configuration, the second advanced sample environments. The millisecond time orchestration for experiments at CoSAXS is comprehensively described in Da Silva et al. (2023).

The SAXS and WAXS detectors are controlled and read out using dedicated detector control units (DCUs) and servers. Detector normalization and energy dependent flat field correction are deployed automatically through the DCUs. The acquisition time, number of images and energy of the detector can be controlled by the user. Data are streamed from the DCUs to the MAX IV fast data storage and combined with meta data from the beamline and sample environment.

Data reduction is carried out using the python implementation of MatFRAIA (Jensen, Christensen, Weninger & Birkedal, 2022) for radial and azimuthal integration at faster than the 500 Hz frame rate of the Eiger®2 X 4 M. PyFAI (Kieffer, Petitdemange & Vincent, 2018) is used for calibration and masking of the SAXS and WAXS detectors. Live displays of the 2D X-ray detector images and 1D integrations, with a 1 Hz refresh rate, allow the user to follow their experiment. The integrated data are normalized to the X-ray beam transmittance, background subtraction and scaled to water absolute intensities and plotted using custom Python programs in Jupyter Notebook.

Above we have outlined how we achieve the desired X-ray beam properties and the possibility to achieve time resolutions down to milliseconds. In the following sections we will describe how we use this potential for a range of experimental methods. First, we will describe the SUrF sample environment before describing how we have achieved time-resolved measurements using commercially available equipment like microfluidic chips and stopped-flow devices. Finally, we will detail the design and synchronization of laser, LED and UV sources with SAXS/WAXS measurements. For each section we will describe the possible information which we gained from these experiments using model proteins. These sample environments are available for general users to use and several publications from their experiments are available (Berntsson, Terry & Plivelic, 2022; Gilbert et al., 2024; Magkakis, Orädd, Ahn, Da Silva, Appio, Plivelic & Andersson, 2024).

3. SUrF—combined SAXS with UV–vis/Raman/fluorescence

Gaining structural information from SAXS is not always straightforward, this can become more challenging as the complexity of the sample increases and has to be monitored. One effective solution to this problem is to combine SAXS with spectroscopic tools. This approach can be compelling, especially if the spectroscopic techniques are familiar to the researcher and readily accessible within their home laboratory. However, collecting and combining the results from data collected at different times, in different sample volumes, in different sample holders and even on different samples, is challenging, making drawing subsequent conclusions difficult; this is mainly the case in biologically relevant samples where all the above factors can affect the structures. One solution can be to combine directly the most relevant techniques in a single experiment, as has been previously described (Bras, Nikitenko, Portale, Beale, van der Eerden & Detollenaere, 2010; David & Pérez, 2009; Caetano et al., 2014; amongst others). The combination of SAXS, UV–vis, and fluorescence in a dedicated sample environment was previously developed at MAXLab, Lund University (Haas, Plivelic & Dicko, 2014) and re-engineered in a joint project for CoSAXS and the Soleil Synchrotron SWING beamline (https://www.synchrotron-soleil.fr/en/beamlines/swing). It enables us even consider time resolved measurements on these samples which are have been well described in terms of concentration, conformation, hierarchical organization and chemical interactions. The latter are highly important to understand and quantify complex systems. It should be noted that the 'r' in the name "SUrF" is for Raman spectroscopy. This will be a future development.

3.1 Design of the multiprobe platform SUrF

The SUrF sample environment allows the coincident illumination of a liquid sample within an exchangeable quartz capillary with X-ray, UV–vis and fluorescence spectroscopy. Fig. 2 shows a photograph of the core assembly (A) and arrangement of the spectroscopic probes perpendicular to the long axis of the capillary, and the path of the incident and transmitted X-ray beam (B); the design is such that the point of incidence of the spectroscopy probes can be offset a short distance from the X-ray beam, either before or after the X-ray beam if there are concerns about either the effect of the X-ray beam or the spectroscopic probes on the sample. The fiber optic coupled UV–vis and fluorescence probes and detectors are detailed in Table 2. Internal fiber optics are permanently fixed to vacuum-compatible feedthroughs on the exterior of the vacuum housing which permits changing the inlet and outlet ports for the spectroscopy signals. During testing of the SUrF equipment, no crosstalk (interference) between the optical spectroscopy sources and probes

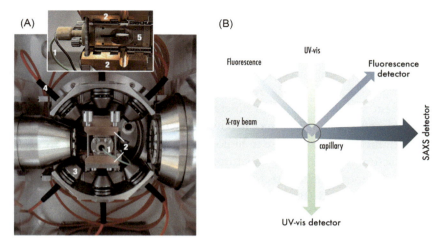

Fig. 2 (A) photograph of the core of the SUrF sample environment and the capillary holder (insert) and (B) schematic representation of the arrangement of the X-ray, UV and fluorescent probes within the core. 1: sample injection tube to the quartz capillary; 2: Peltier heaters for temperature control; 3: collimator pieces for optimizing the fiber optic alignment; 4: fiber optics for spectroscopic probes; 5: quartz capillary inside the PEEK capillary holder. During operation, the core assembly is housed within a vacuum housing (not shown).

was observed. The capillary path length is an essential trade-off for multimodality; since the SUrF is primarily optimized for X-ray scattering, the path length for the other probes is therefore set (for example, 1.5 or 2 mm). The UV–visible and fluorescence optimization is then achieved by carefully focusing via collimators, close to the sample capillary and collecting the light using detectors with a high signal-to-noise ratio, powerful light sources, and appropriate integration time.

This core assembly is housed within a small vacuum box which can be mounted on the sample translation stages at CoSAXS for easy alignment with the X-ray beam. The entire assembly is coupled via flexible bellows along the X-ray beam path to give through-vacuum, for as low background noise as possible. The sample is introduced into the 1.5 mm diameter, 10 μm thick walled, quartz capillary through an externally accessed connection (funnel); the sample is not in exposed to low vacuum, instead the exterior of the quartz capillary is sealed with Viton O-rings to separate the vacuum ($<10^{-2}$ mbar). Peltier heaters are installed above and below the sample capillary to allow rapid temperature changes and the temperature at the sample monitored.

Table 2 Technical specifications for the SAXS, UV–vis and fluorescence measurements in the SUrF sample environment.

SAXS specification

Beamline	CoSAXS
Wavelength	0.99 Å
q-Range	$0.0038 < q < 0.27\,\text{Å}^{-1}$
Beam size at the sample	$200 \times 200\,\mu\text{m}^2$
SAXS detector	EIGER®2 X 4 M

UV–vis specification

Company/model	Ocean Optics HDX
Light source	Deuterium halogen lamp
Wavelength range	200–800 nm
Optics	monochromatic beam
Detector	CCD array

Fluorescence specification

Company/model	Ocean Optics QE65000
Light source	LED Thorlab 280 nm
Excitation wavelength	280 nm
Wavelength range	240–1000 nm
Resolution	$3\,\text{cm}^{-1}$
Detector	CCD array

3.1.1 Data acquisition protocol for SUrF experiments

To successfully perform a "SUrF" experiment, it is necessary to synchronize the illumination and collect and timestamp the data from the various sources and detectors. Data is typically collected as a sequence of time separated measurements, orchestrated through the PandABox. The parameters for the differing lengths of time for the data collection from the X-ray detectors and spectrometers can be specified from the CoSAXS control software. Live viewers, updating at 1 Hz typically, provide a way to monitor changes in the data from the X-ray, UV–vis and fluorescence detector streams. Fig. 3 shows an example sequence for the TTL triggers for each technique.

To highlight the potential of SUrF measurements, an example of measuring the time-dependent acid denaturation of bovine serum albumin (BSA) is described. The SAXS measurements were conducted using a wavelength, $\lambda = 0.99$ Å which, with the SDD of 3.5 m, gave an accessible q-range of $0.0038 < q < 0.27$ Å^{-1}. The X-ray beam was focused on the detector and is approximately 200×200 μm^2 (width x height, FWHM) at the sample position. The data were scaled relative to the absolute differential scattering cross section in cm^{-1} of (Milli-Q) water.

The UV–vis measurements were performed using an *Ocean Optics* HDX Spectrometer, with a range of 200–800 nm at a resolution of 1 nm. The detector is a back-thinned CCD array. In the configuration of SUrF, the UV–vis source optical fiber is mounted from the top and the detection fiber from the bottom at 180° (Fig. 2B).

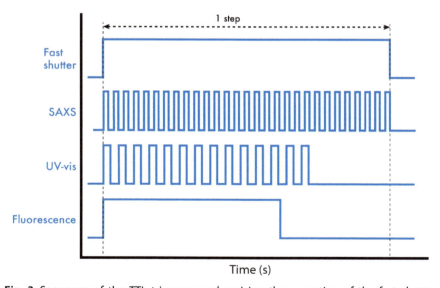

Fig. 3 Sequence of the TTL triggers synchronizing the operation of the fast shutter and the SAXS, UV–vis and fluorescence measurements as implemented for the SUrF experiments at CoSAXS. The fast shutter opens at the start and remains open until the end of the sequence. The SAXS, UV–vis and fluorescence start synchronously with the fast shutter opening. The SAXS and UV–vis are collected as a time-resolved series of individually timestamped data. A single fluorescence spectrum is collected once per sequence.

Fluorescence spectroscopy measurements were performed using an *Ocean Optics* QE65000 Spectrometer, with a range of 240–1000 nm and a back-thinned FFT-CCD detector with a 2D arrangement of pixels. The fluorescence detector optical fiber is aligned with the source light forming a 90° angle (Fig. 2B).

3.2 Time-dependent acid-induced unfolding of BSA with SUrF

The experiment to demonstrate the capability of time-resolved experiments using the SUrF sample environment follows the structural characterization of the unfolding process of BSA. This unfolding process was induced by acidification of the BSA/buffer solution over time using glucono-δ-lactone (GdL). This approach has been utilized previously to follow protein unfolding (Dockal, 2000.; Del Giudice, Dicko, Galantini & Pavel, 2017).

Combined measurements of SAXS, UV–vis and fluorescence were performed on a 4 mg/mL BSA (*Sigma Aldrich*) solution in 20 mM Tris buffer, pH 7.4, the BSA being used directly without further purification and centrifuged at 14 000 rpm for 15 min at 4 °C. The BSA was mixed with GdL at a ratio of 20:1. For the duration of the SUrF experiment, the sample within the X-ray beam was continually refreshed using a flow rate of 0.2 mL/min; this flow rate ensures that each SAXS pattern is collected from a sample not previously exposed to the X-ray beam (although a small layer at the walls of the capillary would be static).

Data collection was performed as a sequence of 10 s steps. Each step consisted of 400 SAXS frames of 20 ms with 10 ms latency time, 100 frames of UV–vis of 70 ms with 10 ms latency time and 1 frame of fluorescence of 5 s with 2 s latency time. The fast shutter did not close between individual frames. A total of 301 steps were recorded meaning that a total of 124000 SAXS curves, 30100 UV–vis absorbance spectra and 301 fluorescence spectra were acquired in 42 min.

Before the experiment, UV–vis dark spectra were collected for the Tris buffer with the deuterium-halogen lamp turned off. In the same way, a fluorescence dark spectrum was measured with the LED light switched off. One step of the data collection sequence was collected for the Tris buffer for use as a background reference. The Tris buffer was injected manually into the capillary.

The integrated time-resolved SAXS data for every frame collected during 42 min is plotted in Fig. 4A; the first data (blue) were collected 2 min after mixing of GdL with the BSA/buffer solution, i.e. at t = 120 s and each line

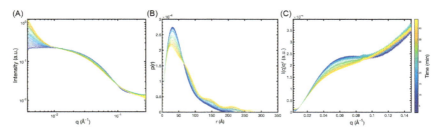

Fig. 4 (A) Integrated time-resolved SAXS data following the acidification of 4 mg/mL BSA in Tris buffer by GdL, mixed at a ratio of 20:1. (B) shows the p(r) obtained by ATSAS and (C) the Kratky plot generated from the SAXS data. Blue represents the start of the SAXS data collection at 120 s after mixing.

represents a further 12 s time delay. The initial SAXS curve is representative of a globular protein in a dilute solution, as is expected for this concentration of BSA in buffer. The intensity at $q < 8 \times 10^{-3}$ Å^{-1} increases significantly during the experiment. This increase indicates large structures forming in solution, likely aggregating protein molecules. However, we are interested in how the individual protein molecules unfold; thus, we choose to estimate the radius of gyration, R_g, which is the root-mean square of the distance of all electrons in the protein from the center of gravity. This was calculated using the Guinier approximation via the AUTORG function from the ATSAS package suite (Petoukhov, Konarev, Kikhney & Svergun, 2007; Manalastas-Cantos et al., 2021). As shown in Fig. 5, the R_g values are stable at ~ 45 Å during the first 5 min of the reaction and then steadily increase to a maximum radius of 65 Å without reaching a plateau until the end of the experiment at 42 min.

Another way to access valuable information on the size and shape of a molecule is to calculate, via indirect Fourier transform of the SAXS data (Glatter, 1977), the pair distance distribution function, p(r), i.e. the distribution of distances between two points in the molecule, weighted by the relative electron densities. The p(r) was calculated using the ATSAS package suite. Fig. 4B shows the p(r) change with time. The initial p(r) (blue curve) indicates a globular protein with a size of 32 Å. As the experiment progressed, the peak shifted to lower distances to ~ 25 Å. The peak also becomes more asymmetric with the Dmax increasing. This increase in Dmax indicates a more elongated particle with the bumps along the curve being due to possible unstructured aggregation.

To understand further we examined the degree of unfolding in the protein molecule by presenting the SAXS data in a Kratky plot (Fig. 4C).

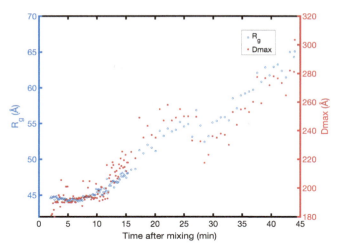

Fig. 5 Increase of R_g (blue) and Dmax (red) with time for the SAXS data shown in Fig. 4A, following the acid denaturation of 4 mg/mL BSA in Tris buffer upon mixing with GdL (mixing ratio 20:1).

The shape of the initial state (blue curve) in the Kratky plot reveals that the proteins are already partially unfolded when we estimate a R_g of 45 Å at time t = 120 s. This degree of unfolding remains approximately the same for the first 5 min before the Kratky plot shows that the protein unfolds further without reaching a fully disordered state during this experiment.

To further disentangle the pH change induced in this reaction, we compared the pH change upon adding GdL in the buffer and the protein solution (Fig. 6), as measured with a pH meter. Both solutions show a dramatic decrease in the pH from an initial value of about 7 to around a pH of 3.5 within the first 5 minutes of the reaction, thus becoming a quasi pH-jump experiment. After that, the pH approaches a constant value for the rest of the experiment. Thus, we might conclude that the immediate pH drop already induces the unfolding in the BSA protein, i.e. a minimal decrease in pH is sufficient to trigger unfolding, as the R_g is already at a much larger value than the native state. The pH is relatively constant at intermediate timescales, and no further change in R_g or unfolding is observed. At longer timescales above 15 minutes, extensive aggregation to form much larger aggregates is observed, possibly because the unfolded proteins might be more unstable and thus self-aggregate with time.

We also examined the changes in the UV–vis spectra. The turbidity of a solution, τ, can be calculated from UV–vis spectra, as described by Haas, Plivelic & Dicko, 2014, using the equation:

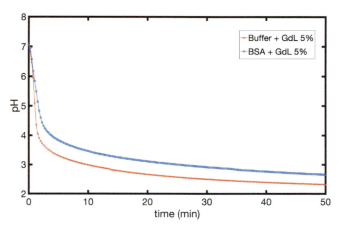

Fig. 6 pH change upon addition of GdL to 4 mg/mL BSA in Tris buffer and to Tris buffer (red), monitored with time using a pH meter. The ratio of BSA/buffer to GdL was 20:1 giving 5 % GdL in the final solution.

$$\tau(\lambda, t) = b(t)\lambda^{-m(t)}$$

where λ is the wavelength, b is a scaling factor, and m is an exponent (also referred to as turbidity). Both b and m are time dependent and were determined by a nonlinear regression of every measured absorbance spectrum in the range $\lambda = 390$ to 420 nm. Fig. 7 compares the time evolution of R_g and turbidity (i.e. m). As R_g increases, the turbidity also increases. Despite both values presenting a time lag at the beginning of the process, the onset of turbidity happens earlier than the change in R_g and is more abrupt than the growth of R_g.

Examining the UV–vis spectra (Fig. 8A) shows a continuous decrease in signal in the aromatic region of the spectra. This was not observed in previous experiments (Del Giudice, Dicko, Galantini & Pavel, 2017) and might be an artefact due to a negative slope in the signal prior to correction for turbidity.

The fluorescence emission spectra show a shift upon acidification which occurs after approximately 10 min, despite the poor signal-to-noise ratio (Fig. 8B). These results are consistent with previous research which demonstrated such a shift as a consequence of protein unfolding (Dockal, 2000; Del Giudice, Dicko, Galantini & Pavel, 2017).

3.2.1 Correlation of structural changes with spectroscopic data upon acidification

The increase of R_g and Dmax with time likely could be a signature of the structural change of the unfolding process due to acidification upon adding

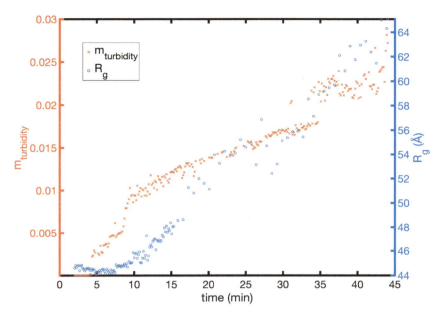

Fig. 7 Increase in turbidity (red) and R_g (blue) with time after mixing GdL with 4 mg/mL BSA in Tris buffer, measured synchronously using the SUrF sample environment. The turbidity is extracted from the UV–vis spectra and the R_g from the SAXS data.

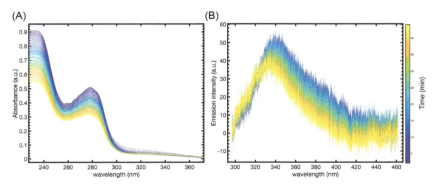

Fig. 8 (A) Evolution of UV–vis and (B) fluorescence emission spectra measured synchronously with SAXS in the SUrF sample environment following the mixing of BSA with GdL. Blue represents the first spectra. The UV spectra show a continuous decrease in signal in the aromatic region of the spectra. The fluorescence over the same time shows a shift within the first 10 min after mixing.

GdL to the BSA/buffer solution (Fig. 5). However, we use this experiment to highlight that care must be taken when carrying out and interpreting SAXS data. The values of R_g and Dmax during the current experiment are larger than previously measured for the same chemical unfolding of BSA measured at the lower flux SWING beamline, Soleil Synchrotron (Del Giudice, Dicko, Galantini & Pavel, 2017). The several orders of magnitude higher flux at CoSAXS could induce significant radiation damage in protein solutions and thus beam-induced aggregation. Thus, there may be two processes occurring, both of which act to increase the R_g and the Dmax, (Del Giudice, Dicko, Galantini & Pavel, 2017).

Strategies to avoid radiation induced aggregation in protein solutions include using free-radical scavengers (Stachowski, Snella & Snella, 2021) or the addition of 5% glycerol to the solutions (Jeffries, Graewert, Svergun & Blanchet, 2015). Neither of these approaches might be suitable when studying time-resolved reactions in radiation sensitive samples. Instead, using the SUrF equipment in these measurements, we kept the protein solution flowing during the SAXS measurements at an appropriate rate to ensure that a fresh sample was introduced for each SAXS data point. Using flow in this way is a common strategy used to prevent radiation damage when measuring the static structure of radiation sensitive samples (Jeffries, Graewert, Svergun & Blanchet, 2015) thus, we are (relatively) confident that the SAXS measurements do not cause radiation-induced aggregation. However, it is not a guaranteed approach especially as the sample at the capillary walls is static. Fortunately for a 1.5 mm diameter sample, the proportion of sample which is static is very small.

Comparing the current and the previous studies (Del Giudice, Dicko, Galantini & Pavel, 2017) revealed the multiprobe approach's power. The UV–vis spectra (Fig. 8A) showed that there were significant changes in the amide bonding during the entire SUrF experiment at CoSAXS, during the first 10 min after the mixing, whilst the SAXS and fluorescence data were not changing (Figs. 4A and 8B). The unfolding processes and radiation-induced damage can complicate the study of individual albumin isomers. This experiment illustrates the importance of employing time-resolved reactions with multiple techniques to extract more comprehensive information. It also highlights the potential for detecting and thus preventing radiation damage, which is a significant issue in synchrotron facilities.

Applying this method to other systems could be valuable for understanding time-dependent processes. Such an approach would enhance our fundamental understanding of protein conformational behavior, for example, and have applications in fields such as examining drug release and enzymatic degradation. Following a scientifically relevant structural change is also the basis for exploiting microfluidic devices with SAXS measurements. In a cross-beamline project at MAX IV, the Adaptocell platform was developed to facilitate the use of microfluidic devices for protein solutions for hard X-ray spectroscopy, SAXS and MX measurements (https://www.maxiv.lu.se/beamlines-accelerators/beamlines/balder/science-at-balder/external-funded-projects/adaptocell/). CoSAXS has benefited from this development and now utilizes commercially produced, and thus highly reproducible, custom microfluidic chips which have been optimized for our beamline.

4. Time-resolved SAXS studies using microfluidic chips

Microfluidic chips are widely used in biological sciences, from lab-on-a-chip incorporating multiple characterization techniques to complex mixing chips for precision manufacturing (see reviews by Beebe, Mensing & Walker, 2002; Nguyen, Wereley & Shaegh, 2019; Tabeling, 2023, for example). At CoSAXS we have performed many user experiments using microfluidic chips, an example of which describes understanding the in-situ formation of lipid nanoparticles in a herringbone mixing chip (Gilbert et al., 2024). In this chapter, we consider how to exploit relatively simple chip designs to follow the structural changes occurring upon diffusion-based mixing on milli- to second timescales.

Microfluidics refer to the flow of liquid samples within narrow channels, typically no more than 100 s of microns in cross-section. Usually, the materials exploited in microfluidic chip manufacture are highly rigid, stable thermoset plastics, although we have utilized silica based etched devices for MX at MAX IV. Additive manufacturing can be used to produce chips but we have found the best reproducibility and lowest background for SAXS from mold-templated manufacturing.

The use of microfluidic chips is ubiquitous in many scientific research areas and their commercialization has meant that they provide a low-cost method to manipulate small (μL) volumes of sample. This latter point is significantly attractive, for example in the study of protein solutions, where obtaining large sample quantities can often be expensive or time-consuming. Thus, the combination of small sample volumes, highly controlled sample manipulation and a high level of reproducibility make microfluidic chips an enticing possibility for SAXS experiments. However, the intrinsically small sample volumes (and sample depth) obtained with microfluidics can also be problematic for SAXS. Typically, quartz capillaries used for biological SAXS measurements are an order of magnitude larger in diameter, and thus the X-ray path length, and SAXS scattered intensity, are also proportionally larger. This problem is to some extent reduced at modern synchrotron beamlines, such as at CoSAXS, which can produce stable and high intensity X-ray beams, thereby overcoming the limited sample exposed to the incident beam. Time-resolution is obtained by mixing the sample at a defined point in the chip and then moving the chip laterally within the beam, whilst simultaneously measuring SAXS at well-defined distances from the mixing point. Thus, the obtained time resolution originates from the lateral position of the beam relative to mixing point and the flow rate of the sample within the chip. Radiation damage is broadly avoided as the sample is continuously flowing within the chip, provided the flow rate is sufficiently high.

A stable X-ray beam and sample environment is necessary to be able to extract the subtle structural changes observed during biological processes or reorganization of the biomolecular ensemble, such as liquid-liquid phase separation. Even though there is an enhanced contribution from air scattering, the ease and speed of replacing the chip and manipulating fluidic and electrical connections to the chip and respective holder outweighs the potential gain to mounting the chips in vacuum; the increased air scattering can be offset by increased measurement time.

However, SAXS measurements with microfluidic devices are relatively complex to perfect. Careful consideration should be paid to the fluidic properties of the various solutions which will be flowed through the chip. A chip that performs well in the lab may not have the desired performance when mounted vertically in the X-ray beam and so we provide a microfluidic testing platform using a horizontally mounted microscope within the Biology Support Laboratory at MAX IV, to optimize experimental conditions prior to installation at the beamline. We also provide two pumping systems, either a pressure driven system with in-line flow meters (*Elveflow Microfluidics*) or 3 Cetoni S syringe pumps (*NEMESYS Ltd Oy Ab*) depending on the requirements for the experiment. Typically, we use the Cetoni syringe pumps more extensively at CoSAXS. Typical microfluidic experiments vary the flow rate and the flow rate ratio during mixing. We are also able to provide simple electrical heating of the microfluidic chip up to 40 °C; it should be noted that the syringe pumps and liquid lines to these devices remain at room temperature. Below we highlight the possibilities of microfluidics coupled SAXS measurements at CoSAXS. These types of experiments are particularly attractive for protein solutions. Indeed, changes to the solution conditions can be simply achieved by mixing with different components, leading to changes in protein structure and protein-protein interactions, which can then be monitored by SAXS. We use BSA as a model protein and induce its unfolding by mixing with SDS.

4.1 Microfluidic chip design

During a typical SAXS experiment a volume of sample is exposed to an incident beam and the angular dependency of the scattered intensity is measured. The scattering from the sample is then obtained by subtracting the measured intensity for the empty cell and any air scattering from this measured intensity. The scattered intensity, for dilute proteins in solution, is proportional to the volume of the sample in the incident beam, such that:

$$I(q) \propto c\, I(0) P(q)\, t\, A\, d\, e^{-\mu(E)d}$$

Where $q = 4\pi\frac{\sin\theta}{\lambda}$, θ is half of the scattering angle, λ is the wavelength, c is the protein concentration, $I(0)$ the forward scattering, $P(q)$ the normalized form factor, t the exposure time, A the beam dimensions at the sample position, d the sample thickness and $\mu(E)$ the linear attenuation coefficient (adapted from Dreiss, Jack & Parker, 2005). We have utilized microfluidic channels with 100 to 400 μm depth, i.e. < 25 % of the sample thickness for conventional BioSAXS measurements. The reduction in scattered intensity for thin samples in microfluidic chips is, to some extent, overcome by the high flux at CoSAXS. Another important consideration is the contribution to the scattering of the microfluidic chip itself which must have a sufficiently small contribution to the transmitted scattering to enable the detection of the inherently weak scattering. We have demonstrated good success using chips manufactured from COC (Cyclic Olefin Copolymer, also commonly known as Topaz®); the thin polymer layers forming the front and back surfaces of the chip are sufficiently robust to not flex during the flow of liquids within the channels. The chips can be manufactured from custom made molds with good reproducibility, for example channel dimensions, low internal channel roughness, and consistent overall chip dimensions.

At CoSAXS we offer users our custom designed, hydrodynamically focusing, microfluidic chips (Fig. 9), with 3 inlet channels with the same dimensions of 100 × 400 μm (width x depth, where depth is the pathlength of the X-ray beam in this orientation) (*microfluidic ChipShop GmbH*). The inlet channels join the main outlet channel at a 45° angle, at the X-junction marked on Fig. 9C, the dimensions of the outlet channel are 300 × 400 μm (width x depth).

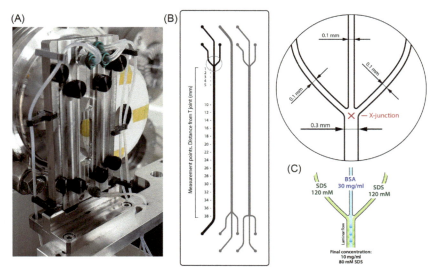

Fig. 9 (A) Photograph of a microfluid chip installed at the CoSAXS beamline. (B) and (C) provide details of the custom microfluidic chips, manufactured in COC, offered for user experiments at CoSAXS. The X-junction marks the nominal mixing point defining time t = 0. Distances are shown for the X-ray beam position below the X-junction as the chip is moved vertically in the beam. Individual Cetoni syringe pumps are connected via tubing to the inlets of the flow channel. The outlet is collected in a waste container.

The side inlets serve a dual purpose. At the X-junction, the flow from the side acts to hydrodynamically focus the flow of the central inlet in the center of the main channel. The width of the central "sample" in our geometry can be calculated considering the ratio of flow rates for the inlets, FR, as the dimensions of the inlets are identical (Merkens et al., 2019).

$$FR = \frac{Flow\ rate_{left\ inlet} + Flow\ rate_{right\ inlet}}{Flow\ rate_{central\ inlet}}$$

And:
width of central sample = $\frac{1}{(FR+1)}$

As we keep the same flow rates in the inlets, FR = 2 and the sample width is 1/3 of the total width of the main channel, i.e. 100 μm wide at the X-junction. By matching the lateral width of the X-ray beam at the microfluidic chip to the width of sample, one can collect information of the structural changes that the sample undergoes with time.

A metal holder allows reproducible (within 100 μm), stable clamping of the microfluidic chip on our sample translation stages. All fluidic and electrical connections are made and tested before translating the chip into the measurement position to avoid any collision or leaks onto our highly fragile vacuum windows and the tubing is carefully clamped out of the way to allow for the translation of the chip. We mount the chip so that the outlet channel flows vertically to avoid a pressure differential on the opposing inlet channels. Individual Cetoni syringe pumps are connected to the inlets of the microfluidic chip, care being taken to ensure the tubing lengths from the syringe to the microfluidic chip connection are known.

4.2 Following surfactant-induced structural changes in a microfluidic chip

BSA, sodium succinate and sodium dodecyl sulfate (SDS) were supplied by *Sigma Aldrich* and used without further purification. Stock solutions of BSA (30 mg/mL) and SDS (120 mM) were prepared in 20 mM succinate buffer with a pH of 5.4. The concentration of BSA was confirmed by measuring the SAXS scattering of a dilution series and UV absorption at 280 nm.

SAXS data were collected with a similar SAXS configuration as for SUrF, $\lambda = 0.99$ Å and q-range was $0.004 < q < 0.3$ Å^{-1}. The X-ray beam was focused at the sample position, the horizontal and vertical FWHM of the approximately gaussian beam profile being $140 \times 120 \, \mu m^2$, respectively. This allows us to achieve good spatial resolution within the microfluidic chip. The position of the X-junction in the chip was determined from the vertical center of the horizontal inlet and the horizontal center of the main channel. To ensure consistency and background subtraction, it is important not to change the chip between these measurements and the subsequent buffer and sample measurements or alter the X-ray beam parameters.

SAXS patterns measured at the X-junction are at time $t = 0$; subsequent measurements at vertical offsets of the X-ray beam in the main channel, relative to X-junction, at a given flow rate, will give the time resolution of the diffusive mixing of the protein solution and buffer. Different vertical distances were used, as shown in Fig. 9B. At each point, 40 SAXS frames of 50 ms acquisition were collected and averaged to give sufficient signal to noise. During these measurements, the sample and buffers were kept under flow.

Four experiments were performed, utilizing the same flow rates on the three syringes of 60, 30, 7, 5 and 3.75 µL/min, which correspond to a flow velocity within the main channel of 12.5, 6.25, 1.56 and 0.77 mm/s, respectively. The total measurement time, including the lateral translation of the chip in the beam, for each flow rate is around 4 to 6 min, meaning that even with only several hundred μL of sample many data points can be collected. The longest time in the channel is investigated at the furthest distance from X of 38 mm and for the slowest velocity of 0.77 mm/s, $t_{max} = 49.35$ s. Only processes that occur faster than that can be followed in these microfluidic chips, with these conditions. The time scale of the positions of the chip for the 4 flow rates are shown in the Table 3. The distances and flow rates can be optimized for different experimental setups, yielding different time-resolutions and sample volume requirements.

Table 3 Calculated time after mixing at vertical distances along the outlet channel of the microfluidic chip shown in Fig. 9.

Distance from the X-junction (mm)	Time (s)			
	60 μL/min	30 μL/min	7.5 μL/min	3.75 μL/min
	12.5 mm/s	6.25 mm/s	1.56 mm/s	0.77 mm/s
0	0.00	0.00	0.00	0.00
1	0.08	0.16	0.64	1.30
2	0.16	0.32	1.28	2.60
3	0.24	0.48	1.92	3.90
4	0.32	0.64	2.56	5.19
5	0.40	0.80	3.21	6.49
10	0.80	1.60	6.41	12.99
12	0.96	1.92	7.69	15.58
14	1.12	2.24	8.97	18.18
16	1.28	2.56	10.26	20.78
18	1.44	2.88	11.54	23.38
20	1.60	3.20	12.82	25.97
22	1.76	3.52	14.10	28.57
24	1.92	3.84	15.38	31.17
26	2.08	4.16	16.67	33.77
28	2.24	4.48	17.95	36.36
30	2.40	4.80	19.23	38.96
32	2.56	5.12	20.51	41.56
34	2.72	5.44	21.79	44.16
36	2.88	5.76	23.08	46.75
38	3.04	6.08	24.36	49.35

The mixing point, X-junction, at time $= 0$ s, is the junction where the horizontal side inlets join the central inlet. The dimensions of inlets are $100 \times 400 \, \mu m^2$ (width × depth) and the outlet channel $300 \times 400 \, \mu m^2$ (width × depth). These times are calculated for the outlet flow rates, the flow rate in each inlet being constant

4.2.1 Diffusion-limited mixing in protein solutions

The flow in the microfluidic channel is laminar for the Reynolds numbers of the protein and buffer solutions, so mixing will occur by diffusion processes only. By Fick's law of diffusion, we can calculate the distance D_y a solute can diffuse with time t (With et al., 2014):

$$D_y = \sqrt{2Dt_D}$$

Where D is the diffusion coefficient and t_D is the mean diffusion time. Using the published diffusion coefficients of water for the buffer in the BSA solution and water and SDS for the buffer in the side inlets, (D_{water} = 1.6×10^{-9} m^2/s, D_{SDS} = 4.5×10^{-10} m^2/s) (Wang, 1965; Weinheimer, Evans & Cussler, 1981), it is possible to estimate the mean diffusion time t, at which the diffusion will have occurred across the channel; as diffusion occurs from both edges of the central sample, this is for a diffusion distance of 50 μm. Thus, the sample should be fully mixed after 4.55 s, independent of the flow rate in the channel. From Table 3, this means that there will be an incomplete mixing of the components for the fastest flows used.

4.2.2 Following the dilution of the SDS-containing buffer in the chip

To perform the background subtraction, microfluidic experiments were performed with SDS-containing buffer in all three inlets, and repeated for the SDS-free buffer. An example of the scattering from a singular position in the microfluidic chip is shown in Fig. 10A. A clear peak at intermediate $q = 0.2$ Å^{-1} can be observed in the SDS-containing buffer. The scattering pattern for just SDS (i.e. after subtraction of buffer scattering) is indicative of the presence of SDS micelles (Fig. 10B). If now the SDS-containing buffer is injected at the side inlets and SDS-free buffer into the central inlet of the microfluidic chip, the concentration of the SDS increases at increasing distances from the X-junction, and we can estimate the relative concentration by scaling the intensity so the form factor of the SDS peak overlaps (Fig. 10C).

4.2.3 Following the surfactant induced structural changes of BSA

After measuring all the backgrounds, we next introduced the BSA-containing sample into the central inlet of the microfluidic chip with SDS-buffer in the side inlets. Fig. 11 shows the time evolution of the SAXS patterns measured at different distances from the X-junction, after subtraction of the SDS-containing buffer. At the initial time points the scattering curve represents that expected of a globular protein in solution.

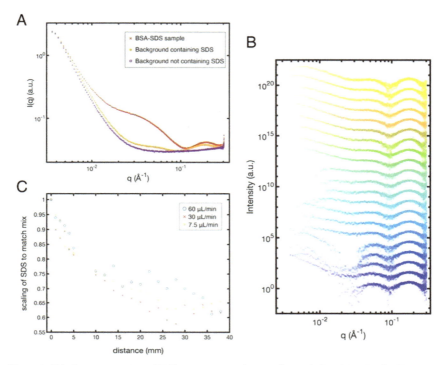

Fig. 10 (A) Representative SAXS scattering data collected for BSA (red), SDs-containing (orange) and SDS-free buffer (purple). (B) SAXS curves for SDS after subtraction of the SDS-free buffer with time corresponding to the vertical position in the microfluidic chip, each curve is offset for clarity. (C) Scaling factor of the scattering intensity to match the form factor peak over distance, for the flow rates indicated.

At increased times after mixing, a peak develops in the scattering at $q = 1.8\,\text{Å}^{-1}$ and becomes progressively more defined. The peak is clear evidence of the formation of the micellar-like aggregates. We note that this peak appears above the scattering for the SDS-containing buffer (Fig. 10A).

The change in protein shape can also be followed at low q by monitoring the time evolution of R_g (Fig. 12) and of p(r) (Fig. 13). The calculated values for R_g are very noisy but we can see that the initial rate of change of R_g and Dmax is independent of the flow rate. The increasing R_g and Dmax observed in the scattering curves indicate that, as previously observed, the protein is expanding in the presence of SDS (Santos, Zanette, Fischer and Itri, 2003).

Finally, we compare the microfluidic SAXS data to BioSAXS measurements from P12, EMBL, Hamburg, of 10 mg/mL BSA collected at

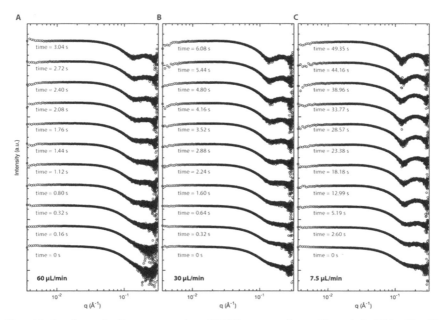

Fig. 11 Small angle X-ray scattering (SAXS) curves from 30 mg/mL BSA, 20 mM sodium succinate buffer, pH 5.4 at 22 °C, for the flow rates (A) 60 μL/min, (B) 30 μL/min and (C) 7.5 μL/min. For clarity curves are offset relative to each other with time being calculated from the flow rate and from the relative position of the X-ray beam in the microfluidic chip after the mixing point.

different SDS concentrations (Fig. 14A). Here we observe a shift in lower q of the scattering peak with increasing SDS concentration. The increase in intensity at lower q is the increasing contribution of the SDS micelle to the scattering. If we subtract the SDS-free buffer from the final mixing point in the slowest flow rate and compare both the scattering curve with the BioSAXS measurement (Fig. 14B) and the peak position (Fig. 14C), we observe that the data is similar to the 70 mM BioSAXS measurement (and of similar signal-to-noise).

The combination of microfluidic chips for sample delivery and processing when coupled with high brilliance synchrotron sources is very powerful for capturing the intricate time evolution of structure and interactions, as we have shown above. High brilliance, small X-ray beams and fast X-ray detectors allow us to capture information directly within the processing window. In the example above we were able to capture information during passive, diffusion-based mixing of two components and thus have access to many concentrations with small sample volumes, if the diffusion process is well enough

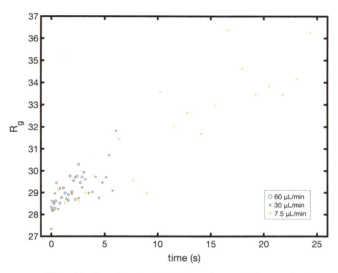

Fig. 12 Increase of R_g with time for the SAXS data shown in Fig. 11 for three flow rates (blue—60 μL/min, red—30 μL/min and orange—7.5 μL/min). The Rg values were obtained using the AUTORG function in ATSAS.

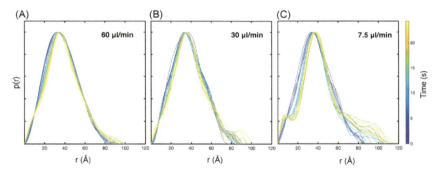

Fig. 13 Evolution in p(r) with time for the SAXS data shown in Fig. 11 for three flow rates, (A) 60 μL/min, (B) 30 μL/min and (C) 7.5 μL/min. The p(r) were calculated in ATSAS. The blue curve in each plot represents time t = 0 s at the X-junction in the microfluidic chip. The data reveal that the Dmax increases with time at each flow rate.

understood. This was crucial to be able to separate the effect due to dilution from the larger structural reorganization caused by the addition of surfactant to the protein solution. As the sample was continually flowing, the same relative time points could be re-interrogated until sufficient signal to noise was achieved. A criticism of this approach using laminar flow, and thus diffusion-driven mixing, is that the sample will be inhomogeneous, and the scattering

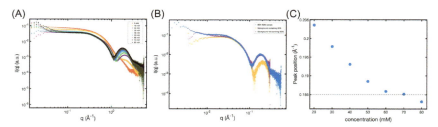

Fig. 14 (A) SAXS data for 10 mg/mL BSA with increasing SDS concentration, data measured in the BioSAXS P12 beamline, Petra III. The data reveal a shift to lower q with increasing SDS concentration. (B) BioSAXS data (blue curve) for BSA with 70 mM SDS in plot A is compared with SAXS data in the microfluidic chip for the final time point at the slowest flow rate (t = 49.35 s) with SDS-containing buffer signal subtracted (orange curve) or SDS-free buffer signal subtracted (purple curve). (C) q-value with SDS concentration for the SDS form factor peak for the data presented in plot A. The dotted line represents the q-value for the peak observed in Fig. 11.

will be an average of an ensemble of different conformations and shapes within the scattering volume. It might be also that more active mixing would help understand the structural changes a protein might undergo during processing. Turbulent/chaotic mixing chips exist, the herringbone mixer used previously at CoSAXS is one such device where we accessed information about the structure of lipid nanoparticles as they are created within the mixing channels and after the outlet from the mixer (Gilbert et al., 2024). In another example of active mixing, we have coupled a commercially available stopped-flow device with SAXS/WAXS at CoSAXS to follow time-resolved processes in proteins.

5. Structural kinetics investigated with stopped-flow

A description of potential time-resolved studies for biological processes would only be complete by considering stopped-flow devices. These were introduced in the 1950s (Chance, 1951; Chance & Legallais, 1951) to study chemical reactions at the milli- to microsecond timescale. Since then, stopped-flow coupled with fast photo-spectrometry techniques has become a standard technique to understand the kinetics of reactions, in biological and chemical sciences. All devices operate under the same basic principle in a highly automated fashion to ensure accurate, fast, and reproducible mixing of two or more solutions. Small volumes of the solutions are rapidly injected in a continuous flow into a highly efficient mixer which initiates the reaction, and the relative ratios of the reactant solutions are carefully controlled during this mixing process. The mixed solution passes into an observation cell and once the desired volumes have been injected through the mixer, the flow is stopped, often by closing a valve called the hard-stop. The changes in the reaction are monitored by the spectroscopic technique chosen. The whole sequence can be repeated as often as required to access the fast time scales. The ability to study the structural evolution in a solution following active mixing has led to stopped-flow devices being installed at many of the SAXS, and even small angle neutron scattering (SANS), instruments at large scale facilities. The achievable time resolution relies on access to sufficiently fast detectors and a highly stable beam. At CoSAXS, we have demonstrated this potential using the 500 Hz operation of our SAXS detector which allows us to achieve 2 ms time resolutions. Below we explore using stopped-flow and SAXS to study the structural changes during the self-association of an intrinsically disordered model protein upon adding divalent cations. For our study, α-casein proved to be a fortuitous choice of protein as it is very radiation resistant. Hence, we did not have to consider methods to avoid radiation damage.

5.1 Structural changes induced by addition of Ca^{2+} ions

α-casein from bovine milk was bought from *Sigma Aldrich* and contains both α_{s1}- and α_{s2}-casein. The protein was dissolved and dialyzed for 48 h in 2 L, 10 mM NaCl, 20 mM Tris pH 8.5 using *Biotech CE* tubing with a molecular weight cut-off of 500–1000 Da, the buffer being changed after 24 h. The protein was subsequently sterile filtered with *Whatman Uniflo* syringe filters with pore size of 0.2 μm (*Cytiva*) and concentrated to 4 mg/mL with Vivaspin Rutbo 15, molecular cut-off 5 kDa (*Sartorius*). Immediately prior to experiments, the protein was centrifuged at 12 000 g for 20 min. The purity of the protein was checked with matrix assisted laser desorption ionization-time of flight mass spectrometry (MALDI-TOF MS) analysis (see Table 4). Some impurities were present, among others β- and κ-casein. The findings in this chapter hence primarily demonstrate the principle and possibilities of the stopped-flow SAXS technique and are not a precise elucidation of α-casein self-association upon addition of $CaCl_2$.

The SFM-4000 stopped-flow mixing system (*BioLogic*) with a 1.5 mm diameter quartz capillary mounted in an X-ray head (*BioLogic*) was used for these experiments. This device can inject from four thermostatically controlled internal syringes. For our measurements, we were mixing two components; (i) the protein solution was loaded in syringe 4, while (ii) the

Table 4 MALDI TOF MS analysis of α-casein from *Sigma Aldrich*.

Protein	Number of matches
α_{s1}-Casein	23
α_{s2}-Casein	17
κ-Casein	3
Protein Shroom3	2
β-Casein	1
Mauriporin	1

In addition to the listed matches, also keratin and trypsin were found which reside in MS sample handling and are hence excluded here.

buffer (10 mM NaCl, 20 mM Tris, pH 8.5 with or without 10 mM $CaCl_2$) was loaded in syringe 3 (Fig. 15). This ensures that the solutions are mixed in the final mixer, situated in the X-ray head, just before the quartz capillary. The measuring point is at the midpoint of the quartz capillary 12.5 mm after the mixer. The solutions were pushed with a combined flow-rate of 4 mL/s, which is a safe operating flow rate for the quartz capillary and gives an estimated deadtime of 5.5 ms arising from the distance from the mixer to the measurement capillary (no delay line was installed between them). About 400 μL total volume was pushed per run, to ensure removal of previous solution and complete filling of the capillary. As for the microfluidic setup, the stopped-flow is mounted in air. A highly tapered nose cone attached to the vacuum vessel ensures that the air path for the scattered intensity was reduced as much as possible and improved the signal-to-noise ratio of the SAXS data.

As α-casein is a relatively large IDP, approximately 20 kDa, and since we are interested in self-association thus forming larger oligomers (also called casein micelles), we decided to use a SDD of 10 m to capture the relevant length scales. Thus, we achieved a q-range of $0.001 < q < 0.097\,\text{Å}^{-1}$, at an energy of 12.4 keV ($\lambda \sim 1\,\text{Å}$). The X-ray beam was focused at 10 m which means that the beam size within the quartz capillary was $\sim 200 \times 200\,\mu m^2$. The SAXS data acquisition was synchronized via TTL signal from the stopped-flow device such that the fast shutter was opened, and the X-ray exposure started 10 ms before the flow was stopped by the hard-stop. The fast shutter has a rise time of approximately 5 ms until the X-ray beam can be fully transmitted along the beamline.

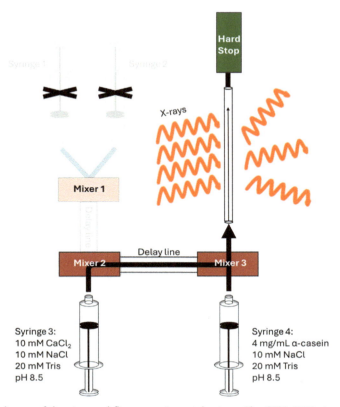

Fig. 15 Scheme of the stopped-flow experimental set-up. The SFM-4000 stopped-flow mixing system has four internal syringes and three mixers, which allow three consecutive mixing steps. In the experiment presented in this chapter, only syringes 3 and 4 are used, with the protein in the latter. This guarantees that mixing the CaCl₂ buffer with α-casein occurs in the final mixer, and after that, the mixture is injected into the capillary for the SAXS measurement. The distance between the mixer and measuring point, results in a deadtime of 5.5 ms, but it could, in principle, be reduced to 2.75 ms by doubling the flow rate to 8 mL/s, the capillary limit.

The SAXS data were acquired with two different acquisition sets; in the first set, 1000 frames of 2 ms were recorded, and repeated for a total of 16 replicates. In the second set, 1200 frames of 100 ms were recorded in duplicate. The first three frames for the first set and the first frame for the second were rejected as the fast shutter was still opening resulting in 'dark' frames and frames with flare from the edge of the shutter in the X-ray beam. Data at identical time intervals from a sequence were averaged together. Background measurements were also measured with the same acquisition protocol and mixing CaCl₂ buffer with a CaCl₂-free buffer loaded into syringes 3 and 4, respectively. All the data are normalized and scaled to absolute intensities following the methods described in the SUrF experiments. The scattering from the buffer measurements were subtracted prior to further analysis.

Fig. 16(A) shows the scattering of α-casein at different time points after mixing with CaCl₂ buffer. For comparison the scattering for the protein in CaCl₂-free buffer at the same concentration is plotted. The change in the scattering signal with time after mixing with CaCl₂ buffer and the corresponding Dmax from the p(r) plots (Fig. 16(C and D)) reveal the overall

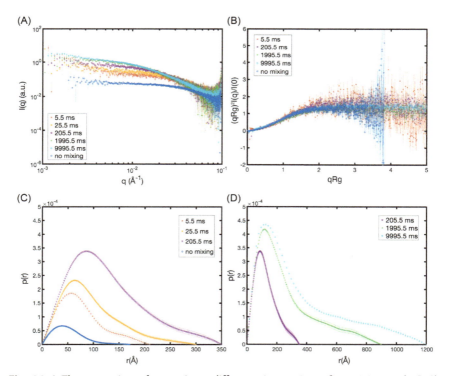

Fig. 16 A The scattering of α-casein at different time points after mixing with CaCl₂ buffer; a measurement without CaCl₂ but with the same concentration of the other buffer components and of the protein itself, is also included for comparison. It is clear that even within the 5.5 ms dead-time, there are changes affecting the overall scattering signal (A). The scattering changes throughout the experiment, with a noticeble shift of the Guinier region from q ≈ 0.005–0.03 Å⁻¹ to q ≈ 0.003–0.01 Å⁻¹ accompanied by an increase in I(0). The effect on the Kratky plot (B) seems to be minimal. The overall increase in size as α-casein oligomerizes is clear from the increase in Dmax in the p(r) plot (C–D) as well as a shift in the peak of the plot to higher r.

increase in size as α-casein oligomerizes. However, there is minimal change in the Kratky plot (Fig. 16(B)). By considering the R_g and $I(0)$ extracted from the scattering patterns (Fig. 17), we are able to follow the self-association of α-casein upon mixing with $CaCl_2$ buffer to a final protein concentration of 2 mg/mL. The final $CaCl_2$ concentration was 5 mM. For the $CaCl_2$-free dataset, the R_g of ~ 39.1 ± 0.4 Å and $I(0)$ remain constant during the initial 10 s of measurement. In contrast, it can be observed that for the $CaCl_2$-mixed α-casein solution, the initial R_g, measured directly after the 5.5 ms deadtime, is already 66.8 ± 9.6 Å, i.e. the $CaCl_2$ already causes self-association within 5.5 ms. The measurement window in the metal surrounding the X-ray head does not allow the X-ray beam to be positioned lower in height and thus closer to the mixer, to reduce the deadtime. Instead, the only way to reduce the deadtime further would have been to increase the flow rate such that doubling the flow rate to 8 mL/s would halve the deadtime to 2.75 ms.

The entire data sequence ran for 2 minutes during which R_g and $I(0)$ can still be seen to increase for α-casein mixed with $CaCl_2$ buffer. Ca^{2+}-induced changes in α-casein structure are known to extend over minute timescales from previous studies (Liu, Jiang, Ahrné & Skibsted, 2022). However, after 10 s, the $CaCl_2$-free α-casein solution also shows increasing R_g and $I(0)$; this is unexpected and is likely an effect of radiation damage. Hence we will only consider the first 10 s for analysis of the Ca^{2+}-induced changes. To access longer timescales, individual frames could be collected widely spaced in time, with the fast shutter being closed between SAXS measurements to prevent radiation damage.

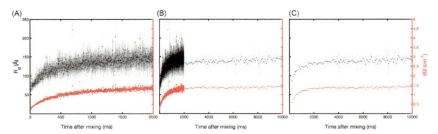

Fig. 17 R_g (black) and $I(0)$ (red) for α-casein as a function of time as calculated by autoRg. (A) shows the first 2000 ms, while (B) and (C) show the full 10 min. Errors are high for R_g, as the Guinier region is automatically estimated, the 2 ms data is somewhat noisy and polydispersity is high during the transition from one stable structural state to another. To improve the statistics, timepoints were averaged together in (C). Error bars show the standard error of the mean.

The data show that 16 repeats of the mixing were sufficient to generate data with good signal to noise, i.e. it is possible to interpret 32 ms data for a 2 mg/mL α-casein/buffer solution measured in a 1.5 mm diameter capillary measured in air. For the analysis, R_g and $I(0)$ were calculated with AUTORG, however, uncertainties are high when the Guinier region is not selected manually. Further averaging of frames together to reduce the time resolution can allow us to investigate which range of time evolution we should concentrate the analysis on. If the automatic procedure fails to calculate R_g and $I(0)$, individual 2 ms frames can be fitted manually.

Ca^{2+} stimulates micelle formation and oligomerization of caseins, this mechanism depending on interactions with the phosphorylated serines, natively present in α_{s1}-, α_{s2}- and β-caseins (Yoshikawa, Sasaki & Chiba, 1981; Gerber & Jost, 1986; Sato, Noguchi & Naito, 1983; Berrocal, Chanton, Juillerat, Favillare, Scherz & Jost, 1989; Müller-Buschbaum, Gebhardt, Roth, Metwalli & Doster, 2007). Our data suggests the formation of larger oligomeric states, as $I(0)$ increases almost 20 times within 500 ms of mixing with $CaCl_2$ buffer. The formation of unstable oligomeric micellar states of α-casein when exposed to $CaCl_2$ is consistent with previous reportings (Müller-Buschbaum, Gebhardt, Roth, Metwalli & Doster, 2007; Vogel, Brokx & Ouyang, 2002; Liu, Jiang, Ahrné & Skibsted, 2022).

The rate of increase of both R_g and $I(0)$ is initially very fast, the R_g doubling within the first 60 ms. For times greater than 500 ms, the rate of increase of R_g and $I(0)$ slows but continues to increase for the remaining 9.5 s. The p(r) also shows that there is an accumulation of larger structures with time. Eventually the $CaCl_2$ containing α-casein will become opaque, i.e. the oligomers are above 1 µm in diameter (Liu, Jiang, Ahrné & Skibsted, 2022).

In the above sections we have described how external perturbations can influence the chemical surroundings of proteins, for example, by introducing excipients or surfactants to modify a protein's intramolecular structure and shape or changing the electrostatic screening in the buffer to promote interactions between proteins. The methods used necessarily increased the dilution of the protein solution. There are other physical parameters to consider as well, for example, temperature can influence the association dynamics between oligomeric states of proteins and/or reveal the intrinsic flexibility of the protein. The experimental methods discussed so far allow us to control the temperature of the sample. For the stopped-flow equipment, we have the possibility to separately control the temperature of the observation head and the internal syringes. In addition, temperature jump experiments are possible. It is also possible to explore the response of a protein/excipient interactions to a defined heating or cooling ramp using a Linkam THMS 600 hotstage (*Linkam Scientific Instruments*) with SAXS/WAXS (Bogdanova, Lages, Phan-Xuan, Kamal, Terry, Fureby & Kocherbitov, 2023). We will now discuss the development of a novel sample environment at CoSAXS using tunable lasers to probe the response of a solution of biologically relevant molecules to temperature jumps and light. We will also present another form of analysis where instead of interpreting every individual scattering pattern, we consider the differences between sequential X-ray data with and without laser illumination and relate those to the changes in the structure and interactions. This technique is commonly called X-ray Solution Scattering (Berntsson et al., 2017) to differentiate it from the more typical BioSAXS presented above.

6. Time-resolved X-ray solution scattering, TR-XSS

Laser pump-probe techniques have been developed to study the perturbation and subsequent dynamics in biological important molecules. Such techniques exist at highly specialized synchrotron beamlines (ID09 at the European Synchrotron Radiation Source—Extremely Brilliant Source (ESRF-EBS) (Wulff, Plech, Eybert, Randler, Schotte & Anfinrud, 2003), and BioCARS 14 ID at the Advanced Photon Source (APS) (Henning, Kosheleva, Srajer, Kim, Zoellner & Ranganathan, 2024), as well as at free electron lasers, like SPB/SFX, European X-ray Free Electron Laser (EU-XFEL) (Blanchet et al., 2023). In these pump-probe experiments, the laser induces a rapid change in the molecules under investigation and the response can be measured by the X-ray beam, often by diffraction and scattering methods. The timing of the experiment is inherent to the time structure within the X-ray beam, i.e. the beam is not a continuous source of X-rays, instead its structure is that of pulses of ultra-short (in time) bunches of X-ray photons and the X-ray detection can be synchronized on that timescale.

A different, simpler, approach was adopted to exploit the existing infrastructure at CoSAXS, which avoids the use of a rotating mechanical "chopper" to provide short X-ray pulses (Cammarata et al., 2009). At CoSAXS we achieve temporal resolution using the fast acquisition capabilities of the X-ray detectors which is currently limited to 500 Hz, i.e. dynamics and motions faster than 2 ms are not accessible by this method. The detector acquisition is precisely synchronized to the fast shutter and the sample environment and laser systems via TTL pulses orchestrated by the

PandABox controllers at CoSAXS. In a typical TR-XSS experiment at CoSAXS (Berntsson, Terry & Plivelic, 2022; Magkakis, Orädd, Ahn, Da Silva, Appio, Plivelic & Andersson, 2024), a fresh sample is loaded into a capillary and the flow is stopped. SAXS/WAXS data are collected at 500 Hz for a certain duration, alternating between with and without laser illumination (laser-on/laser-off). The differences in the signals between laser-on and laser-off are compared. The sample is replenished and the experiment automatically repeated until a sufficient signal-to-noise ratio is achieved.

There is a wide diversity in the type of laser-induced changes we might hope to study in biological samples. The absorption of an infrared laser (IR) by water molecules in an aqueous solution will introduce a temperature jump in that solution, and the extent of the temperature jump will depend on the fluence of the laser (both the flux and the length of the laser pulse). For soluble (Cammarata et al., 2008) and for intrinsically disordered proteins (Orädd & Andersson, 2021), the thermal perturbation can cause transitions between different oligomeric states (Rimmerman et al., 2018) or reveal the intrinsic flexibility of the proteins (Thompson et al., 2019). For light sensitive proteins, such as rhodopsin (Andersson et al., 2009) and phytochromophores (Takala et al., 2014), Ultraviolet (UV) lasers in the visible range can excite transient states and structural rearrangements. More recently, indirect laser activation of proteins can liberate caged compounds such as ions and neurotransmitters into the protein solution (Klan et al., 2013; Ravishankar et al., 2020). This wide range of laser induced transitions require that we offer suitable wavelength lasers to enable their study. For this reason, CoSAXS provides two laser sources: (i) a continuous wave IR laser, $\lambda = 1470$ nm, 50×10^3 mJ output power (*LuOceanP2, Lumics*) and (ii) a nanosecond, Q-switched, Nd:Yag laser, 4–6 ns pulse duration, 10 Hz repetition rate (*Surelite II-10 Continuum*). The principal wavelengths for the nanosecond laser are 1064 nm (690 mJ), 532 nm (300 mJ) and 355 nm (100 mJ) and can be tuned with an optical parametric oscillator (OPO) from 675 to 2500 nm (20–140 mJ). A cross-cylindrical lens array (*Nr.18–00142, SUSS MicroOptics, Switzerland*) and a plano-convex spherical lens (*LA1608, Thorlabs, United States*) are used to homogenize and couple high-energy laser pulses into a multimode fiber (*FT1500UMT, Thorlabs, United States*). For easing laser safety considerations, both sources are fiber coupled to the sample environment.

6.1 Design of the TR-XSS experimental setup

The sample environment is designed such that the laser fiber is mounted perpendicularly to a quartz capillary with a 10 μm thick wall that holds the sample (Fig. 18). Different capillaries can be chosen with inner diameters between 0.3 to 1.5 mm diameter, depending on the illumination depth within the sample and the amount of sample available. A peristaltic pump connected with silicon tubing from a reservoir to the measurement capillary, delivers an aliquot of the sample solution at the start of the experiment, the volume being sufficient to completely fill the capillary and then the flow is stopped. The sample can be re-circulated back into the reservoir to reduce the amount of sample required for these measurements. The capillary is glued into a machined aluminum block or to allow for the range of capillary diameters, into a 3D printed, polyethylene terephthalate glycol holder, and sandwiched between two thermostatically controlled copper blocks to allow baseline temperature stabilization of the sample from 5 to 60 °C. The capillary holder, with the laser fiber optic coupling, is mounted on the X/Y stages at CoSAXS to permit alignment of the capillary to the X-ray beam.

To be able to study thermally driven conformation changes, it is necessary to correlate precisely the increase in thermal energy to the X-ray data by measuring the instantaneous temperature of the solvent/buffer upon laser illumination. Equally for light activated protein reactions the thermal effect must be removed, or at least considered, from the effect of the optical triggering. The position of the water scattering peak in the WAXS regime ($1.7 < q < 2.5 \text{ Å}^{-1}$) is sensitive to temperature changes and so can be used to calibrate the magnitude of the temperature jump. In order to access this high q range, a second WAXS detector, MYTHEN®2 X 1 K, is mounted in air on the nose cone of the vacuum vessel, at a SDD of about 0.25 m; the

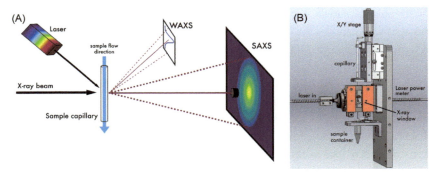

Fig. 18 (A) Schematic showing the orientation of the sample capillary to the X-ray and laser beams and the SAXS and WAXS X-ray detectors at CoSAXS. (B) Details of the flow cell mounted on a X/Y stage for TR-XSS.

MYTHEN®2 X 1 K with 1 kHz operation replaces the slower PILATUS®3 X 2 M detector for TR-XSS measurements, see Table 1. A calibration curve comparing the positional change in the WAXS scattering from Milli-Q water during laser illumination was compared to reference scattering from the same sample collected at elevated temperatures using the thermostatically controlled copper blocks in the sample environment. As can be seen in Fig. 19B, there is an almost linear relationship which permits the magnitude of the temperature jump to be calibrated. In a further step, it was also possible to use the same principle of the thermal shift in the position of the water peak to map the temperature distribution in the quartz capillary by scanning the capillary through a small X-ray beam (Fig. 19D) and relate this to simulations of the heat disipation of an IR laser in a water filled quartz capillary. This approach has allowed us to follow the thermally-induced conformational changes in a model protein, in this case lysozyme, below its denaturation temperature (Berntsson, Terry & Plivelic, 2022).

6.2 Thermal response of lysozyme to laser induced temperature jumps

Lysozyme (hen egg-white) was purchased from *Sigma Aldrich* and used without further purification. The protein was dissolved in a buffer solution (HEPES 20 mM, pH 7.2, 5 % v/v glycerol) to a final concentration of $\sim$ 40 mg/mL and filtered through a 0.2 μm filter before use. Glycerol in the solution mitigates radiation damage which is an important consideration for TR-XSS measurements where the sample is not flowing during X-ray exposure.

The IR laser, coupled to the TR-XSS sample environment, was used to induce temperature jumps of between 20 and 40 °C above a baseline temperature held constant using the attached water bath. SAXS data were collected on the EIGER®2 X 4 M at a SDD of 2.08 m, X-ray energy of 12.4 keV, giving a q-range of $6 \times 10^{-4} < q < 0.6 \text{ Å}^{-1}$. WAXS data were recorded using the MYTHEN®2 X 1 K placed in air, at a SDD of 0.253 m from the sample, and covers a q-range of $1.4 < q < 2.5 \text{ Å}^{-1}$. Each scan consisted of alternating laser-off and laser-on 4 s duration steps. Fig. 20 shows the triggering scheme for the synchronisation of data acquisition and the laser. The detectors are read out at 500 Hz (1.9 ms acquisition, 100 ms readout) throughout each 4 s step, resulting in 2000 frames per step. The fast shutter was opened at the start of the data acquisition and closed whilst the sample was replenished between steps. During each laser-on step a T-jump was triggered, 1 s after data acquisition commenced, by a 2 ms-long IR laser pulse. The T-jump is then maintained by a train of secondary

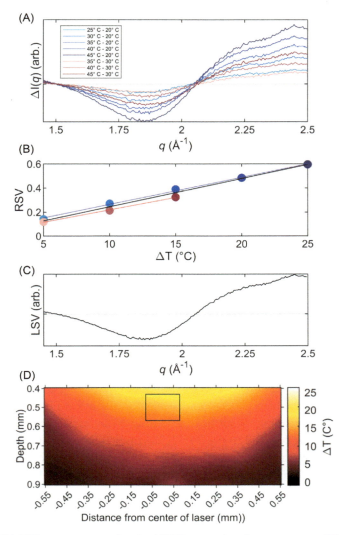

Fig. 19 (A) Difference spectra for the WAXS scattering from water at different temperatures. (B) and (C) are the results of a SVD analysis, (B) the first RSV against the temperature difference and (C) the first LSC. (D) The spatial profile of the temperature jump as seen by the X-ray beam. The gray box marks the position of the X-ray for the regular SAXS data acquisition.

pulses, each 50 ms long, fired at 90 Hz, for 3 s. The SAXS data were radially integrated and the intensities normalized to the WAXS scattering at $q = 1.5\,\text{Å}^{-1}$. In the data reduction, the SAXS contribution from the neighboring laser–off steps are averaged and subtracted from the laser–on

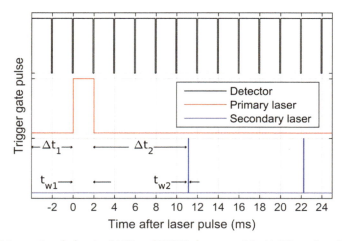

Fig. 20 Trigger signals for the SAXS and WAXS detectors (black). When the trigger gate pulse is high, the detector acquires data or the laser fires. The triggers for the primary laser pulse (red) to initiate the temperature jump and the secondary laser pulses (blue) to maintain the temperature are offset for clarity, and are separate pulses sent to the same laser. *Reproduced with permission from Berntsson, Terry and Plivelic (2022).*

step, i.e. the averaged steps m and m + 2 (laser-off) are subtracted from step m + 1 (laser-on). As the SAXS data will also depend on the structural change in the solvent, it is necessary to subtract the background contribution from the solvent, collected under the same temperature jump conditions, the intensities scaled at $q = 2\,\text{Å}^{-1}$ to avoid oversubtraction.

It is important to consider whether the temperature of the aqueous solution affected the structure of lysozyme or whether the changes might be due to the IR radiation being absorbed by the protein, thus the experiments were also performed with lysozyme dissolved in a buffer prepared with deuterated water, since this is virtually transparent to IR light of $\lambda = 1470$ nm (Berntsson, Terry & Plivelic, 2022). This showed that the IR radiation absorbed directly by the protein does not have a pronounced effect on the structure of the protein. Instead the structural changes observed are caused by the temperature change of the solvent. In these styles of experiments, it is also important to consider whether X-ray radiation damage is occurring. Over the 4 s data collection, the same radiation damage would have occurred for the laser-on and laser-off steps, and assuming that any proteins which have undergone radiation damage will not respond to the temperature jump, any radiation damage might be effectively canceled out (Berntsson, Terry & Plivelic, 2022).

TR–XSS difference data are shown in Fig. 21 for a 13.1 °C temperature jump above the starting baseline temperature of 20 °C. The signal develops over time and can be attributed to large-scale structural changes in the protein as the proteins expands. Fitting the SAXS difference data, via a singular value decomposition (SVD), allows us to quantify the q-range and the rate over which the structural changes occur, the left and right singular vectors, respectively (Fig. 21). Interestingly, data collected for the same

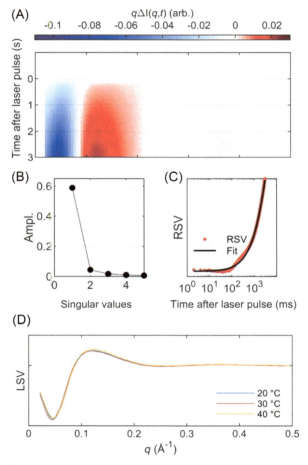

Fig. 21 Data from a T-jump TR-XSS experiment on lysozyme. (A) Heat subtracted data in the SAXS range for a T-jump in lysozyme starting at 20 °C. (B–D) SVD analysis of lysozyme data. (B) Relative amplitude of the first five singular values from an SVD analysis for the data shown in plot A. (C) The time evolution of the first RSV at 20 °C. (D) The LSV at three temperature jumps.

approximate temperature jumps for different initial temperatures, 20 °C, 30 °C and 40 °C, revealed that lysozyme undergoes the same structural reorganization but at a different rate (Berntsson, Terry & Plivelic, 2022). As the denaturation temperature of lysozyme is neared, the magnitude of the difference is reduced as more of the protein has already adopted the higher temperature structure and so is not influenced by the temperature jump. A more detailed structural analysis is also possible using conventional methods to interpret the SAXS data.

The TR-XSS methodology has been widely applied to light-sensitive proteins, yet these types of proteins constitute only a relatively small subset of proteins. Of significant interest would be to follow the intermediate conformations which dictate a protein's biological function, for example, how a compound like adenosine 5'-triphosphate (ATP) influences transient structures in an enzymatic reaction. At CoSAXS, it was possible to follow such dynamical structural changes induced by the release of ATP from a caged compound with the laser in the TR-XSS setup and thus triggering an ATP activated reaction in adenylate kinase enzyme (Magkakis, Orädd, Ahn, Da Silva, Appio, Plivelic & Andersson, 2024).

6.3 Laser-induced activation of ATP binding in adenylate kinase

The sample preparation is fully described previously (Orädd et al., 2021). A final adenylate kinase concentration of 12.5 mg/mL was used for these measurements, with 10 mM nitrophenylethyl ester (NPE)–caged ATP. In a separate measurement, a UV-fluorescent dye was exploited to show that thermal expansion induced by the laser did not contribute to the difference scattering in the q-range specific to protein dynamics. The configuration of the beamline and detectors was the same as the TR-XSS measurements described above, with a 0.3 mm diameter sample capillary being mounted vertically. The release of the caged ATP was activated using the nanosecond laser with a wavelength of 355 nm and 12 mJ/mm^2 fluence. Up to 1000 repetitions of laser-on/laser-off cycles were made, the time-resolved SAXS/WAXS data being collected for 50 ms at 500 Hz and the sample replenished between consecutive repeats.

The SAXS difference data were averaged into 8 ms time intervals and show, with time, a growth in the difference intensity at 0.1 Å^{-1} and a decrease at neighboring q-ranges (Fig. 22A). It is possible to estimate that at least 1.7 % of the protein in the sample was activated. SVD analysis allowed us to determine a two-state sequential model which broadly

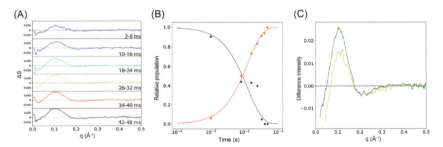

Fig. 22 (A) Difference X-ray scattering plots for adenylate kinase protein upon heating in the TR-XSS setup. The difference profiles were averaged for 8 ms. (B) Temporal evolution of the relative proportions of early (black) and late (red) states obtained from a two-state sequential model. (C) Difference spectra at CoSAXS compared with data collected at ID09 ESRF-EBS under the same conditions.

explains the kinetic data. Fig. 22B shows the relative evolution of two structural populations. The first component contained the major positive peak at $q = 0.1 \, Å^{-1}$ observed at all time points. The second component contributed a positive peak at $q < 0.05 \, Å^{-1}$, where the first component showed a negative feature. The temporal evolution of relative populations of the early (black) and late (red) states from a 2-state sequential kinetic model is presented in Fig. 22B. The intermediate state is formed with a 13.7 ms rise-time. Interestingly, the data generated with the continuous mode at CoSAXS show great similarity to measurements in single-pulse mode at ID09, ESRF-EBD, under similar conditions (Orädd et al., 2021) (Fig. 22C). The observed differences between the spectra are most apparent at low q, which results from CoSAXS being a dedicated small-angle beamline capable of resolving such spectral features.

7. Future perspectives

Above we have introduced some of the insights that modern high brilliance SAXS/WAXS beamlines can bring to the scientifically complex questions raised by biological sciences through enabling time-resolved structural studies. As with all experiments, there is continual development of our techniques and so the methods described are a snapshot of the current possibilites for samples in solution at CoSAXS to inspire you in thinking of what these methods will bring to your research interests.

There is a still a mismatch of these experimental methods to the timescales of MD simulations, for example, which are limited to below microsecond timescales for all-atom simulations unless coarse-grained (reaching milliseconds) or mesoscale methods are employed. Beamlines like CoSAXS have sufficient X-ray flux to push to sub-millisecond data acquisition and so we are exploring methods to improve the data acquisition mode of the X-ray detectors which is now the limiting factor and avoiding dead-time in our sample environment to be able to capture the onset of interactions and structural changes. Such sub-millisecond dynamics hold information on the structural origins of biological processes and many protein targets can be envisioned (Colletier, Bourgeois, Sanson, Fournier, Sussman, Silman & Weik, 2008).

Fig. 23 represents the timescales that CoSAXS can achieve currently in relation to the timescales of simulations and the structural processes occuring in proteins.

Biologically relevant samples can be highly complex and we have shown the advantage of combining structural characterization with

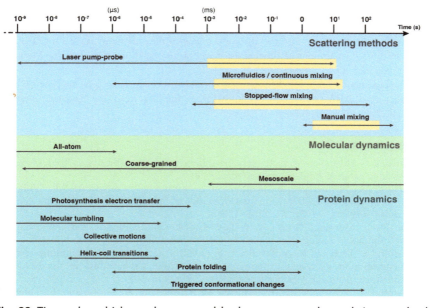

Fig. 23 Timescales which can be accessed by laser pump-probe and time-resolved structural methods for biology compared with those explored by simulations. For comparison, the timescale of structural dynamics in proteins is included. The highlighted area in each method shows the timescales that CoSAXS can achieve.

spectroscopic methods in SurF, the advantage of exploring parameter space like concentration in situ with sufficiently small samples using microfluidics, the advantage to be able to repeat the experiments with controlled volumes of components and accurate timing from the start of an interaction with stopped-flow, and the advantage of external stimuli like light and temperature to drive perturbation in TR-XSS. One dream would be to combine all of these advantages in a single multiprobe experiment but perhaps we still have to dream a while longer.

Acknowledgments

We acknowledge MAX IV Laboratory for time on the CoSAXS beamline under proposals 20200104, 20200221, 20210924, 20220421, 20221045, 20230785 and 20240930. Research conducted at MAX IV, a Swedish national user facility, is supported by the Swedish Research Council under contract 2018–07152, the Swedish Governmental Agency for Innovation Systems under contract 2018–04969, and Formas under contract 2019–02496.

The SUrF sample environment was jointly funded by the COSAXS-LU-SOLEIL project. The microfluidics experiments are supported by the Adaptocell platform, Stiftelsen för strategisk forskning (SSF), ITM-0375. VF and SJL acknowledge funding from the VILLUM FONDEN by the Villum Young Investigator Grant "Protein superstructure as Smart Biomaterials (ProSmart)' 2018–2023 (project number: 19175), the Novo Nordisk foundation (projects NNF20OC0065260 and NNF22OC0080141) and HALOS project UCPH-009.

The stopped-flow measurements were jointly funded by the Carl Tryggers Stiftelse and the Crafoord foundation, and the mass-spectrometry measurements were carried out by Katja Bernfur (Department of Biochemistry and Structural Biology, Center for Molecular Protein Science, Lund University).

Development of the TR-XSS setup was funded through Röntgen-Ångström Cluster (contract No. 2015–06099). MA acknowledges funding from the Swedish Research Council (2020–03840).

References

AdaptoCell, https://www.maxiv.lu.se/beamlines-accelerators/beamlines/balder/science-at-balder/external-funded-projects/adaptocell/.

Andersson, M., Malmerberg, E., Westenhoff, S., Katona, G., Cammarata, M., Wöhri, A. B., ... Neutze, R. (2009). Structural dynamics of light-driven proton pumps. *Structure (London, England: 1993), 17*, 1265–1275.

Beebe, D. J., Mensing, G. A., & Walker, G. M. (2002). Physics and applications of microfluidics in biology. *Annual Review of Biomedical Engineering, 4*(1), 261–286.

Berntsson, O., Terry, A. E., & Plivelic, T. S. (2022). A setup for millisecond time-resolved X-ray solution scattering experiments at the CoSAXS beamline at the MAX IV Laboratory. *Journal of Synchrotron Radiation, 29*(2), 555–562.

Berntsson, O., Diensthuber, R. P., Panman, M. R., Björling, A., Gustavsson, E., Hoernke, M., ... Westenhoff, S. (2017). Sequential conformational transitions and alpha-helical supercoil- ing regulate a sensor histidine kinase. *Nature Communications, 8*, 284.

Berrocal, R., Chanton, S., Juillerat, M. A., Favillare, B., Scherz, J.-C., & Jost, R. (1989). Tryptic phosphopeptides from whole casein. II. Physicochemical properties related to the solubilization of calcium. *Journal of Dairy Research, 56*(3), 335–341.

Bizien, T., Durand, D., Roblina, P., Thureau, A., Vachette, P., & Pérez, J. (2016). A brief survey of state-of-the-art BioSAXS. *Protein and Peptide Letters, 23*(3), 217–231.

Blanchet, C. E., Round, A., Mertens, H. D. T., Ayyer, K., Graewert, M., Awel, S., ... Svergun, D. I. (2023). Form factor determination of biological molecules with X-ray free electron laser small-angle scattering (XFEL-SAS). *Communications Biology, 6*, 1057.

Bogdanova, E., Lages, S., Phan-Xuan, T., Kamal, M. A., Terry, A. E., Fureby, A. M., & Kocherbitov, V. (2023). Lysozyme-sucrose interactions in the solid state: Glass transition, denaturation, and the effect of residual water. *Molecular Pharmaceutics, 20*(9), 4664–4675.

Bras, W., Nikitenko, S., Portale, G., Beale, A., van der Eerden, A., & Detollenaere, D. (2010). Combined time-resolved SAXS and X-ray Spectroscopy methods. *Journal of Physics, Conference Series, 247*, 012047.

Brosey, C. A., & Tainer, J. A. (2019). Evolving SAXS versatility: Solution X-ray scattering for macromolecular architecture, functional landscapes, and integrative structural biology. *Current Opinion in Structural Biology, 58*, 197–213.

Caetano, B. L., Meneau, F., Santilli, C. V., Pulcinelli, S. H., Magnani, M., & Briois, V. (2014). Mechanisms of SnO2 nanoparticles formation and growth in acid ethanol solution derived from SAXS and combined Raman–XAS time-resolved studies. *Chemical Materials, 26*(23), 6777–6785.

Cammarata, M., Levantino, M., Schotte, F., Anfinrud, P. A., Ewald, F., Choi, J., ... Ihee, H. (2008). Tracking the structural dynamics of proteins in solution using time-resolved wide-angle X-ray scattering. *Nature Methods, 5*, 881–886.

Cammarata, M., Eybert, L., Ewald, F., Reichenbach, W., Wulff, M., Anfinrud, P., ... Polachowski, S. (2009). Chopper system for time resolved experiments with synchrotron radiation. *Review of Scientific Instrumentation, 80*, 015101.

Chance, B. (1951). Rapid and sensitive spectrophotometry. I. The accelerated and stopped-flow methods for the measurement of the reaction kinetics and spectra of unstable compounds in the visible region of the spectrum. *Review of Scientific Instruments, 22*(8), 619–627.

Chance, B., & Legallais, V. (1951). Rapid and sensitive spectrophotometry. II. A stopped-flow attachment for a stabilized quartz spectrophotometer. *Review of Scientific Instruments, 22*(8), 627–634.

Colletier, J. P., Bourgeois, D., Sanson, B., Fournier, D., Sussman, J. L., Silman, I., & Weik, M. (2008). Shoot-and-trap: Use of specific x-ray damage to study structural protein dynamics by temperature-controlled cryo-crystallography. *Proceedings of the National Academy of Sciences of the United States of America, 105*(33), 11742–11747.

Coutinho, T. M., Cuni, G., Fernandez-Carreiras, D. F. C., Klora, J., & Pascual-Izarra, C. (2011). Sardana: The software for building SCADAS in scientific environments. *Proceedings of ICALEPCS'11, Grenoble, France, WEAAUST01*, 607–609.

Da Silva, J., Appio, R., Mota-Santiago, P., Plivelic, T. S., Terry, A. E., & Herranz, F. Design of an in-line sample viewer for SAXS/WAXS experiments at the CoSAXS beamline, MAX IV. (*in preparation*).

Da Silva, V., Appio, R., Freitas, A., Ahn, B., Alcocer, M., Lindberg, M., ... Terry, A. E. (2023). DAQ system based on Tango, Sardana and PandABox for millisecond time resolved experiment at the CoSAXS beamline of MAX IV Laboratory. *Proceedings of ICALEPCS '23, Cape Town, South Africa.*

David, G., & Pérez, J. (2009). Combined sampler robot and high-performance liquid chromatography: A fully automated system for biological small-angle X-ray scattering experiments at the Synchrotron SOLEIL SWING beamline. *Journal of Applied Crystallography, 42*, 892–900.

Del Giudice, A., Dicko, C., Galantini, L., & Pavel, N. V. (2017). Time-dependent pH scanning of the acid-induced unfolding of human serum albumin reveals stabilization of the native form by palmitic acid binding. *Journal of Physical Chemistry B, 121*(17), 4388–4399.

Deng, H., Zhadin, N., & Callender, R. (2001). Dynamics of protein ligand binding on multiple time scales: NADH binding to lactate dehydrogenase. *Biochemistry, 40*(13), 3767–3773.

Dockal, M. (2000). Conformational transitions of the three recombinant domains of human serum albumin depending on pH. *Journal of Biological Chemistry, 275*(5), 3042–3050.

Dreiss, C. A., Jack, K. S., & Parker, A. P. (2005). On the absolute calibration of bench-top small-angle X-ray scattering instruments: A comparison of different standard methods. *Journal of Applied Crystallography, 39*, 32–38.

Gerber, H. W., & Jost, R. (1986). Casein phosphopeptides: Their effect on calcification of in vitro cultured embryonic rat bone. *Calcified Tissue International, 38*, 350–357.

Gilbert, J., Sebastiani, F., Arteta, M. Y., Terry, A. E., Fornell, A., Russell, R., Mahmoudi, N., & Nylander, T. (2024). Evolution of the structure of lipid nanoparticles for nucleic acid delivery: From in situ studies of formulation to colloidal stability. *Journal of Colloid and Interface Science, 660*, 66–76.

Glatter, O. (1977). A new method for the evaluation of small-angle scattering data. *Journal of Applied Crystallography, 10*(5), 415–421.

Graewert, M. A., & Svergun, D. I. (2013). Impact and progress in small and wide angle X-ray scattering (SAXS and WAXS). *Current Opinion in Structural Biology, 23*(5), 748–754.

Haas, S., Plivelic, T. S., & Dicko, C. (2014). Combined SAXS/UV–VIS/Raman as diagnostic and structure resolving tool in materials and life sciences applications. *Journal of Physical Chemistry B, 118*(8), 2264–2273.

Henning, R. W., Kosheleva, I., Srajer, V., Kim, I. S., Zoellner, E., & Ranganathan, R. (2024). BioCARS: Synchrotron facility for probing structural dynamics of biological macromolecules. *Structural Dynamics, 11*, 014301.

Janvier, N., Clement, J. M., Fajardo, P., & Cuni, G. (2013). IcePAP: An advanced motor controller for scientific applications in large user facilities. *Proceedings of ICALEPCS 2013, San Francisco, USA.*

Jeffries, C. M., Graewert, M. A., Svergun, D. I., & Blanchet, C. E. (2015). Limiting radiation damage for high-brilliance biological solution scattering: Practical experience at the EMBL P12 beamline PETRAIII. *Journal of Synchrotron Radiation, 22*, 273–279.

Jensen, A. B., Christensen, T. E. K., Weninger, C., & Birkedal, H. (2022). Very large-scale diffraction investigations enabled by a matrix-multiplication facilitated radial and azimuthal integration algorithm: MatFRAIA. *Journal of Synchotron Radiation, 29*, 1420–1428.

Khodadadi, S., & Sokolov, A. P. (2015). Protein dynamics: From rattling in a cage to structural relaxation. *Soft Matter, 11*(25), 4984–4998.

Kieffer, J., Petitdemange, S., & Vincent, T. (2018). Real-time diffraction computed tomography data reduction. *Journal of Synchotron Radiation, 25*, 612–617.

Kirby, N. M., & Cowieson, N. P. (2014). Time-resolved studies of dynamic biomolecules using small angle X-ray scattering. *Current Opinion in Structural Biology, 28*, 41–46.

Klan, P., Solomek, T., Bochet, C. G., Blanc, A., Givens, R., Rubina, M., ... Wirz, J. (2013). Photoremovable protecting groups in chemistry and biology: Reaction mechanisms and efficacy. *Chemical Reviews, 113*, 119–191.

Le Ferrand, H., Duchamp, M., Gabryelczyk, B., Cai, H., & Miserez, A. (2019). Time-resolved observations of liquid–liquid phase separation at the nanoscale using in situ liquid transmission electron microscopy. *Journal of the American Chemical Society, 141*(17), 7202–7210.

Levantino, M., Yorke, B. A., Monteiro, D. C., Cammarata, M., & Pearson, A. R. (2015). Using synchrotrons and XFELs for time-resolved X-ray crystallography and solution scattering experiments on biomolecules. *Current Opinion in Structural Biology, 35*, 41–48.

Li, Y., Beck, R., Huang, T., Choi, M. C., & Divinagracia, M. (2008). Scatterless hybrid metal–single-crystal slit for small-angle X-ray scattering and high-resolution X-ray diffraction. *Journal of Applied Crystallography, 41*, 1134–1139.

Liu, X.-C., Jiang, Y., Ahrné, L. M., & Skibsted, L. H. (2022). Temperature effects on calcium binding to caseins. *Food Research International, 154*, 110981.

Magkakis, K., Orädd, F., Ahn, B., Da Silva, V., Appio, R., Plivelic, T. S., & Andersson, M. (2024). Real-time structural characterization of protein response to a caged compound by fast detector readout and high-brilliance synchtron radiation. *Structure (London, England: 1993), 32*, 1–9.

Manalastas-Cantos, K., Konarev, P. V., Hajizadeh, N. R., Kikhney, A. G., Petoukhov, M. V., Molodenskiy, D. S., ... Franke, D. (2021). ATSAS 3.0: Expanded functionality and new tools for small-angle scattering data analysis. *Journal of Applied Crystallography, 54*, 343–355.

Merkens, S., Vakili, M., Sánchez-Iglesias, A., Litti, L., Gao, Y., Gwozdz, P. V., ... Trebbin, M. (2019). Time-resolved analysis of the structural dynamics of assembling gold nanoparticles. *ACS Nano, 13*(6), 6596–6604.

Müller-Buschbaum, P., Gebhardt, R., Roth, S., Metwalli, E., & Doster, W. (2007). Effect of calcium concentration on the structure of casein micelles in thin films. *Biophysical Journal, 93*(3), 960–968.

Nguyen, N. T., Wereley, S. T., & Shaegh, S. A. M. (2019). *Fundamentals and applications of microfluidics*. Artech House.

Orädd, F., & Andersson, M. (2021). Tracking membrane protein dynamics in real time. *Journal of Membrane Biology, 254*, 51–64.

Orädd, F., Ravishankar, H., Goodman, J., Rogne, P., Backman, L., Duelli, A., ... Andersson, M. (2021). Tracking the ATP-binding response in adenylate kinase in real time. *Scientific Advances, 7*, eabi5514.

Petoukhov, M. V., Konarev, P. V., Kikhney, A. G., & Svergun, D. I. (2007). ATSAS 2.1—Towards automated and web-supported small-angle scattering data analysis. *Journal of Applied Crystallography, 40*, S223–2S8.

Plivelic, T. S., Terry, A. E., Appio, R., Theodor, K., & Klementiev, K. (2019). X-ray tracing, design and construction of an optimized optics scheme for CoSAXS, the small angle x-ray scattering beamline at MAX IV laboratory. *AIP Conference Proceeding, 2054*, 030013.

Ravishankar, H., Pedersen, M. N., Eklund, M., Sitsel, A., Li, C., Duelli, A., ... Andersson, M. (2020). Tracking Ca(2+) ATPase intermediates in real time by x-ray solution scattering. *Scientific Advances, 6*, eaaz0981.

Rimmerman, D., Leshchev, D., Hsu, D. J., Hong, J., Abraham, B., Kosheleva, I., ... Chen, L. X. (2018). Insulin hexamer dissociation dynamics revealed by photoinduced T-jumps and time-resolved X-ray solution scattering. *Photochemical and Photobiological Sciences, 17*, 874–882.

Sato, R., Noguchi, T., & Naito, H. (1983). The necessity for the phosphate portion of casein molecules to enhance Ca absorption from the small intestine. *Agricultural and Biological Chemistry, 47*(10), 2415–2417.

Stachowski, T. R., Snella, M. E., & Snella, E. H. (2021). A SAXS-based approach to rationally evaluate radical scavengers—Toward eliminating radiation damage in solution and crystallographic studies. *Journal of Synchrotron Radiation, 28*(5), 1309–1320.

SWING beamline, Soleil Synchrotron, https://www.synchrotron-soleil.fr/en/beamlines/swing.

Tabeling, P. (2023). *Introduction to microfluidics*. Oxford University Press 2023.

Takala, H., Björling, A., Berntsson, O., Lehtivuori, H., Niebling, S., Hoernke, M., ... Westenhoff, S. (2014). Signal amplification and transduction in phyto- chrome photosensors. *Nature, 509*, 245–248.

Tango, https://tango-controls.org.

Thompson, M. C., Barad, B. A., Wolff, A. M., Sun Cho, H., Schotte, F., Schwarz, D. M. C., ... Fraser, J. S. (2019). Temperature-jump solution X-ray scattering reveals distinct motions in a dynamic enzyme. *Nature Chemistry, 11*, 1058–1066.

Vogel, H. J., Brokx, R. D., & Ouyang, H. (2002). Calcium-binding proteins. *Calcium-Binding Protein Protocols, 1*, 3–20.

Wang, J. H. (1965). Self-diffsion coefficients of water. *Journal of Physical Chemistry, 69*(12), 4412.

Weinheimer, R. M., Evans, D. F., & Cussler, E. L. (1981). Diffusion in surfactant solutions. *Journal of Colloid and Interface Science, 80*(2), 357–368.

With, S., Trebbin, M., Bartz, C. B. A., Neuber, C., Dulle, M., Yu, S., ... Förster, S. (2014). Fast diffusion-limited lyotropic phase transitions studied in situ using continuous flow microfluidics/microfocus-SAXS. *Langmuir: The ACS Journal of Surfaces and Colloids, 30*(42), 12494–12502.

Wulff, M., Plech, A., Eybert, L., Randler, R., Schotte, F., & Anfinrud, P. (2003). The realization of sub-nanosecond pump and probe experiments at the ESRF. *Faraday Discussion, 122*, 13–26.

Yoshikawa, M., Sasaki, R., & Chiba, H. (1981). Effects of chemical phosphorylation of bovine casein components on the properties related to casein micelle formation. *Agricultural and Biological Chemistry, 45*(4), 909–914.

Zhang, S., Abiven, Y. M., Bisou, J., Renaud, G., Thibaux, G., & Ta, F. (2017). PandABox: A multipurpose platform for multi-technique scanning and feedback applications. *Proceedings of ICALEPCS'17, 143*–150 (Barcelona, Spain).